Hugh Craig

The Animal Kingdom based upon the Writings of the Eminent

Naturalists

Audubon, Wallace, Brehm, Wood and Others

Hugh Craig

The Animal Kingdom based upon the Writings of the Eminent Naturalists
Audubon, Wallace, Brehm, Wood and Others

ISBN/EAN: 9783337033415

Printed in Europe, USA, Canada, Australia, Japan

Cover: Foto ©berggeist007 / pixelio.de

More available books at **www.hansebooks.com**

DUGONG

...THE...
ANIMAL KINGDOM

Based upon the Writings of the Eminent Naturalists,

AUDUBON, WALLACE, BREHM, WOOD, AND OTHERS

.... Edited by ....
HUGH CRAIG, M. A.,
Trinity College, Cambridge.

...WITH...
SIXTY-FOUR FULL-PAGE ILLUSTRATIONS

Accurately and Beautifully Executed in

EIGHT COLORS AND TINTS.

VOLUME TWO.

NEW YORK:
JOHNSON & BAILEY,
114 and 116 Nassau Street.

CONTENTS OF VOLUME TWO.

SIRENIA.

THE SEA COWS.

The Order Sirenia (397)—Mermaids (397)—The Family Manatidæ (398)—The Manatees of America (399)—Their Voracity and Laziness (399)—Modes of Capture (399)—Tame Specimens (399)—The Florida Manatee (400)—The African Lamantin (400)—The Eastern Dugong (400)—The Australian Dugong (401)—The Northern Sea Cows (401)—Steller's Description (401)—Extinct since 1768 (403).

UNGULATA.

CHAPTER I.

HOOFED ANIMALS.

The Order Ungulata (407)—The Numerous Families (407)—The Ruminants (407)—Their Peculiar Stomach (408)—Horns (408)—Antlers (408)—Extinct Species (408)—The Original Horse Protohippus (409)—Gradual Development (409)—The Family Equidæ (409)—The Genus Equus (409)—The Horse (410)—The Tarpan or Wild Horse of Tartary (410)—The Mustang or Wild Horse of America (411).

CHAPTER II.

THE ARAB AND THE BARB.

Early Domestication of the Horse (416)—The Horse in Egypt (416)—Assyria—Judæa (416)—Greece—Persia (417)—Bits and Stirrups (417)—Chariot Races (417)—The Arab Horse (418)—Exaggerated Pedigrees (419)—The Best Arabs (419)—Their Training (419)—Attachment of the Arab for his Mare (420)—Speed and Endurance (421)—The Barb (422)—The Same Horse as the Arab (422)—Abd-el-Kader on the Horse (422).

CHAPTER III.
THE RACE-HORSE AND TROTTING HORSE.

The Race-Horse (425)—The English Turf (426)—The American Turf (427)—Imported Horses (427)—The Trotting Horse (428)—Flora Temple (431)—Steve Maxwell (432)—St. Julien and Maud S (432)—The Narragansett Pacers (432)—Pocahontas (432).

CHAPTER IV.
EUROPEAN HORSES.

The Hunter (434)—The Hackney (434)—The Russian Horse (436)—The Austrian Horse (437)—The Holstein Horse (438)—The French Horse (438)—The Italian Horse (440)—The Races at Rome (440)—The Spanish Horse (440)—The Shetland Pony (441)—The Carriage Horse (443)—The Cart Horse (443)—The Percheron Horse (443).

CHAPTER V.
THE WILD AND THE COMMON ASS.

The Wild Asses (445)—The Kulan or Dzigguetai (445)—Their Speed (446)—Domestication (446)—The Wild Ass of the Bible (447)—The African Wild Ass (448)—The Common Ass (448)—Its Patience—Its Intelligence (449)—The Egyptian Ass (450).

CHAPTER VI.
THE ZEBRAS.

The Zebras or Tiger-Horses (452)—The Quagga (452)—The Dauw, or Burchell's Zebra (453)—Harris's Description of it (454)—The Zebra Proper (454)—Hunting the Zebra (455)—Cross-Breeds (456)—The Mule (456)—The Hinny (456)—Instances of their Fertility (457)—Darwinism (457).

CHAPTER VII.
THE TAPIRS.

The Family Tapiridæ (458)—The American Tapir (458)—Its Trunk (459)—Its Habits (459)—The Tapir as a Domestic Animal (460)—A Tapir Hunt (461)—Peculiar Marks of the Young Tapir (461)—The Malay Tapir (462)—Its Trunk (462)—Its Color (462)—Discovery of the Animal (462)—Chinese Account (463)—The Pinchaque (463)—Baird's Tapir (463).

CHAPTER VIII.
THE RHINOCEROS.

The Family Rhinocerotidæ (464)—General Description (464)—The Horn—Peculiar Structure of the Horn (465)—Known to the Ancients (466)—Woodcut by Albert Durer (466)—Arab Superstitions (466)—Haunts of the Rhinoceros (466)—A Nocturnal Animal (467)—Its Food—Its Habits (467)—Its Senses (468)—Its Fits of Rage (468)—Maternal Affection (469)—Its Friends the Small Birds (469)—Captive Rhinoceroses (470)—Uses of its Hide (470).

CHAPTER IX.
THE ASIATIC RHINOCEROSES.

The One-horned Rhinoceroses (470)—The Indian Rhinoceros (470)—Its Thick Hide (470)—Mode of Hunting (473)—The Wara or Javanese Rhinoceros (473)—The Emperor Baber (474)—

The Two-horned Rhinoceros or Badak of Sumatra (474)—The Fire-eating Rhinoceros (476)—The Rough-eared Rhinoceros (476).

CHAPTER X.
THE AFRICAN RHINOCEROS.

The Borele or Little Black Rhinoceros (477)—The Sword-Hunters of Abyssinia (479)—The Keitloa (479)—Their Fierceness (480)—The Mohogoo or White Rhinoceros (481)—Hunting Adventure of Mr. Oswell (482)—The Kobaoba (484)—Probability of its Extinction (484).

CHAPTER XI.
THE HIPPOPOTAMUS.

The Hippopotamus or River Horse (485)—Description (486)—Habits (486)—Favorite Haunts (487)—Food (487)—Violence when Provoked (488)—Maternal Affection (488)—Modes of Hunting (489)—Pitfalls and Downfalls (489)—Harpooning (489)—The Hippopotamus in Captivity (491)—The Small or Liberian Hippopotamus (492).

CHAPTER XII.
THE PECCARIES.

The Swine Family (493)—General Characteristics (493)—The Peccaries (494)—The Collared Peccary (494)—Its Courage and Fierceness (495)—The White-lipped Peccary (495)—Its Habits (495)—Methods of Hunting the Peccary (496)—Flesh of the Peccary (497).

CHAPTER XIII.
THE TRUE SWINE.

The Genus Sus (498)—Religious Prohibitions (498)—The Boar of Valhalla (499)—The Boar's Head (499)—The Wild Boar of Europe (499)—Hunting the Wild Boar (500)—The Wild Hog of India (501)—The Domestic Hog (502)—Anecdotes of the Hog (502)—Breeds of Hogs (504)—The Berkshire (504)—Trichiniasis (504).

CHAPTER XIV.
THE RIVER-HOGS, BABYROUSSA, AND WART-HOGS.

The River Hogs (506)—The Pencilled Hog (506)—The Bush Hog, or Bosch Vark (507)—Edwards' River-Hog (508)—The Babyroussa (508)—Its Peculiar Tusks (508)—The Wart-Hogs (509)—Hideous Appearance (510)—The African Wart-Hog, or Vlacke Vark (510)—The Wart-Hog of Ælian or Engallo (511).

CHAPTER XV.
THE CAMEL.

The Ruminants (512)—The Camelidæ (512)—The Camels of the Old World (513)—The Arabian Camel, or Dromedary (514)—The Camel in the Bible (515)—The Camel in Europe (515)—The Camel in Africa (515)—Its Food 516—Its Powers of Resisting Thirst (516)—Its speed (517)—Mode of Riding (517)—Its Behavior when Loading (518)—Its Vices (519)—Anecdote of Latif Pacha (520)—Its Value (521)—The Two-humped Camel of Bactria (522).

CHAPTER XVI.
THE LLAMAS.

The American Camelidæ (524)—The Genus Auchenia (524)—The Guanaco (525)—Its Habits (526)—The Llama (527)—Its Use as a Beast of Burden (527)—The Alpaca or Paco (528)—Its Wool (528)—The Vicuna (529)—Indian Hunts (530).

CHAPTER XVII.
THE MOUSE DEER.

The Tragulidæ or Hornless deer (532)—Disputes of Naturalists (532)—The Kanchil (532)—Its Appearance and Habits (533)—Attempts to introduce it to Europe (534).

CHAPTER XVIII.
THE DEER.

The Cervidæ (535)—Their Antlers (535)—The Process of Growth of the Antler (536)—The Shedding of the Velvet (536)—Habits of the Cervidæ (538)—The Various Genera (538)—The Elk of the Old World or the Moose of the New World (539)—The Elk of Sweden (539)—The Moose of Canada (541)—Habits—Modes of Hunting (541).

CHAPTER XIX.
THE REINDEER AND THE CARIBOU.

The Reindeer (544)—Its Life in Northern Europe (545)—Its Life in Siberia (546)—Its Life when Domesticated (547)—Its Value (547)—The Caribou (548)—Modes of Hunting it (548).

CHAPTER XX.
THE TRUE DEER.

The True Deer (550)—The Wapiti (550)—The Red Deer of Europe (552)—The Virginian Deer or Carcajou (554)—The Persian Deer (556)—The Indian Species (556)—The Barasinga (556)—The Axis Deer (557)—The Sambur (557)—The Maned Stag (557)—The Hog Deer (558)—The South American Species (558)—The Pampas Deer (558)—The Red Deer or Guasupita (559).

CHAPTER XXI.
THE FALLOW DEER, ROE DEER, AND MUSK DEER.

The Genus Dama (560)—Fallow Deer (560)—Genus Capreolus (562)—Roe Deer (562)—Genus Cervulus (564)—Muntjak or Kidang (564)—Genus Moschus (565)—Musk Deer (565)—Its Abode—Habits—The Musk (566).

CHAPTER XXII.
THE GIRAFFE.

The Camelopardalidæ or Giraffes (569)—Its Size and Appearance (569)—Its Habitat (570)—Its adaptation to its Location (570)—Its Movements (570)—Its Food (571)—Its Senses (572)—Giraffes in London and Paris (572)—Modes of Hunting (572)—Meaning of the Word "Giraffe" (573).

CHAPTER XXIII.
THE HOLLOW-HORNED RUMINANTS.

The Bovidæ (574)—The Thirteen Sub-families (574)—The Bovinæ (575)—The Genus Bos (575)—The Domestic Ox (575)—The Wild Cattle (576)—The Cattle of the Pampas (577)—Cattle of Africa (578)—Domestic Cattle (579)—The Highland Cattle (582)—The Durham (582)—The Alderney (582).

CHAPTER XXIV.
THE BISONS.

The Bonassus or European Bison (584)—Called also the Aurochs (584)—The Real Aurochs Extinct (584)—The Forest or Bialowicz (584)—Description of the Bonassus (585)—The Bison of the Caucasus (586)——The American Bison or Buffalo (586)—Enormous Numbers (586)—Terrible Destruction (587)—Estimate of Numbers Killed (588)—The Mountain Buffalo (589)—Death of a Bull (590).

CHAPTER XXV.
EASTERN CATTLE.

The Domestic Cattle of India (591)—The Zebu (591)—The Wild Cattle of India (592)—Genus Bibos (593)—The Gayal (593)—The Gaur (594)—The Banteng (595)—Genus Poephagos (595)—The Yak (595)—The Plough Yak (596)—Hunting the Yak (597)—Genus Anoa (597)—The Chamois Buffalo or Celebes (597)—Its Fierceness (598).

CHAPTER XXVI.
THE BUFFALOES.

The Genus Bubalus (599)—The Cape Buffalo (599)—Drayson's Account (600)—Buffalo Shooting (602)—The Indian Buffalo (602)—Buffalo and Tiger Fights (603)—Williamson's Account (604)—The Kerabau (605)—The Domesticated Buffalo (605)—Its Habits—Its Uses (606).

CHAPTER XXVII.
THE ANTELOPES.

The Antelopes (607)—The Eland (607)—The Koodoo (609)—The Bosch-bok (610)—The Nylghau (611)—The Passan (613)—The Beisa (614)—The Sabre Antelope (614)—The Addax (614)—The Sable Antelope (615)—The Blau Bok (616).

CHAPTER XXVIII.
THE GAZELLES.

The Gazelle (617)—Its Beauty and Grace (617)—The Ariel Gazelle (618)—The Jairou (619)—The Spring-bok (620)—Its Immense Numbers (620)—The Dseren (622)—The Sasin (623)—The Pallah (624)—The Saiga (624)—The Sub-family Antilocaprinæ (625)—The Prong Horn (625).

CHAPTER XXIX.
THE LESSER ANTELOPES.

The Ourebi (627)—The Klippspringer (628)—The Water Buck (628)—The Blue Buck (630)—The Musk Antelope (629)—The Duyker Bok (630)—The Rhoode Bok (631)—The Chickara

(631)—The Hartebeest (632)—The Sassaby (632)—The Gnu (633)—The Chamois (633)—The Goral (635)—The Mountain Goat of the Rocky Mountains (638).

CHAPTER XXX.
GOATS AND IBEXES.

The Genus Capra (637)—The Goats (637)—The Bezoar Goat or Paseng (639)—The Cashmere Goat (639)—The Angora Goat (640)—The Mamber Goat (641)—The Markhor and Tahir (641)—The Egyptian Goat (641)—The Ibexes (642)—The Alpine Ibex (642)—The Pyrenean Ibex (643)—The Arabian Ibex (644).

CHAPTER XXXI.
THE SHEEP AND THE MUSK-OX.

The Aoudad (646)—The Moufflon (647)—The Argali (647)—The Katshkar (648)—The Big Horn (648)—Its Habits (649)—Fat-tailed Sheep (649)—The Cretan Sheep (650)—The Southdown (651)—The Leicester (651)—The Merino (652)—The Highland Sheep (653)—The Genus Ovibos (653)—The Musk-ox of North America (654).

PROBOSCIDEA.

CHAPTER I.
ELEPHANTS IN GENERAL.

The Order Proboscidea—Derivation of Name (657)—The Family Elephantidæ (657)—Fossil Elephants—The Mammoth (657)—The Mastodon (658)—The Elephant (659)—Its Trunk—Its Tusks (660)—The Elephant in History (661)—In the East—In Rome—In Modern Times (663).

CHAPTER II.
THE ASIATIC ELEPHANT.

The Asiatic Elephant (665)—Its Use (666)—Mode of Capture in Ceylon (666)—Points of a Good Elephant (669)—White Elephants (670)—Funeral of a White Elephant (670)—The Dwarf Elephant (671).

CHAPTER III.
THE ELEPHANT.

The African Elephant—Difference from the Indian Elephant (672)—Hunting the Elephant (672)—Delegorgue (672)—Gordon Cumming (673)—The Abyssinian "Hock-cutters" (674)—Captive Elephants (676)—Baby Elephants (676)—Anecdotes of Elephants (677).

HYRACOIDEA.

THE ROCK RABBITS.

The Order Hyracoidea (681)—The Genus Hyrax (681)—Its Characteristics (682).

RODENTIA.

CHAPTER I.
RATS AND MICE.

The Order Rodentia (687)—The Family Muridæ (688)—Rats and Mice (688)—The Black Rat (688)—The Brown Rat (688)—The Mouse (689)—The Harvest Mouse (689)—The Barbary Mouse (690)—The Hamster (690)—The Musk Rat (692)—The Water Rat (693)—The Field Mouse (693)—Wilson's Meadow Mouse (694)—Le Conte's Mouse (694)—The Cotton Rat (692)—The Lemming (695).

CHAPTER II.
MOLE RATS, POUCH RATS, AND BEAVERS.

The Mole Rat (696)—The Jerboa (697)—The Alactaga (697)—The Cape Leaping Hare (697)—The Hudson Bay Jumping Mouse (698)—The Fat Dormouse (698)—The Common Dormouse (699)—The Pouched Rats (699)—The Beavers (701)—The American Beaver (702)—The European Beaver (704).

CHAPTER III.
THE SQUIRRELS AND MARMOTS.

The Family Sciuridæ (707)—The European Squirrel (707)—The Javanese Squirrel (708)—The Hare Squirrel (708)—The Black Squirrel (708)—The Gray Squirrel (708)—The Northern Gray Squirrel (709)—The Red Squirrel (709)—The Long-haired Squirrel (710)—The Flying Squirrel (710)—The American Flying Squirrel (711)—The Taguan (711)—The Chipmuck (712)—The Leopard Marmot (713)—The Marmot (714)—The Babac (715)—The Woodchuck (715)—The Prairie Dog (716).

CHAPTER IV.
THE SEWELLELS, PORCUPINES, AND CAVIES.

The Family Haploodontidæ (718)—The Family Chinchillidæ (718)—The Chinchillas and Viscachas (719)—The Octodontidæ (720)—The Hutia Conga (720)—The Degu (721)—The Tuko-tuco (722)—The Gundy (722)—The Coypu (723)—The Ground Pig (723)—The Canadian Porcupine (724)—The Tufted-tailed Porcupines (726)—The Agouti (726)—The Sooty Paca (727)—The Capybara (727)—The Guinea Pig (728)—The Mara (728)—The Pikas (729).

CHAPTER V.
HARES AND RABBITS.

The Family Leporidæ (730)—The American Hares (730)—The Polar Hare (730)—The Northern Hare (731)—The Wood Hare (731)—The Jackass Rabbit (731)—The African Hares (731)—The Sand Hare (732)—The Common Hare (732)—The Alpine Hare (733)—The Rabbit (733)—The Wild Rabbit (734)—The Domestic Rabbit (734).

EDENTATA.

CHAPTER I.
THE SLOTHS AND ARMADILLOS.

The Edentata (737)—The Sloths (737)—The Two-toed Sloth (738)—The Ai or Three-toed Sloth (738)—The Spotted Sloth (739)—The Scaly Ant-eaters (739)—The Phatagin (739)—The

Pangolin (740)—The Tatouhon (740)—The Giant Armadillo (740)—The Tatouay (741)—The Armadillo (741)—The Apar (741)—The Pichengo (742).

CHAPTER II.
THE AARD VARK AND ANT-EATERS.

The Aard Vark of the Cape (743)—The Great Ant-eater or Tamanoir (744)—The Tamandua (745)—The Little Ant-eater (746).

MARSUPIALIA.

CHAPTER I.
THE OPOSSUMS AND BANDICOOTS.

The Marsupials (749)—The True Opossum (749)—The Virginia Opossum (750)—Merrian's Opossum (750)—The Crab-eating Opossum (750)—The Yapock (750)—The Pouched Mouse (751) The Tasmanian Devil (751)—The Native Cat (751)—The Zebra Wolf (752)—The Native Ant-eater (752)—The Striped Bandicoot (752)—The Chæropus (753).

CHAPTER II.
THE KANGAROOS, PHALANGERS, AND WOMBATS.

The Kangaroo (754)—The Woolly Kangaroo (755)—The Wallabee (755)—The Rock Kangaroo (755)—The Tree Kangaroo (756)—The Kangaroo Hare (756)—The Jerboa Kangaroo (756) —The Potoroo (757)—The Koala (757)—The Sooty Phalangist (757)—The Valpine Phalangist (758)—The Cuscus (758)—The Taguan (758)—The Great Flying Phalanger (759)—The Sugar Squirrel (759)—Opossum Mouse (759)—The Wombat (760).

MONOTREMATA.

THE DUCK MOLE AND AUSTRALIAN HEDGEHOG.

The Monotremata (763)—The Family Ornithorhynchidæ (763)—The Duck Mole (763)—The Family Echidnidæ (765)—The Native Hedgehog (766)—The Tasmanian Species (766)—Conclusion.

LIST OF ILLUSTRATIONS.

VOLUME TWO.

Plate	Order		
XXX.....	VI.	Sirenia...	Frontispiece
XXXI.....	VII.	Ungulata.. To face p.	416
XXXII.....	VII.	Ungulata..	428
XXXIII.....	VII.	Ungulata..	442
XXXIV.....	VII.	Ungulata..	458
XXXV.....	VII.	Ungulata..	472
XXXVI.....	VII.	Ungulata..	486
XXXVII.....	VII.	Ungulata..	504
XXXVIII.....	VII.	Ungulata..	512
XXXIX.....	VII.	Ungulata..	526
XL.....	VII.	Ungulata..	535
XL.....	VII.	Ungulata..	544
XLII.....	VII.	Ungulata..	560
XLIII.....	VII.	Ungulata..	570
XLIV.....	VII.	Ungulata..	582
XLV.....	VII.	Ungulata..	588
XLVI.....	VII.	Ungulata..	598
XLVII.....	VII.	Ungulata..	605
XLVIII.....	VII.	Ungulata..	618
XLIX.....	VII.	Ungulata..	632
L.....	VII.	Ungulata..	648
LI.....	VIII.	Proboscidea..	666
LII.....	VIII and IX.	Proboscidea and Hyracoidea...................	672
LIII.....	X.	Rodentia..	688
LIV.....	X.	Rodentia..	696
LV.....	X.	Rodentia..	704
LVI.....	X.	Rodentia..	710
LVII.....	X.	Rodentia..	716
LVIII.....	X.	Rodentia..	722
LIX.....	X.	Rodentia..	728
LX.....	X.	Rodentia..	734
LXI.....	XI.	Edentata..	744
LXII.....	XII.	Marsupialia..	750
LXIII.....	XII.	Marsupialia..	756
LXIV.....	XIII.	Monotremata..	764

ORDER VI.

SIRENIA.

42. MANATIDÆ - - - MANATEES, SEA COWS.

THE SEA COWS.

THE ORDER SIRENIA—MERMAIDS—THE FAMILY MANATIDÆ—THE MANATES OF AMERICA—THEIR VORACITY AND LAZINESS—MODES OF CAPTURE.—TAME SPECIMENS—THE FLORIDA MANATEA—THE AFRICAN LAMANTIN—THE EASTERN DUGONG—THE AUSTRALIAN DUGONG—THE NORTHERN SEA COWS—STELLER'S DESCRIPTION—EXTINCT SINCE 1768.

THE SIRENIA, or Sea Cows, as they are commonly called, are regarded by some authorities as intermediate between the Eared Seals and the Whales, while others consider them as constituting a sub-family of Cetacea. But many parts of their structure exhibit so close an alliance with the thick-skinned, or pachydermatous mammalia, that modern investigators place them in a separate order, preceding immediately the orders which are sometimes united under the common appellation Pachydermata.

The name of Sirenia, which is given to this group of animals, is chiefly owing to the peculiar form and habits of the Dugong, which has a curious custom of swimming with its head and neck above the surface of the water, so that it bears some grotesque resemblance to the human form, and might have given rise to the poetical tales of mermaids and sirens which have prevailed in the literature of all Eastern countries. When the female Dugong is nursing her child, she carries it in one arm, and takes care to keep the head of her offspring, as well as her own, above the surface of the water, and thus presents a strangely human aspect. If alarmed, she immediately dives below the waves, and flinging her fish-like tail into the air, corresponds in no inadequate degree with the popular notions of mermaid form.

The Sea Cows are herbivorous animals, living on the coasts or in the great rivers of several parts of the globe. The nostrils are placed at the extremity of the muzzle, and are never used as blow-holes. They possess only the two fore-limbs, which are developed into fin-like flippers; but the skin so completely covers the fingers, that all separate movement of the joints is impossible, traces of nails being the only external indication of the internal division of the hand. The tail, representing the hinder-limbs, ends in a fin. One striking peculiarity exists in the form of the female Sirenia; this peculiarity is the position of the mammae, which are placed on the breast and between the fore-limbs, and which are more prominent than in the other marine mammals. It requires, however, very great powers of imagination to see in these clumsy and awkward creatures any resemblance to the human form divine; and perhaps it is the same power of imagination which has led to the assertion that these creatures live in strict monogamy.

The Sirenia are much more sea-animals than the seals are; they very seldom protrude their unwieldy bodies above the surface of the water, while their movement on dry land—on to which they never emerge voluntarily—is slow, and requires great exertion, as the fore-limbs are too weak to bear the weight of the body, which is much less flexible than that of the seals. They swim and dive excellently, but avoid deep water, preferring places where the marine plants, weeds and grasses, on which they feed, can be found. They are very voracious, and like all voracious animals, heavy, lazy, and dull. They do nothing but eat and sleep, and may therefore be described as peaceful and harmless. Both sexes display great mutual affection, and the female is a devoted mother, clasping the little one to her bosom with one of her fore-flippers. In danger or in pain, tears roll from their eyes, and they utter a weak, dull moan, which somewhat resembles that of a human being in pain.

The Sirenia constitute only one family, MANATIDÆ, which is divided into *three* genera.

I.—GENUS MANATUS.

The *two* species of this genus inhabit both shores of the Atlantic; one ranging from the gulf of Mexico to North Brazil, and ascending the Amazon River; while the other is found on the West Coast of Africa. The genus is distinguished from the following one by the shape of the tail-fin, which is round, and by the thick fleshy dirk which terminates the muzzle. The body is covered with short, thin hairs which become bristles on the muzzle; traces of four small nails can be detected on the flippers.

The MANATEE, *Manatus Australis* (Plate XXX), attains the length of nine feet, on the coast of Brazil. Our first accurate knowledge of the animal is due to the great traveler, Humboldt, who dissected one captured in the lower Orinoco. He observed that the Manatees prefer to linger in parts of the sea where fresh water springs arise, as in the Bay of Jagua in Cuba, and are very common in the Amazon, the Orinoco, and its tributaries. As all these southern rivers are rich in quiet nooks where water-plants of all kinds grow, they have no need to swim to any great distances; they eat ravenously, and when their hunger is satisfied, they lie in shallow places with the snout out of the water; they are thus saved the trouble of diving and rising, and sleep tranquilly for hours. During their waking time, they only come to the surface to breathe, but this emergence from the water occurs very often.

In all places where the Manatee is found, it is eagerly pursued. Its flesh is compared by Humboldt to pork, but is said to be unwholesome, and to produce fever. Salted and dried, it can be kept for a whole year, and like the flesh of the Cetacea, it is allowed as an article of food in Lent. The mode of taking the animal is very simple. A canoe approaches the feeding-grounds, and waits till one rises to take breath. As it appears, arrows with light wooden buoys fastened to them by cords are discharged at it, or it is harpooned. In the latter case an ingenious method of getting the body on board is adopted. The boat is filled two-thirds with water, and pushed under the dead Manatee, after which the water is baled out again by a calabash. The end of the inundations is a favorable season for its capture, and the Jesuits used to organize hunts on a large scale. The oil obtained from the carcass has not the offensive odor of train oil; the hide is manufactured into whips, from which the luckless Indians of the Missions used to suffer.

The Manatee is susceptible of domestication. The old traveler of the sixteenth century, Peter Martyr, writes: "A cacique in San Domingo has a little fish named Manato, which is quite tame, comes when called, eats bread out of the hand, allows itself to be stroked, and will carry people on its back across the pond in which it is kept." Gomara adds that it lived twenty-six years, crawled on dry land up to the house for food, and then back to the lake accompanied by boys, whose singing pleased it. It once carried ten people on its back across the water.

After a Spaniard had pricked it with his spear to see if its hide was as thick as people said, it would never come near any one in the Spanish dress. A planter in Surinam a few years ago trained a Manatee; he fed it on milk and bananas; it became quite tame, but displayed little intelligence, and its powers of sight and hearing were weak. If its owner stepped into the water, it would come and snuffle at his legs as a dog does, or if he sat down would climb on to his lap. It died after seventeen months of captivity. Dr. Cunningham has had, since 1867, two Manatees in the Public Gardens at Rio de Janeiro, where they were kept in company of some alligators and a number of waterfowl. One of them became very fond of a swan, which reciprocated the attachment, and they were always in company. This Manatee would eat grass out of the hand, and come halfway out of the water to reach the herbage which grew on the bank of the pool in which it was confined.

Some writers raise the Florida Manatee to the dignity of a species, *Manatus latirostris*; it seems to be, however, only a large variety, sometimes measuring fifteen feet or upwards. They used to abound in Tampa Bay, but are now rare. The one that was kept for some time in the Central Park, New York, was of this variety.

Columbus states that he saw three sirens dancing on the waters at Saint Domingo. Their lack of beauty, however, made him think that "they regretted their absence from Greece."

The LAMANTIN, *Manatus Senegalensis*, found on the African coast, is about eight feet long, having a conical head, round and very small eyes, with the iris of a very deep blue; a cylindrical muzzle, fleshy and thick lips, horizontal tail, and thick skin of an ashy-lilac color.

II.—GENUS HALICORE.

The Chinese and Arabians have for centuries been acquainted with this branch of the family. The old Greek writer Megasthenes speaks of creatures in the Indian seas which resemble women; an early Portuguese surgeon professes to have dissected a "mermaid"; and the Dutch traveler Valentyn describes "sea-wives"; in all of which cases one of the two species of *Halicore* is doubtless meant.

The DUGONG, *Halicore Dugong* (Plate XXX), is found in all parts of the Indian Ocean, and of the seas and straits connected with it; it abounds in the southern parts of the China Seas, in the Straits of Banda and Sunda, and extends northward into the southern half of the Red Sea. It is always found near the coast where marine vegetation abounds; it enters occasionally the mouths of rivers, but does not sojourn in the rivers themselves; it does not come up on the land, and any dugongs that may have been seen lying on the shore, have doubtless been left high and dry by the receding tide. It rises to take breath once a minute, showing its muzzle, or even half his body, and then it sinks slowly and steadily back again into the water.

The Dugongs are said by the Eastern fishers to live in pairs; but this is not generally true, as they have been seen in schools in the Indian Ocean, as well as in the Red Sea. Their form is not well adapted for moving through the water; the snout is of an obtuse truncated character; and the tail is proportionately greater than in the Cetacea, its breadth being rather more than one-third of the length of the body; they do not possess the blood-reservoirs which enable the seals to survive for a long time beneath the water; and they are distinguished from the Manatee by their flippers, which have no nails, while the two external incisor teeth of the upper-jaw are elongated into a sort of tusks.

They are caught either in nets, or by spears. The natives on the Malay coast spear them at night, their presence being known by the snuffling noise they utter. When caught, the tail is raised out of the water, in which position the animal is quite powerless. The flesh is said to be good eating, with a peculiar sweet taste, and its skin is manufactured into various useful articles.

The Dugongs of the Red Sea are regarded by the German naturalist Ruppell as constituting a separate species, which he calls *Halicore tabernaculi*, from a notion that the Jews used its hide for covering the tabernacle. The AUSTRALIAN DUGONG, *Halicore Australis*, is undoubtedly a distinct species, and it is eagerly hunted for the sake of the oil which it yields, to which are attributed the same virtues as to cod-liver oil.

III.—GENUS RHYTINA.

This genus is supposed to be now extinct. The celebrated naturalist Steller was stranded, in November, 1741, on Behring Island, and spent ten weary months there. He writes: "Along the whole coast of the island, especially where streams enter the sea and sea-weeds grow, there is found at all times of the year the animal called by the Russians *Morskaja Korowa*, or Sea-Cow. As food from other quarters began to be scarce, we resolved to catch some of these animals. I made my first attempt with a great iron hook provided with a strong and long rope. The hook, however, was too blunt, and the creature's hide too tough. I then tried harpooning. We repaired our jolly-boat, and sent it out with a harpooner; the harpoon was connected to a rope which was passed ashore, and held there by forty men of our crew. The boat was rowed quietly up to the animals as they were feeding near the shore. When one of them had been struck, we drew it gradually to land, stabbing it with bayonets and knives till it lost nearly all its blood, and at high water we made it fast on the strand. When the water had receded we cut off the flesh and preserved it in barrels. The largest of these animals measured four to five fathoms, with a girth of three and a half fathoms. The skull, when stripped of the flesh, was, in its general conformation, not unlike that of a horse; but when the hide was on it, resembled more that of a buffalo, especially in the lips. In place of teeth there were on each side two broad, long, smooth, spongy bones provided with protuberances and furrows which formed sharp angles. The lips were furnished with strong bristles, nearly as thick as the quills of chickens' feathers; the eyes were as large as a sheep's; and had no eyelids; the aperture of the ears was very small, and invisible till the skin was removed; there was no external ear. The head is united to the body by a very short neck; the fore-limbs consist of two joints, the extremity somewhat resembling the hoof of a horse; no nails or fingers could be seen. With these fore-limbs it tears up the sea-weed from the rocks. Below them the breasts are situated, the nipples being black, and two inches long, and abundantly supplied with lacteal ducts. When pressed, these teats discharge a great quantity of milk, which surpasses that of terrestrial animals in sweetness and richness. The back is like that of an ox; the tail is horizontal, as in the whale."

"These animals live in herds in the sea; usually a male and female go together, with the young one before them. The back and half of the body is frequently seen above water. They eat like cows as they slowly advance, scratching the weeds up with their fore-feet, and chewing unceasingly, but they do not ruminate. They care for nothing but eating. While eating, they move their heads like oxen, and after every few minutes, raise their heads from the water, and snort like a horse. When the tide ebbs, they retire into the sea, but return as it rises, approaching the land so closely that they can be struck from the shore. They have no fear of man, and have little trace of intelligence, their most admirable quality being their affection for each other, which is so great, that when one was struck, the others tried to rescue it. Some formed a line, and attempted to keep their wounded comrade from the shore; others tried to upset the jolly-boat, while others exerted themselves to knock the harpoon out, a feat they accomplished successfully in several cases. We observed that a male came for two days in succession to the shore where his dead wife was lying, as if to ascertain her condition. They sometimes rest their heads on the land, leaving the body floating like a log on the water.

"They are found at all times of the year in great abundance around Behring Island, and all the inhabitants of the East Coast of Kamtschatka are thus supplied with plenty of meat and fat. The skin of the Sea-Cow is peculiar; the exterior layer is black or black-brown, an inch thick and firm; it consists of vertical fibres lying close to each other. This exterior layer, which can be easily scaled off, seems to me to be formed of hairs modified in an extraordinary manner. The inner skin is somewhat thicker than a cow's, strong, and white in color. Beneath these two skins the whole body is enveloped with a layer of fat four fingers deep, then comes the flesh. The fat is not oily, but hard; and, after exposure to the sun, becomes as yellow as the best butter; the tail is a mere mass of fat. The flesh of the calves is like veal, that of the old ones like beef; it has the peculiar quality that it can remain, even in summer, exposed to the open air for the space of two weeks without giving forth any bad smell, although it may be covered with flies and maggots; the flesh is very red in color, almost as if it had been rubbed with saltpetre. It was very nutritious, and soon banished the scurvy, from which some of our men were suffering."

We have given Steller's description at some length, as no other exists or will ever be given, for, since the year 1768, no trace of this animal in a living condition has ever been recorded. There can be no doubt, however, that considerable numbers once existed; and these have all fallen a prey to the rapacity of the Aleutian fishermen, and hunters of the sea-otter.

Attracted by the reports of the Russian exploring expedition in which Steller was engaged, whalers and adventurers of all kinds flocked to Behring Island, and commenced such a terrible butchery among the defenceless creatures, that not a single specimen of the animal can now be discovered. All attempts have been in vain to recover even a fragment of the Sea-cow of Steller. Every ship which set sail for the Island of Behring was requested to keep a look-out for them, but no news was ever brought back. Steller regarded this Sea-cow as identical with the Lamantin of Hernandez. It is clear, from his description, that the animal seen by him was a creature perfectly different from any of the Sirenia mentioned by earlier writers. It is well for science that Steller was among the unfortunate voyagers who were shipwrecked on Behring Island, and still more fortunate that he left so perfect a description of his discovery. His narrative was first published in 1749, at St. Petersburg, after his death. The animal seems not to have been abundant even at the time of Steller's voyage, and our destructive race, without giving naturalists an opportunity of unraveling its structure, have swept it from its native shores, and well-nigh obliterated all traces of its existence.

This sea-cow deservedly bears the name of the naturalist who discovered it, and is called RHYTINA STELLERI; it has, however, been described under the generic appellations of *Manatus* and *Trichechus*, while in the British Museum Catalogue of Cetacea it is alluded to under the name of *Rhytina gigas*. Like the Dodo and the Great Auk, Steller's Sea-cow is one of the animals which civilized man has destroyed in the mere wantonness of power.

ORDER VII.
UNGULATA.

43. EQUIDÆ - - - - - - - Horses.
44. TAPIRIDÆ - - - - - - - Tapirs.
45. RHINOCEROTIDÆ - - - - Rhinoceros.
46. HIPPOPOTAMIDÆ - - - - Hippopotamus.
47. SUIDÆ - - - - - - - - Swine.
48. CAMELIDÆ - - - - - - - Camels.
49. TRAGULIDÆ - - - - - - Chevrotains.
50. CERVIDÆ - - - - - - - Deer.
51. CAMELOPARDALIDÆ - - Giraffes.
52. BOVIDÆ - - - - - - - - Cattle, Sheep, Antelopes.

CHAPTER I.

HOOFED ANIMALS.

THE ORDER UNGULATA—THE NUMEROUS FAMILIES—THE RUMINANTS—THEIR PECULIAR STOMACH—HORNS—ANTLERS—EXTINCT SPECIES—THE ORIGINAL HORSE PROTOHIPPUS—GRADUAL DEVELOPMENT—THE FAMILY EQUIDÆ—THE GENUS EQUUS—THE HORSE—THE TARPAN OR WILD HORSE OF TARTARY—THE MUSTANG OR WILD HORSE OF AMERICA.

WITH the exception of the Australian region, in which this order of mammalia is almost entirely wanting, the UNGULATA are well distributed over the world; they are the dominant vegetable feeders of the great continents, and are of larger size and less activity than the CARNIVORA. Among them are the most valuable and most domestic animals which man possesses, as the horse and the camel, the ox and the sheep. Among them are the strongest, as well as the most timid, the rhinoceros and the hippopotamus, the giraffe and the tapir, the antelope and the deer.

The order UNGULATA, from the Latin *unguis*, "a hoof," embraces *ten* extensive families. Many subdivisions have been proposed; one method is based on the number of toes, and whether they are odd and even; by it the families *Equidæ, Tapiridæ, Rhinocerotidæ* are classed as *Perissodactyla* or "odd-toed"; the remaining families, as *Artiodactyla*, or "even-toed." By another classification those animals that "divide the hoof and chew the cud" are taken from the UNGULATA and formed into an order RUMINANTIA. Among these latter are those domesticated animals which are especially adapted for human food.

With the exception—in some instances—of the *Suidæ* or Swine, all the families in this order are herbivorous, and the molar teeth have hard crowns adapted for grinding vegetable substances. In the Ruminant animals, the typical dentition consists in the absence of incisor and canine teeth in the upper jaw, while the lower jaw holds six incisors and two canines, which are all similar in size and form. There are six molars on each side of each jaw. The lower incisors bite against a callous pad of gum. But, more remarkable than the arrangement of the teeth, is the structure of the stomach in the cud-chewing families. They first swallow their food unmasticated, and then bring it up again to chew it. This regurgitation is effected as follows: the stomach is divided into compartments, of which the largest, lying to the left, is called the *rumen* or "paunch," and receives the food. Here it remains soaking for some time, and is then passed into the *reticulum* or "honey-comb" bag, where it is made into little balls or pellets which are returned to the mouth by a reverse action of the muscles. In the mouth the food is thoroughly chewed and then swallowed a second time, passing not into the paunch, but into the *psalterium* or "manyplies." From the "manyplies" a wide aperture leads into the *abomasum* or fourth stomach, where the gastric juice is secreted, and digestion completed.

Among the ruminants alone are found animals which possess those appendages that are usually called horns. But these horns are of two distinct kinds. The true Deer bear on their forehead two solid, bony antlers, which, except in the Reindeer, are confined to the males. These antlers are deciduous, and are shed annually; they increase in size every time they are reproduced. Oxen, sheep, goats, and antelopes have true horns, consisting of a horny sheath surrounding a central bony axis; these horns are persistent, that is, they are not shed. In the antelope, the horns are compact, without cells; in the goat, ox, and sheep they have cells which communicate with the nose.

Wallace remarks that the present distribution of the *Ungulata* is utterly unintelligible without reference to the numerous extinct forms of existing and allied families; he adds, that we have good evidence that their wide range over the globe is a comparatively recent phenomenon. Tapirs and Llamas have probably not long inhabited South America, while Rhinoceroses and Antelopes were once perhaps unknown in Africa, although abounding in Asia and Europe. Swine are the most ancient types in both hemispheres, and their great hardiness, omnivorous diet, and powers of swimming, have led to their wide distribution. The sheep and goats, on the other hand, are perhaps the most recent development of the Ungulata, and seem to have arisen in the Northern region of the Eastern hemisphere, when the climate approximated to that which now prevails in the same regions.

The animals belonging to this order being usually of larger size, and accustomed to travel in herds, are liable to wholesale destruction by floods, bogs, precipices, drought, or hunger. Hence their remains are exceedingly numerous in the older geological strata. Such fossil remains are especially abundant in America. The horse is peculiarly interesting. When Columbus and the Spaniards landed, it was entirely unknown on this continent; but in the earlier ages of the world's history, horses of all kinds must have roamed over our plains. Among these was the *Protohippus*, only two feet and a half high, with the lateral toes not externally developed, the *Mesohippus* and *Anchitherium*, about the size of a sheep with three toes used for locomotion, but still unmistakably equine; and in the deposits of Utah and Wyoming, species have been discovered about the size of a fox, with four toes in front and three behind. These form the genus *Orohippus*, the oldest ancestral horse known. America thus possesses a perfect series of forms which, beginning with this minute ancient type, is gradually modified by gaining increased size and increased speed, by concentration of the limb-bones, by elongation of the head and neck, by the canine teeth decreasing in size, and by the molars becoming larger and being coated with cement till we at last come to animals hardly distinguishable, specifically, from the living horse with which we are all acquainted.

The family EQUIDÆ, as befits the beauty and utility of the animals embraced in it, stands at the head of the order. It contains *one* genus, which is divided into *eight* species; among which are *four* species of Asses, and *three* of Zebras.

GENUS EQUUS.

The members of this genus are often styled *Solipedes* or "solid-footed," as they have only one apparent toe on each foot, which is enclosed in a single hoof. They have, however, under the skin, the rudiments of lateral toes. They have six incisor teeth in each jaw, small canines exist in the males, but are wanting in the females. Between the canines and the molars there is a space where the bit is placed, an arrangement by which alone man has been able to subdue these vigorous animals. They all attain a moderate size, and possess a graceful figure with strong limbs; their head is long, the eyes vivacious, the nostrils expanded, the ear small, and the hair is short and thick, forming a mane on the neck.

THE HORSE.

The HORSE, *Equus caballus* (Plate XXXI), the noble companion of man in the battle and the chase, in the labors of agriculture and commerce, is distinguished by the uniformity of his color, and by his tail being ornamented with long hair throughout its entire length. Wild horses exist both in Tartary and America, where they live in troops, conducted and defended by an old male. We know that in the latter, these herds are all descended from imported ancestors, and many authorities consider that those that roam over the steppes of Asia are likewise the offspring of domesticated parents.

THE TARPAN.

The TARPAN, as the wild horse of Tartary is called, is considered by the Cossacks and Tartars to be a wild animal in the proper sense of the word. It is small, with powerful legs, long pasterns, a long thin neck, a thickish head, pointed ears inclined forward, and small, lively, wicked looking eyes. A light or yellowish brown is the prevailing color of the summer-coat, but in winter the hair becomes lighter, at times even white. The mane is bushy, and, like the tail, is uniformly dark. Dappled or black horses are rare.

The Tarpans live in herds, numbering several hundreds; usually the larger herds are divided into smaller family groups, each under an old stallion; they live in the open steppes, but are not found in the desert of Gobi, or in the highlands north of India. They migrate from place to place, traveling against the wind, are very watchful and shy, and recognize in good time the approach of danger. The leading stallion is the autocrat of the troop; at the sight of anything suspicious he raises his head, advances, and when he thinks it advisable, neighs the signal for flight. These herds of horses have no dread of beasts of prey, they attack wolves without fear, and strike them down with their fore-feet. In the combats between the males, teeth as well as feet are used, and the young ones have to establish their right to found a family by a series of desperate duels with their seniors of the troop.

The appearance of one of these herds as they sweep over the plains, and career around any intruder on their domain, is best described in the well-known lines of the poet Byron:

> "A trampling troop! I see them come
> In one vast squadron they advance!
> I strove to cry—my lips were dumb.
> The steeds rush on in plunging pride.
> But where are they the reins who guide?
> A thousand horse and none to ride!
> With flowing tail and flying mane,
> Wide nostrils—never stretch'd by pain—
> Mouths bloodless to the bit or rein,
> And feet that iron never shod,
> And flanks unscarr'd by spur or rod;
> A thousand horse, the wild, the free,
> Like waves that follow o'er the sea.
> On came the troop....
> They stop—they start—they snuff the air,
> Gallop a moment here and there,
> Approach, retire, wheel round and round,
> Then plunging back with sudden bound,
> They snort, they foam, neigh, swerve aside,
> And backward to the forest fly."

The inhabitants of the steppes dread the Tarpans, on account of the injury they do to their haystacks, and because the male Tarpans run away with their mares. The Cossacks are accustomed to hunt the wild horses, partly to keep up their own stock, and partly for food. A species of vulture is sometimes made use of in this affair. The bird pounces upon the poor animal, and fastens itself on his head or neck, fluttering his wings, and perplexing and half-blinding him, so that he becomes an easy prey to the Tartar. The young horses are generally tamed without much difficulty; they are, after a little while, coupled with a tame horse, and grow gentle and obedient. The wild horses thus reclaimed are usually found to be stronger and more serviceable than any which can be bred at home, in the Tartar villages.

Pallas, the great Russian traveler, considers these wild horses to be descended from those employed at the siege of Azof in 1657, which were turned loose by the army, owing to want of forage.

THE MUSTANG.

The MUSTANG (Plate XXXI) is the wild horse of the American prairies and pampas. The city of Buenos Ayres was abandoned by the Spanish colonists soon after its settlement in 1535, and about half-a-dozen horses were left behind. The city was again occupied in 1580, and the new-comers saw, to their astonishment, that the neighborhood was swarming with wild horses. The descendants of these Spanish derelicts are now spread in every part of the pampas, and have been seen in troops of ten thousand. They possess much of the form of the Spanish horse, but are not speedy; they are capable of enduring immense fatigue, and are frequently ridden sixty or seventy miles without drawing bit, while they have been known to be urged on by the cruel spur of the Gauchos more than a hundred miles a day, at the rate of twelve miles an hour. They know no pace between the walk and the gallop, and at the end of the day's journey are turned loose on the plains. The mares are never ridden, and when the Gaucho or native Indian of the South American plains wants a horse, he sets out armed with his lasso, mounted on a horse that has been used to the work. He gallops alongside a troop of the wild horses, and as soon as he comes sufficiently near his prey, the lasso is thrown round the two hind-legs, and as the Gaucho rides a little on one side, the jerk pulls the entangled horse's feet laterally, and throws him on his side, without endangering his knees or his face. Before the horse can recover the shock, the rider dismounts, and snatching his *poncho* or cloak from his shoulders, wraps it round the prostrate animal's head. He then forces into his mouth one of the powerful bridles of the country, straps a saddle on his back, and bestriding him, removes the poncho; upon which the astonished horse springs on his legs, and endeavors by a thousand vain efforts to disencumber himself of his new master, who sits quite composedly on his back, and, by a discipline which never fails, reduces the horse to such complete obedience, that he is soon trained to lend his whole speed and strength to the capture of his companions.

When the Gauchos have a grand breaking-in, a whole herd is driven into the corral. A young horse is lassoed by the neck and dragged out; some men on foot lasso his fore-legs and throw him; in an instant a Gaucho is seated on his head, and, with his long knife, in a few seconds cuts off the whole of the horse's mane, while another cuts the hair from the end of his tail; this is a mark that the horse has been once mounted. They then put a piece of hide into his mouth to serve for a bit, and a strong hide halter on his head. The Gaucho who is to mount arranges his spurs, which are unusually long and sharp, and while two men hold the horse by his ears, he puts on the saddle, which he girths extremely tight. He then catches hold of the horse's ear, and in an instant vaults into the saddle; upon which the man who holds the horse by the halter throws the end to the rider, and from that moment no one takes any further notice of him.

"The horse instantly began," Sir F. Head writes, "to jump in a manner which made it very difficult for the rider to keep his seat, and quite different from the kick or plunge of an English horse; however, the Gaucho's spurs soon set him going, and off he galloped, doing everything in his power to throw his rider.

"Another horse was immediately brought from the corral; and so quick was the operation, that twelve Gauchos were mounted in a space which I think hardly exceeded an hour. It was wonderful to see the different manner in which different horses behaved. Some would actually scream while the Gauchos were girding the saddle upon their backs; some would instantly lie down and roll upon it; while some would stand without being held, their legs stiff and in unnatural positions, their necks half bent towards their tails, and looking vicious and obstinate; and I could not help thinking that I would not have mounted one of those for any reward that could be offered me, for they were invariably the most difficult to subdue.

"It was now curious to look around and see the Gauchos on the horizon in different directions, trying to bring their horses back to the corral, which is the most difficult part of their work, for the poor creatures had been so scared there that they were unwilling to return to the place. It was amusing to see the antics of the horses; they were jumping and dancing in different ways, while the right arm of the Gauchos was seen flogging them. At last they brought the horses back, apparently subdued and broken in. The saddles and bridles were taken off, and the young horses trotted off towards the corral, neighing to one another."

In the *Estancias*, or establishments dedicated to the breeding of cattle on the pampas, large numbers of these horses are required, and in order to secure a good supply of colts, to make good the wear and tear of the year, herds of brood mares are kept, which yield an annual supply. The small herds are called *manadas*, and consist of from twenty-five to thirty mares, and over them presides the father of the family. They are parted off from larger herds, and are shut up in the corral for the night, where at first they seem to feel their separation from their old friends and companions. At sunrise they are let out, and a peon, or Indian servant is appointed to assist the horse to keep them together. Presently one jade bolts away at full speed, and tries to regain her old accustomed herd. Away goes the horse in chase, and as he overhauls her, with his ears laid back, and his nose to the ground, he compels her return quicker than she went away. Before he has got well *breathed*, another starts off in the opposite direction. After her goes the horse, and the jilt is brought back at the top of her speed. This continues nearly all the day, and at night they are again locked up. Next day the same game goes on, the horse still assisted by a well-mounted peon, until, at the end of eight or ten days, their acquaintance is made. If any remain refractory, they are thrown down, the axe applied to one of the hoofs of the hind-legs; it is cut to the quick, and, thus punished, she is compelled to *limp it* for fifteen or twenty days, until the hoof grows again. By that time she becomes reconciled, and the *manada* is established, and keeps together. They are then conducted to a quiet part of the ground, apart from the other herds, in order to prevent collision between the different families.

Apart from these herds of horses already described, there are small troops of horses accustomed to go together in company with a mare selected for that purpose, and called the *madrina*. These are reserved for particular service, and are called *tropillas*. The horses, which generally consist of ten or twelve, are selected of one color, and the mare, to render her more distinguishable, is as different as possible from the horses. For instance, a tropilla of black horses will generally have a white mare; she wears a bell suspended from her neck, so that the horses at night may hear the sound, and prevent them from parting company.

Such is the life of the wild steeds that run over the wide plains of the Argentine Republic, Paraguay, and Central Brazil. The Mustangs north of the Isthmus of Panama are derived from Mexican horses which have escaped into the woods and savannahs, and roamed northward to the Rocky Mountains. The Indians have learned to capture them, and employ them in transporting their families from place to place. The highest ambition of the young Indian is to possess a good horse; and to steal a horse is almost as glorious as to scalp an enemy. The Indian pony, as it is called, is barely fourteen hands high, rather light built, with good legs, straight shoulders, short, strong back, and full barrel. He has no appearance of "blood" except sharp, nervous ears, and intelligent eyes. He is never stabled, washed, rubbed, shod, nor fed; in winter he is an animated skeleton, sustaining a bare existence on cottonwood branches; but when spring brings out the tender grass, he is soon in condition " worthy to be trusted to the death." After endurance, the best quality of the Indian pony is his sureness of foot. He climbs rocks like a mule, he plunges down precipices like a buffalo, and crosses swamps like an elk. The amount of work got out of him by the Indians is considerable, and in Indian hands and with Indian bits he is tractable. But he does not receive civilization any better than his master. According to General Dodge, " he is either a morose, ill-tempered brute, hard to manage, and always dangerous, or he degenerates into a fat, lazy, short-breathed cob, fit only for a baby or an octogenarian. Prosperity spoils him, and his true character and value are best displayed in adversity." Among the Indians a " pony " is the standard of value by which the price of wives and other chattels is fixed.

CHAPTER II.

THE ARAB AND THE BARB.

EARLY DOMESTICATION OF THE HORSE—THE HORSE IN EGYPT—ASSYRIA—JUDEA—GREECE—PERSIA—BITS AND STIRRUPS—CHARIOT RACES—THE ARAB HORSE—EXAGGERATED PEDIGREES—THE BEST ARABS—THEIR TRAINING—ATTACHMENT OF THE ARAB FOR HIS MARE—SPEED AND ENDURANCE—THE BARB—THE SAME HORSE AS THE ARAB—ABD-EL-KADER ON THE HORSE.

NO historian tells us what nation first subdued the horse. Pictures of the horse are not found on the monuments of Egypt till the eighteenth century before the Christian era, when the Pharaohs of the "New Kingdom" employed them to draw their war-chariots. As these animals are represented well-trained and harnessed, they must have been domesticated for several generations previously, unless we suppose that they were introduced from Asia by the nomad tribes who visited the delta of the Nile. But from whatever source they derived their horses, the Egyptians were the first to give systematic attention to improving the breed: the symmetry, beauty, and color of the steeds represented on the ornaments show that they were designed from high-bred types. Grooms are represented as rubbing their joints, and sedulously attending to their comfort. If we turn from the sculptures of Egypt to the bas-reliefs of Assyria, we find the horse again serving for warlike purposes, and harnessed to the chariot. The Jews, according to the Book of Deuteronomy, were prohibited from breeding or importing horses. At all events, they seem not to have used them for five hundred years after their migration from Egypt; Solomon, however, possessed fourteen hundred chariots, and stabling for forty thousand horses, most of which were imported from Egypt. In the Homeric poems, the oldest Greek records, we again find the horse, used solely for drawing chariots. Mounted cavalry is not heard of till comparatively a late period. The Persians were the first to become celebrated for their

MUSTANGS   ARABIAN HORSE   SHETLAND PONY

PLATE XXXI UNGULATA

mounted troops, and every Persian boy was taught "to ride, to shoot, and to speak the truth." The provinces of the Persian Empire contained many excellent breeds of horses, Armenia and Cappadocia being the most famous in this respect. Cyrus, the founder of the Persian power, devoted himself to the improvement of native horses, and declared it ignominious for a Persian to be seen in public except on horseback. The Persian horseman mounted on the right side, and his horse was carefully taught to have a high action, and was ridden by a halter. We first see the bit in some of the Grecian sculptures, and this Greek bit is the common snaffle of the present day. To their great credit, the Greeks never rode with a curb. The Romans invented a cruel bit with sharp projections like the teeth of a wolf. No mention is made by ancient writers of saddles such as are used in modern times; the back of the horse was covered with a cloth or a skin, which was kept in its place by a surcingle. Stirrups, too, were unknown; in fact, they are first mentioned in the year 1158 A. D., but are represented in the Bayeux tapestry executed nearly a century earlier, and representing William the Conqueror's invasion of England. The ancient heroes trusted chiefly to their own agility in leaping on their horses' backs; sometimes they climbed up by the aid of the loop of their lance, sometimes the horse was taught to kneel; wealthy men were assisted to mount by their slaves. Shoes of metal were not fixed to the hoof by nails, but both the Greeks and Romans were accustomed to fasten on the foot a sort of shoe of leather, which was sometimes strengthened with a plate of iron, or adorned with silver or gold.

Next to war, the race was the chief field on which the ancients exercised their horses. The favorite race was the chariot race, each chariot having four horses attached to it; the circus was about one-third of a mile in length, and six laps formed the course. In the later Roman Empire and the Empire of Constantinople, the passion for the circus reached such a point that the history of the city became a history of the struggle between the *blue* and *green* parties, as they were called, from the colors of their drivers. This was at a time, too, when the use of chariots in war had fallen into disuse, and when cavalry was the important part of the army.

The Romans were the first to introduce the custom of employing horses in ordinary agricultural or mercantile pursuits, and they have from that time, in every part of Europe and North America, gradua' become one of the most useful of the servants of man. In the East the practice does not prevail, and the Arabs say that a horse is degraded if employed in tillage or carrying burdens. Abd-el-Kader, the great Algerian chief, who for so many years withstood the French and baffled the French generals, has written on the horse, and tells us: " My father —may God receive him in mercy—was accustomed to say, ' No blessing on our country since we have changed our coursers into beasts of burden and tillage. Has not God made the horse for the race, the ox for the plough, and the camel for the transport of merchandise? Nothing is gained by changing the ways of God.'"

M. Toussenel exclaims: "Tell me what sort of horse a people have, and I will tell you the manners and institutions of that people. The history of the horse is the history of the human race, for the horse is the personification of the aristocracy of blood. Every revolution which exalts the people, abases the horse. The horse is one of the first conquests of man, and is one of the pivots on which the very existence of a patriarchal tribe depends. Some day or other the tribe of horsemen conquers its horseless fellows, and organizes itself so as to ensure a firmer settlement on the conquered territory. The first step is to ennoble the horse; in other words, to establish a feudal system. The highest functionary in a feudal kingdom was the Constable, the 'Count of the Stable,' or the Marshal, the 'Horse-servant'; the lesser nobles are *chevaliers*, that is, horsemen (from the French *cheval*, 'a horse'), or *ritters*, the men who ride. But gunpowder has killed both the horse and the feudal system." The remarks of M. Toussenel are decidedly fanciful, but he states the truth when he says, "There is only one horse in the world—that is, one real horse—the Arab."

THE ARAB HORSE.

The ARAB HORSE (Plate XXXI). The horse of the Egyptian monuments closely resembles the Arab, but no ancient writer celebrates any breed of steeds belonging to that peninsula. In the seventh century of our era the Arabs had few horses, and it is said that when the Prophet attacked the Koreish near Mecca he had only two in his whole army, nor does he enumerate a horse among the spoils of his bloody campaign. It is only since the thirteenth century that the Arab horse began to obtain its unrivaled celebrity. There are now three breeds recognized by the Arabs, the *Attechi*, an interior breed which is found wild in some places, the *Kadischi*, or mixed breed, and the *Kochlani*, or pure breed. Till the last century, the pedigrees of these horses used to be traced to the five which bore Mohammed and his companions on the flight to Medina; at present, no Arab chief is happy unless he can trace the descent of his favorite steed from one of the stud of King Solomon. The Bedouins themselves, however, never make use of any written pedigree, and, according to Abd-el-Kader, the Arabs of Algiers care nothing for written registers, the genealogy of the blood-horses being as well known as that of their masters. The province of Nedjed adjoining the Syrian desert is said to produce the noblest horses of the present day, the pasture there is rich and succulent, and unrivaled for its nutritious and aromatic properties, and it is a mistake to suppose that the Arabian is reared in an arid desert, and owes his powers of endurance to early hardships. Another error is to suppose that horses are numerous in Arabia; Burckhardt estimates the total number at fifty thousand. Still fewer, comparatively, are those of perfect quality and beauty, probably not two hundred in the whole desert. A true Arab of Arabia never mounts a stallion, and never parts with his mare.

The owner of a mare bestows great pains in seeking out for her a mate of unblemished descent. The mare and foal live in the tent with the Bedouin and his children, who roll about with her and her foal; no accident ever occurs, and the animal becomes a loving friend. At the end of a month the foal is weaned, and is fed for three months on camel's milk; then a little wheat is allowed for another three months; at the expiration of this time, the young animal is allowed to graze near the tent, and some barley is given it. The kindness with which the Arab is treated from her very birth gives her an affection for her master, a wish to please, a pride in exerting every energy, and an apparent sagacity, which is seldom seen in other breeds.

When the rider falls from his mare, and is unable to rise, she will immediately stand still, and neigh until assistance arrives. If he lies down to sleep, as fatigue sometimes compels him, in the midst of the desert, she stands watchful over him, and neighs and rouses him if either man or beast approaches. The Arab horses are taught to rest occasionally in a standing position; and a great many of them never lie down.

The Arab loves his horse as truly and as much as the horse loves him; and no little portion of his leisure time is often spent in talking to and caressing his faithful steed.

An old Arab had a valuable mare that had carried him for fifteen years in many a rapid weary march, and many a hard-fought battle; at length, eighty years old, and unable longer to ride her, he gave her and a scimitar that had been his father's to his eldest son, and told him to appreciate their value, and never lie down to rest until he had rubbed them both as bright as a mirror. In the first skirmish in which the young man was engaged, he was killed, and the mare fell into the hands of the enemy. When the news reached the old man, he exclaimed that "life was no longer worth preserving, for he had lost both his son and his mare, and he grieved for one as much as the other." He immediately sickened, and soon afterward died.

The following anecdote of the attachment of an Arab to his mare has often been told: "The whole stock of an Arab of the desert consisted of a mare. The French consul offered to purchase her, in order to send her to his sovereign, Louis XIV. The Arab would have rejected the proposal, but he was miserably poor; he had scarcely a rag to cover him, and his wife and his children were starving. The sum offered was great it would provide him and his family with food for life. At length, and reluctantly, he yielded. He brought the mare to the dwelling of the consul, dismounted, and stood leaning upon her; he looked now at the gold, and then at his favorite. 'To whom is it,' said he, 'I am going to yield thee up? To Europeans, who will tie thee close, who will beat thee, who will render thee miserable. Return with me, my beauty, my jewel, and rejoice the hearts of my children.' Thus speaking, he sprang upon her back, and was out of sight in an instant."

The Arab horse would not be acknowledged by every judge to possess a perfect form. The head, however, is inimitable. The broadness and squareness of forehead, the smallness of the ears, the prominence and brilliance of the eye, the shortness and fineness of the muzzle, the width of nostril, the thinness of the lower jaw, and the beautifully developed course of the veins, are always striking characteristics. The body is too light, and the chest too narrow, but the barrel is roomy, the neck is long and arched, and beautifully joined to the chest, the withers are high, the shoulder-blade well sloped, and though covered with plenty of muscle, has no appearance of heaviness.

The fineness of the legs and the oblique position of the pasterns might be supposed by the uninitiated to lessen his apparent strength, but the leg, although small, is deep, and composed of bone of the densest character. The tendons are sufficiently distinct from the bone, and the starting muscles of the forearm and the thigh indicate that it is fully capable of accomplishing many of the feats that are recorded of it.

The young Arab commences its career with a severe trial. Probably the filly has never before been mounted. Her owner springs on her back, and goads her over the sands and rocks of the desert for fifty or sixty miles without one moment's respite. She is then forced, steaming and panting, into water deep enough for her to swim. If, immediately after this, she will eat as if nothing had occurred, her character is established, and she is acknowledged to be a genuine descendant of the *Koehlani* breed. The Arab does not think of the cruelty which he thus inflicts; he only follows an invariable custom.

We may not, perhaps, believe all that is told us of the speed and endurance of the Arabian. It has been remarked that there are on the deserts which this horse traverses no mile-stones to mark the distance, or watches to calculate the time; and that the Bedouin is naturally given to exaggeration, and, most of all, when relating the prowess of the animal that he loves as dearly as his children. But it cannot be denied that, at the introduction of the Arabian into the European stables, there was no horse comparable to him. The mare in her native deserts will travel fifty miles without stopping; she has been urged to the almost incredible distance of one hundred and twenty miles, and occasionally, neither she nor her rider has tasted food for three whole days.

Our horses would fare badly on the scanty nourishment afforded the Arabian. The mare usually has but two meals in twenty-four hours. During the day she is tied to the door of the tent, ready for the Bedouin to spring, at a moment's warning, into the saddle; or she is turned out before the tent ready saddled, the bridle merely being taken off, and she is so trained that she immediately gallops up at her master's call. At night she receives a little water; and with her scanty provender of five or six pounds of barley or beans, and sometimes a little straw, she lies down content, if she is accustomed to lie down at all, in the midst of her master's family.

Burckhardt relates a story of the speed and endurance of one of them, which shows with what feelings an Arab regards his quadruped friend: "A troop of Druses on horseback attacked, in the summer of 1815, a party of Bedouins, and pursued them to their encampment; the Bedouins were then assisted by a superior force, and becoming the assailants in their turn, killed all the Druses except one, who fled. He was pursued by some of the best mounted Bedouins, but his mare, although fatigued, could not be overtaken. Before his pursuers gave up the chase they called to him, and begged to be permitted to kiss his excellent mare, promising him safe conduct for her sake. He might have taken them at their word, for the pledge of an Arab, in such circumstances, might have been relied on; he, however, refused. They immediately left the pursuit, and, blessing the noble beast, cried out to the fugitive, 'Go and wash the feet of your mare, and drink off the water.' This expression is often used by the Bedouins to show the regard they have for their mares."

A writer of eminence observes: " Taking the comparative excellence of the different races, Nejed, between the desert of Syria and Yemen, and now in the possession of the Wahabes, is generally reckoned to produce the grandest, noblest horses. Hejaz (extending along the Red Sea, from Mount Sinai to Yemen, and including in it Medina and Mecca) the handsomest; Yemen (on the coast of the Red Sea and the Indian Ocean, and the most fertile part of Arabia) the most durable; Syria the richest in color; Mesopotamia the most quiet; Egypt the swiftest; Barbary the most prolific; and Persia and Koordistan the most warlike."

THE BARB.

By the Barb is meant the breed of horse found in the north of Africa, from the coast to the Desert of Sahara. The common horse of the country is a miserable creature, but the tribes of the desert possess some horses of superior breed, form, and power. Their general points are those of the Arab. Abd-el-Kader refutes the notion that there is any difference between the Barb and the Arab. "The people of Barbary," he writes, "are of Arab origin, who entered the country before Islam, and no one doubts that the Arabian horses have spread in the same way as the Arab families." The Arabs of Africa, contrary to the habits of the Arabs of Arabia, always ride stallions. Abd-el-Kader, in his reply to the questions of General Daumas respecting the horse of the Sahara, says that an Arab horse ought to be able to travel sixty miles a day, without fatigue, for three or four months, without resting a single day. A horse, if necessary, ought to accomplish one hundred and eighty-seven miles in a day. "We have seen," he adds, "a great number of horses perform this distance. Nevertheless, the horse which has accomplished this distance ought to be spared the next day." He then describes one of his own forays: "We completed our excursion thither and back, in twenty or twenty-five days. During this interval our horses had no barley to eat except what their riders could carry, eight feeds apiece Our horses went without drinking one day or two, and on one occasion three days. The horses of the desert do much more than this." The Arabs commence to educate their foals early. In the first year, continues the Emir, they teach the horse to be led with the *resenn*, a sort of bridle; they call him *djeda*, and begin to tie him up. When he is two years old he is ridden for a mile, then for two, then for a parasang (nearly four miles). The Arabs have a proverb, "In the first year tie him up, for fear of accident; in the second, ride him till his back bends; in the third, tie him up again." Respecting the feeding of the horse, he writes: "Offer it to the horse saddled and bridled, as the proverb says: 'Water with the bridle, barley with the saddle.' 'To give drink at sunrise makes a horse lean; in the evening, fat; in the middle of the day, it keeps him in condition.' In the great heats they give water only every other day." Abd-el-Kader concludes with a string of proverbs. "Horses are birds without wings." "For horses, nothing is distant." "He who forgets the beauty of horses for the beauty of women will never prosper": and finally drops into poetry:

> "Love horses, care for them,
> Spare no trouble for them,
> By them comes beauty, by them honor."

He repeatedly insists on the necessity of keeping the horse free from all servile employment, and relates the following story: "A man was riding on a horse of pure race. He was met by his enemy, also mounted on a noble courser. One pursued the other; he who gave chase was distanced by him who fled. The former shouted, 'I ask you, in the name of God, has your horse ever worked on the land?' 'He has worked for four days.' 'Mine has never worked! By the head of the Prophet, I am sure of catching you.' Toward the end of the day the pursuer began to gain. He soon succeeded in fighting with the man whom he had given up all hopes of reaching."

While the horse of Arabia has only become famous recently, the horse of Barbary has been famed since the light Numidian cavalry of

Hannibal swept from end to end of Italy. They were light men on light horses, and rode without rein or bit.

The following are the characteristic points of a true Barb, and especially from Morocco, Fez, and the interior of Tripoli: The forehand is long, slender, and ill-furnished with mane, but rising distinctly and boldly out of their withers; the head is small and lean; the ears well-formed, and well-placed; the shoulders light, sloping backward, and flat; the withers fine and high; the loins straight and short; the flanks and ribs round and full, and with not too much band; the haunches strong; the croup, perhaps, a little too long; the quarters muscular and well-developed; the legs clean, with the tendons boldly detached from the bone; the pastern somewhat too long and oblique; and the foot sound and good. They are rather lower than the Arabian, seldom exceeding fourteen hands and an inch, and have not his spirit, or speed, or continuance, although in general form they are probably his superior.

The training is adapted to the exigencies of desert warfare. All that is required of the best-taught and most valuable Barbary horse is thus to gallop and to stop, and to stand still, all the day if it is necessary, when his rider quits him. As for trotting, cantering, or ambling, it would be an unpardonable fault were he ever to be guilty of it. A Barbary horse is generally broken in in a far severer way, and much earlier than he ought to be, usually in his second year, and therefore he usually becomes unfit for service long before the Arabian. The usual food of the Barb is barley and chopped straw, and grass while it is to be found, but of the provision for winter food in the form of hay they are altogether ignorant.

The Barb has chiefly contributed to the excellence of the Spanish horse; and, when the improvement of the breed of horses began to be systematically pursued in Great Britain, the Barb was very early introduced. The Godolphin Arabian, as he is called, who was the origin of some of our best racing blood, was a Barb; and others of our most celebrated turf-horses trace their descent from African mares.

CHAPTER III.

THE RACE-HORSE AND TROTTING-HORSE.

THE RACE-HORSE—THE ENGLISH TURF—THE AMERICAN TURF—IMPORTED HORSES—THE TROTTING HORSE—FLORA TEMPLE—STEVE MAXWELL—ST. JULIEN AND MAUD S.—THE NARRAGANSETT PACERS—POCAHONTAS.

"AS breeders of horses," Brehm remarks, "the English of the present day stand on a level with the Arabs." The horse, indeed, is curiously connected with the history of England, the leaders of the Anglo-Saxon invaders of Britain bearing, according to tradition, the very equine names of Hengest (Stallion), and Horsa (Horse). Lappenberg, the historian, supposes that these names do not indicate individuals, but the fact that the expedition was undertaken in obedience to an oracle derived from the sacred horses which were kept for purposes of divination. Tacitus describes these sacred horses as being white; the White Horse was a favorite symbol of the Saxon, and on the Chiltern Hills in Berkshire, England, by removing the green turf an enormous white horse is still visible, cut on the slopes of the chalky hill. The figure has most probably been unchanged from the time of the Saxon kings, and is periodically renewed and cleaned with great ceremony. The "White Horse" figures conspicuously in the coats of arms of the Princes of the House of Hanover and Brunswick, and the sovereigns have always affected on state occasions the use of cream-colored horses in their royal chariots. The horses used by the Queen of England when she opens Parliament, are descendants of a North German race, and may probably be derived from those mentioned by the Roman historian as dedicated to the gods, and endowed with the powers of prophecy. Most of the superstitions connected with the horse have died away, but that of the horseshoe remains as vigorous as ever, and this mysterious symbol is seen everywhere nailed over doors, and has lately spread into the domain of personal ornament.

THE RACE-HORSE

The first improvement in the breed of the original English horse was effected by William the Conqueror, who imported some Spanish horses, thus introducing Barb blood; and down to the time of Henry VIII Spanish and Italian horses were frequently imported. While the use of heavy armor in war prevailed, heavy, stout animals were necessary, and the real amelioration of the English horse commenced when the musket superseded the battle-axe and lance.

Ever since men possessed horses, they have been fond of trying the speed of their favorites; but it was not till the reign of James I. that regular race-courses were laid out, and rules for the sport laid down. James did his best to improve the stock by purchasing an Arab for the enormous sum, in those days, of five hundred pounds. But, strangely, this Arab was disapproved of by the celebrated writer on horsemanship, the Duke of Newcastle, who described it as a little, bony horse, of ordinary shape. James bought another Eastern horse, the White Turk, and the Helmsley Turk, and Fairfax's Morocco Barb were imported about the same time. No attention was paid to the genuine Arabian till Mr. Darley, in Queen Anne's time, purchased the famous "Darley Arabian," whose figure contained every point that could be desired in a turf-horse. This horse was the sire of "Flying Childers" and "Bartlett's Childers," and through them his blood has passed into most of our racers. Twenty years after Darley had established the value of Arabian blood, Lord Godolphin bought, out of the shafts of a cart in Paris, the celebrated "Godolphin Arabian," which in all probability, however, was a Barb. The most illustrious of the English racers of the last century was "Eclipse," a grandson of Bartlett's Childers, which never was beaten, nor ever paid forfeit.

Races originally were merely meetings at which men could test the speed of horses which were on ordinary occasions used for other purposes, such as war, or the chase. But the establishment of regular race-meetings with valuable prizes to be won by the victors, soon introduced great changes. It was at once seen that the ordinary hunter or roadster or charger had no chance in a competition with horses bred for the sole purpose of racing, and therefore men ambitious of distinction or desirous of making money on the turf, turned their attention to producing a breed of animals which were racers and nothing else. The racers, down to the middle of the last century, were powerful animals; the course was usually four miles, and the race was run in heats. The object of all breeders of race-horses being to win, they did their best to gain speed, and very soon stoutness became a secondary consideration; horses soon showed that they could not run the distances their ancestors performed, heats became unfashionable, the length of the course was diminished, and younger horses entered. At present, the length of the "Derby" course is a mile and a half; that of the "St. Leger" nearly two miles. With young horses and short courses, the result of a race depends in a very great degree on the riders, if the animals are at all fairly matched, and hence has arisen the system of betting on "mounts."

The ENGLISH RACE-HORSE (Plate XXXII) is the result of long-continued efforts to obtain the maximum of speed, and it is to the English Race-horse that other nations turn, when they desire to improve their native stock. By numerous crossings with English thorough-breds, the Prussians have produced the Trakehner, a powerful and useful animal, and the French have employed it with advantage in improving their Norman stock.

English horses of pure blood were introduced into the colonies at a very early date, and race-horses were kept and trained, especially in those States which had not been settled by the Puritans. Virginia and Maryland took a decided lead in this pursuit. Spark, grandson of the Darley Arabian, a stallion presented by Frederick, Prince of Wales, to Lord Baltimore, was one of the earliest importations, and in 1750 he was followed by Selima, a daughter of the Godolphin Arabian. From Virginia the love for racing spread to the Carolinas, and northward to New York, where Colonel Delancey of King's Bridge imported Wildair and Lath. Within thirty years of the close of the Revolutionary War, Bedford, Citizen, Diomed, Messenger, Saltram, Shark, Sir Harry, Spread Eagle, and many other sires of fame were brought over, and the good breed thus introduced was continued by the importation of such horses as Priam, Barefoot, Margrave, Rowton, St. Giles, Squirrel,—six Derby and St. Leger winners—Glencoe, Riddlesworth, Monarch, Sovereign, Leamington, and others. American-bred horses are now entered in the English races. Mr. Ten Brock led the way, and his example has been followed in the last two years by Mr. Sanford, Mr. Lorillard, and Mr. J. Keene. Mr. Lorillard's Parole especially distinguished itself, winning both the Metropolitan and City and Suburban stakes at Epsom, while Keene's Don Fulano gained a brilliant victory at Newmarket.

THE TROTTING-HORSE.

The TROTTING-HORSE (Plate XXXII) is pre-eminently the American horse, and the pace attained during the last few years is almost incredible, but is well authenticated.

In England, trotting matches have never been fashionable, but some celebrated ones are recorded. In 1822, there was a match of nine miles between Mr. Bernard's mare and Captain Colston's horse, near Gerrard's Cross, for five hundred guineas. It was won easily by the mare, who performed the distance in twenty-seven minutes and forty-six seconds. The horse went the same distance in twenty-seven minutes forty-nine seconds, which is nearly at the rate of nineteen and a half miles an hour. The race was on an ordinary road.

This, however, had been equalled or excelled some years before. Sir Edward Astley's Phenomenon mare, when twelve years old, trotted seventeen miles in fifty-six minutes. There being some difference about the fairness of the trotting, she performed the same distance a month afterward in less than fifty-three minutes, which was rather more than nineteen miles an hour. Her owner then offered to trot her nineteen and a half miles an hour; but, it being proved that in the last match she did one lap of four miles in eleven minutes, or at the rate of more than twenty-one and a half miles an hour, the betting men would have nothing more to do with her.

Mr. Osbaldeston had a celebrated American trotting-horse, called Tom Thumb. He matched him to trot one hundred miles in ten hours and a half. It seemed to be an amazing distance, and impossible to be accomplished; but the horse had done wonders as a trotter; he was in the highest condition; the vehicle did not weigh more than one hundred pounds, nor the driver more than one hundred and forty-three pounds. He accomplished his task in ten hours and seven minutes; his stoppages to bait, etc., occupied thirty-seven minutes—so that, in fact, the hundred miles were done in nine hours and a half. He was not at any time distressed; and was so fresh at the end of the ninetieth mile, that his owner offered to take six to four that he did fourteen miles in the next hour, but no one was bold enough to take the bet.

An English-bred mare was afterward matched to accomplish the same task. She was one of those animals, rare to be met with, that could do almost anything as a hack, a hunter, or in harness, and on one

RUNNING HORSE
CLYDESDALE CART HORSE
TROTTING HORSE

occasion, after having, in following the hounds, and traveling to and from cover, gone through at least sixty miles of country, she fairly ran away with her rider over several ploughed fields. She accomplished the match in ten hours and fourteen minutes—or, deducting thirteen minutes for stoppages, in ten hours and a minute's actual work; and thus gained the victory. She was a little tired, and, being turned into a loose box, lost no time in taking her rest. On the following day she was as full of life and spirit as ever. These are matches which it is pleasant to record, particularly the latter one; for the owner had given positive orders to the driver to stop at once, on her showing decided symptoms of distress, as he valued her more than anything he could gain by her enduring actual suffering. These matches, it will be seen, are all for long distances, and are trials of endurance rather than speed; the system of matches against time had not yet been introduced.

In England, as we have observed, the efforts of breeders and trainers were turned to producing a swift gallop; the sportsmen of America, especially in the North and West, directed all their energies to developing the trot. Frank Forester writes in explanation of the difference in the sporting tastes of the two countries: "I do not think I ever knew, or heard tell of such a thing, in my life, in England, as of two gentlemen going out to take a drive in a light carriage. In England, every man who can keep a horse for pleasure, keeps it with a view of occasionally taking a run with hounds. In America, every farmer keeps his wagon and driving-horse, and, as it costs no more to keep a good horse than a bad one, he keeps one that enables him to combine business and pleasure. Trotting in America is the popular pastime, and the trotting-course is open to all."

The trotting-horse is not a distinct breed, and his qualities as a trotter cannot be ascribed to his origin or connection with any one blood. Some trotters of first-rate powers have come from the Canadian or Norman stock, some from the Vermont stock, some from the Indian pony, and some entirely from the thorough-bred. It is, however, beyond doubt that "the best type of American trotter descends from the English thorough-bred horse Messenger, imported into this country toward the end of the last century. Mr. C. J. Foster writes that "when the old gray, Messenger, came charging down the gang-plank of the ship which brought him over, the value of not less than one hundred million dollars struck our soil." The estimate appears at first sight to be extravagant, but when it is found upon investigation that the Messenger blood crops up in almost all the best American trotters of the present century, the figures are not too high.

The first Kentucky sire which seems to have done great good to the trotting stock of that sporting State was Abdallah, who was a grandson of Messenger. From Abdallah's loins sprang by far the best trotting sire that the United States have hitherto produced, whose name was Rysdyk's Hambletonian; and in the "List of Trotters with Records of 2:25 or Better," in the agricultural paper the *National Live Stock Journal*, published at Chicago, it appears that Hambletonian is to the trotting turf what Touchstone is to the running course. Hiram Woodruff believes that the Messenger blood existed in the most historical horse ever bred in Hindoostan, whose name was Lylee, and who was the favorite of Runjeet Singh, "the Lion of the Punjab." It is known that Runjeet spent enormous sums upon his stud; that his bridles and saddles were inlaid with gold and studded with precious stones; and that the Maharajah himself was a desperately hard rider. In order to get possession of his incomparable gray stallion Lylee, Runjeet Singh used to boast that he had spent six hundred thousand pounds and the lives of twelve thousand men. When the fame of Lylee first reached Runjeet's ears, the horse was the property of Yan Mohammed Khan, one of the Punjabee princes, who had his capital at Peshawur. Runjeet opened negotiations to get hold of Lylee, and having failed, went to war for that purpose. After a long contest the arms of the Maharajah prevailed, and the first condition upon which he offered peace was that Lylee should be ceded to him by his vanquished foe. After an infinite number of evasions and subterfuges resorted to by Mohammed Khan, the horse became the property of Runjeet, but he had to fight another war in order to retain him. Lylee, who was believed to be the son of an English thoroughbred, was seen in 1839 by some English officers, and was "a flea-bitten gray, very old, standing sixteen hands high," and with all the characteristics of the Messenger blood.

From Messenger are descended Abdallah, Hambletonian, Volunteer, Mambrino Chief, Edward Everett, Alexander's Abdallah, Conklin's Abdallah, Dexter, and a host of famous trotting-mares. But, richly though our sportsmen are indebted to the Messenger blood, it would be injustice to deny that they owe still more to the skill, patience, and persistency lavished upon training and bringing the trotter to perfection by such men as Hiram Woodruff and Dan Mace. The system of training, teaching, driving, and riding the trotting-horse of the United States is an art of itself, peculiar to the country which has bred St. Julien, Maude S., Rarus, and Goldsmith Maid. These horses are superior by at least forty seconds in the mile to any trotters that Europe can boast. "The English had the stock all along," says Hiram Woodruff, "but it is our method of cultivation and perseverance that has made the difference between their fast trotters of a mile in three minutes and ours in two minutes and twenty seconds, or less."

The development of the trotting-pace and the establishment of the trotting-course is comparatively recent. The first American match against time was in 1818; the wager was, that no horse could be produced to trot a ?e in three minutes. The horse named at the post was Boston Blue, who won easily. He afterward became the property of Thomas Cooper the tragedian. The same horse afterward trotted eighteen miles within the hour. In 1824, the Tredwell mare trotted a mile in two minutes thirty-four seconds, but for many years afterward a two-forty horse was considered extra fast. In 1826, the first authenticated record of authorized trotting is found. The New York Trotting Club opened its course near the Jamaica Turnpike, about a mile below the Union Course, Long Island, with a series of two-mile and three-mile heats. The first two miles was done in five minutes thirty-six seconds; the second, in five minutes thirty-eight seconds. In 1827, the horse Whalebone trotted fifteen miles in harness within the hour, performing the last mile in three minutes five seconds. In 1828 the best time on record was made by Screwdriver, at Philadelphia, by winning a three-mile heat in eight minutes two seconds, beating the celebrated Topgallant. In 1833, Paul Pry was backed to do seventeen and three-quarter miles within the hour; he won with the greatest ease, going eighteen times round the Long Island Trotting Course, covering eighteen miles thirty-six yards in fifty-eight minutes fifty-two seconds. He was ridden "by a boy named Hiram Woodruff." Topgallant, Columbus, Collector, Lady Jackson, and the best trotters of whom Hiram Woodruff makes mention as having flourished about 1830, could not "knock off" their mile in less than 2:50: but in 1834, Edwin Forrest beat the record by trotting a mile in two minutes thirty-one and a half seconds.

Coming down to more recent times, Flora Temple began her performances in 1850, and in 1853 she accomplished a mile in two minutes twenty-seven seconds, and in 1856 made the best time on record, two minutes twenty-four and a half seconds, beating Tacony for a stake of one thousand dollars. She won this race in one heat, distancing Tacony, and the time was one second less than ever made before. She was driven by her favorite driver, Hiram Woodruff, who declared after the race that she could beat a locomotive. But the time, both in mile and two-mile races, has gradually been shortening; we pass over the performances of Goldsmith Maid, Rarus, Hopeful, and Lulu, all of whom have accomplished a mile in 2:15 or less, and come at once to the running of this year. At Rochester, Steve Maxwell trotted two miles in four minutes forty-eight and a half seconds, beating by two seconds Flora Temple's best record at that distance; and both St. Julien and Maud S. trotted a mile in the extraordinary time of *. * minutes eleven and three-quarter seconds. And even this has been surpassed, the former being credited with accomplishing the distance in two minutes eleven and a quarter seconds, the latter with covering the same distance in two minutes ten and three-quarter seconds. Such are the results of careful breeding and skilful training, and there seems to be no reason why a mile may not be done in two minutes ten seconds. We must observe that, respecting these times and distances, there can be no dispute, while, in earlier races, the distance traversed was often inaccurately estimated, and the time never given with anything like precision.

THE NARRAGANSETT PACER.

This beautiful animal, according to Frank Forester, has entirely ceased to exist. The pace, it may be explained, is a gait in which both legs on one side are raised together; while in the trot, one of the fore-legs and the opposite hind-leg are lifted at the same time. In the celebrated Elgin marbles from the Parthenon of Athens, two horses are represented as pacing; and in the famous "Horses of St. Mark" at Venice, the attitude is the same. A writer in the middle of the last century describes Rhode Island as producing fine horses, remarkable for swift pacing, and adds: "I have seen some of them pace a mile in a little more than two minutes, and a great deal less than three." The original genuine Narragansett Pacers are said to pace naturally, but the name soon became applied to all pacers. The most famous of modern pacers was the magnificent mare Pocahontas, an animal described by a eulogist as "the most sumptuous," as well as the fastest of her day. She was a rich chestnut, sixteen hands in height, good crest, high and thin withers; her pedigree was excellent, and her greatest triumph was her defeat of Hero, whom she distanced in the first heat to wagons in the then unparalleled time of two minutes seventeen seconds.

The original breed is said to have been introduced by Governor Robinson of Rhode Island from Spain in the last century, and large numbers of them were produced in New England for exportation to the West Indies, where they were in great demand for the wives and daughters of the planters. But change in our modes of traveling has extinguished the pacer. While our roads were all bad, and horseback-riding the only method of locomotion, pacers were highly-prized luxuries. Now they are superseded by the trotter, and for riding-horses of mere pleasure, the present day requires speed, style, and action, rather than an easy gait which can be kept up at a slow pace for a considerable time. Any animals possessing this gait at the present time have fallen into it by accident, or been taught to pace. As far as is known, there is no breed of horses in Spain or elsewhere, to which the gait is native, and there has always been considerable doubt as to whether the claim put forward for the Narragansett Pacer could be allowed.

CHAPTER IV.

EUROPEAN HORSES.

THE HUNTER—THE HACKNEY—THE RUSSIAN HORSE—THE AUSTRIAN HORSE—THE HOLSTEIN HORSE—THE FRENCH HORSE—THE ITALIAN HORSE—THE RACES AT ROME—THE SPANISH HORSE—THE SHETLAND PONY—THE CARRIAGE HORSE—THE CART HORSE—THE PERCHERON HORSE.

THE HUNTER.

THE best of all hunters is a thorough-bred horse with bone, standing about fifteen hands and a half. A lofty forehand, a good oblique shoulder, and a clear high action are indispensable. The body should be compact, the barrel round, the loins broad, the quarters long, the thighs muscular, and, above all, the hunter's temper should be good. In other words, he must have wind and bottom, plenty of jumping power, and be able to go at a good rapid pace.

THE HACKNEY.

The perfect roadster is more difficult to find than even the hunter. He must have good fore-legs, and good hinder ones too; he must be sound in his feet, even-tempered, no starter, quiet in all situations, not heavy in hand, and never disposed to fall on his knees. Safety in this latter respect depends entirely on the foot being placed at once flat on the ground. His height should rarely exceed fifteen hands and an inch. He is more valuable for the pleasantness of his paces and his safety, good temper, and endurance, than for his speed. It is the roadster which furnishes most of the recorded instances of intelligence and fidelity in the horse. Their memory is often remarkable. A horse was ridden thirty miles into a quite new district by a road very difficult to find. After an interval of two years, during which the animal had never been in that direction, the gentleman had occasion to make the same journey.

He was benighted on a common where it was so dark that he could scarcely see his horse's head. He threw the reins on its neck, and in half an hour was safe at his friend's gate. Another gentleman, riding through a wood in a dark night, struck his head against the branch of a tree and fell, stunned, to the ground. The horse returned to the house they had lately left, and paused at the door till some one arose and opened it. He then turned about and led the man to the place where his master was lying senseless.

In 1809, the inhabitants of the Tyrol captured fifteen horses from the Bavarian troops, on which they mounted their own men. An encounter afterward took place between the hostile forces; but at the commencement of it the Bavarian chargers, which had changed their masters, recognized their former trumpet-call and the uniform of their old regiment, and in an instant darted off at full gallop, in spite of all the efforts of their riders, whom they bore in triumph into the midst of the Bavarian ranks, where the Tyrolese were at once made prisoners.

An orchard had been repeatedly stripped of its best and ripest fruit, and the marauders had laid their plans so cunningly that the strictest vigilance could not detect them. At last the depredators were discovered to be a mare and her colt which were turned out to graze among the trees. The mare was seen to go up to one of the apple-trees and to throw herself against the trunk so violently that a shower of ripe apples came tumbling down. She and her offspring then ate the fallen apples, and the same process was repeated at another tree. Another mare had discovered the secret of the water-butt, and whenever she was thirsty, was accustomed to go to the butt, turn the tap with her teeth, drink until her thirst was satisfied, and then to close the tap again. Two animals are said to have performed this feat, but one of them was not clever enough to turn the tap back again, and used to let all the water run to waste.

A careless groom was ordered to prepare a mash for one of the horses placed under his care, and after making a thin, unsatisfactory mixture, he hastily threw a quantity of chaff on the surface, and gave it to the horse. The animal tried to push away the chaff and get his nose into the mash, but was unable to do so, and when he tried to draw the liquid into his mouth, the chaff flew into his throat and nearly choked him. Being baffled, he paused awhile, and then pulled a lock of hay from the rack. Pushing the hay through the chaff, he contrived to suck the liquid mash through the interstices until the hay was saturated with moisture. He then ate the piece of hay, pulled another lock from the rack, and repeated the process until he had finished his mash.

THE RUSSIAN HORSE.

Russia supplies a magnificent race, which combines elegance of proportion, height, size, vigor, and suppleness. Many of this breed are remarkable for their speed in trotting, and they all much resemble the celebrities of the American trotting turf. They are, in all probability, derived from Cossack blood, but improved by stallions from Poland, Holstein, and England, with some Turkish and Arab blood.

The Cossack horse is remarkable for its combination of speed and endurance, and it was long supposed to be unrivaled in the possession of these qualities. But the Cossack horse was beaten by horses of English blood in a race which fairly tested its powers.

On the 4th of August, 1825, a race of forty-seven miles was run between two Cossack and two English horses. The English horses were Sharper and Mina, well-known, yet not ranking with the first of their class. The Cossacks were selected from the best horses of the Don, the Black Sea, and the Ural.

On starting, the Cossacks took the lead at a moderate pace; but before they had gone half a mile, the stirrup-leather of Sharper broke, and he ran away with his rider, followed by Mina, and they went more than a mile, and up a steep hill, before they could be held in.

Half the distance was run in an hour and fourteen minutes. Both the English horses were then fresh, and one of the Cossacks. On their return, Mina fell lame, and was taken away, and Sharper began to show the effects of the pace at which he had gone when running away, and was much distressed. The Calmuck was completely used up, his rider was dismounted, a mere child was put on his back, and a Cossack on horseback on either side dragged him on by ropes attached to his bridle, while others at the side supported him from falling. Ultimately Sharper performed the whole distance in two hours and forty-eight minutes—sixteen miles an hour for three successive hours and the Cossack was brought in eight minutes after him. The English horse carried fully forty pounds more than the Cossack.

In Southern and Western Russia, and also in Poland, the breeding of horses and cattle has lately occupied the attention of the great landed proprietors, and has constituted a very considerable part of their annual income. There is scarcely now a seigniorial residence to which there is not attached a vast court, in four large divisions, and surrounded by stables. In each of the angles of this court is a passage leading to beautiful and extensive pasture-grounds, divided into equal compartments, and all of them having convenient sheds, under which the horses may shelter themselves from the rain or the sun. From these studs a larger kind of horse than that of the Cossacks is principally supplied, which are more fit for the regular cavalry troops, and for pleasure and parade, than common use. The remounts of the principal houses in Germany are derived from this source; and from the same source the great fairs in the different states of the German empire are supplied.

THE AUSTRIAN HORSE.

The Austrian cavalry has long been the most famous in the world, and most of the horses are bred in the vast plains of Hungary, where there has always been a strong infusion of Turkish and Arab blood. In 1790, an Arabian named Turkmainath was imported, and in 1819, the Archduke Maximilian purchased several valuable stallions in England. Of late years, some of the best sires and dams that could be procured from the English racing stables have been imported by several of the Hungarian nobles for their breeding establishments.

But the improvement of the breed of horses is not left to private enterprise alone; the imperial government maintains noble studs in many places, on which neither care nor money is spared.

The following account is given by the Duke of Ragusa of the imperial establishment for the breeding of horses at Mesohagyés, near Carlsburg in Austria: "This is the finest establishment in the Austrian monarchy for the breeding and improvement of horses. It stands on forty thousand acres of land of the best quality, and is surrounded in its whole extent, which is fifteen leagues, by a broad and deep ditch, and by a broad plantation sixty feet wide. It was formerly designed to supply horses to recruit the cavalry: at present its object is to obtain stallions of a good breed, which are sent to certain dépôts for the supply of the various provinces. To produce these, one thousand brood mares and forty-eight stallions are kept; two hundred additional mares and six hundred oxen are employed in cultivating the ground. The plain is divided into four equal parts, and each of these subdivided into portions, resembling so many farms. At the age of four years the young horses are all collected in the centre of the establishment. A selection is first made of the best animals to supply the deficiencies in the establishment, in order always to keep it on the same footing. A second selection is then made for the use of the others; none of these, however, are sent away until they are five years old; but the horses that are not of sufficient value to be selected are sold by auction, or sent to the army to remount the cavalry, as circumstances may require.

"The whole number of horses at present here, including the stallions, brood mares, colts, and fillies, is three thousand. The persons employed in the cultivation of the ground, the care of the animals, and the management of the establishment generally, are a major-director, twelve subaltern officers, and eleven hundred and seventy soldiers."

THE HOLSTEIN HORSE.

The horses of Holstein and Mecklenburg, and some of the neighboring districts, are on the largest scale. Their usual height is sixteen, or seventeen, or eighteen hands. They are heavily made; the neck is too thick; the shoulders are heavy; the backs are too long, and the croups are narrow compared with their fore-parts; but their appearance is so noble and commanding, their action is so high and brilliant, and their strength and spirit are so evident in every motion that their faults are pardoned and forgotten, and they are selected for every occasion of peculiar state and ceremony.

THE FRENCH HORSE.

In France, as in Hungary, the breeding of horses is an affair of state. Before the creation of the *Administration des Haras*, there existed in Normandy a race of horses which for many years furnished carriage animals to the great lords of olden time. These were of Danish origin; but the present race is the result of a cross between the Norman or Danish mares and the English thorough-bred; the results show the characteristics of both stocks. They are bred in two districts in Nor-

‎mandy: one, the plain of Caen, comprising the grassy meadows of Calvados and La Manche; the other is situated in that part of the Department of Orne which bears the name of Merlerault.

The French horses, however, are generally of a very miscellaneous character. In addition to the Percheron which we speak of under the head of the "cart-horse," there are the Boulonais and Flemish breeds, both of which are in repute as draught-horses. The most remarkable is the horse of the district named the Camargue. In England he would be denominated a pony, for he is small, his height measuring from thirteen to thirteen hands and a half; it is but seldom that he is tall enough to reach the limit for a light cavalry charger. His coat is always of a grayish-white. Although the head is large, and sometimes "Roman-nosed," it is generally squarely made, and well set on; the ears are short and widely separated, the eyes are lively and well opened, the crest is straight and slender, but sometimes ewe-necked; the shoulder is short and upright, but yet the withers are of a sufficient height; the back is prominent, the reins wide, but long, and badly set on; the croup is short and drooping, the haunches are poor, the hocks narrow and close, but yet strong; the foot is very sure and naturally good, but wide, and sometimes even flat. The Camargue horse is active, abstemious, mettlesome, high-spirited, and capable of enduring both bad weather and fasting. For centuries he has maintained the same type, notwithstanding the state of distress to which he is sometimes reduced by carelessness and neglect.

These small horses are kept in the marshes and wild meadows which stretch away from Arles to the sea. They live in perfect freedom, in small droves, together with semi-wild oxen. In harvest time these horses are used for threshing out the grain; they are led in upon the threshing-floors, and are made to stamp upon the sheaves to beat out the corn from the ears. Their hard, but elastic hoof forms an excellent flail. When they have done their allowance of work they are permitted to return to their independent existence, to roam and feed over the wide expanse of uncultivated districts which surround their homes.

The breed of Camargue horses is, as a rule, but little valued, even in the south of France. The best of them are, however, occasionally sent into the market. It is stated that these horses are the descendants of some of those left by the Moors in one or the other of the frequent descents and incursions made by them on the south coast of France during the early years of history.

THE ITALIAN HORSE.

We need say little respecting the horse in Italy. During long ages of misgovernment and oppression, the Italian horse, once celebrated for the beauty of his form and his paces, sadly deteriorated. The Neapolitan horses were particularly remarkable for their size and majestic action, but, with few exceptions, have degenerated. A rapid improvement has now taken place, and is still in progress.

As a striking contrast between our style of racing, in which the rider or driver is so important, we subjoin an account of the peculiar method in which the races were run during the Roman Carnival.

The horses—termed Barberi, because the race was at first contested by Barbs—are brought to the starting-post, their heads and their necks gayly ornamented; while, to a girth which goes round the body of each, are attached several loose straps, having at their ends small balls of lead thickly set with sharp steel points. At every motion these are brought in contact with the flanks and nelhes of the horses, and the more violent the motion, the more dreadful the incessant torture. On their backs are placed sheets of thin tin or stiff paper, which, when agitated, will make a rustling, rattling noise. A rope is placed across the street to prevent them getting away, and a groom holds each horse. When all is ready for starting, a troop of dragoons gallops through the street in order to clear the way. A trumpet sounds, the rope drops, the grooms let go their hold, and the horses start away like arrows from a bow. The harder they run, the more they are pricked; the cause of this they seem scarcely able to comprehend, for they bite and plunge at each other, and a terrible fight is sometimes commenced. Others, from mere fright or sulkiness, stand stock-still, and it is by brute force alone that they can again be induced to move. A strong canvas screen is passed along the bottom of the street. This is the goal. It has the appearance of a wall; but some of the horses, in the excess of their agony and terror, dart full against it, tear through it, or carry it away.

THE SPANISH HORSE.

The common breed of Spanish horses have nothing extraordinary about them. The legs and feet are good, but the head is rather large, the forehand heavy, and yet the posterior part of the chest deficient, the crupper also having too much the appearance of a mule. The horses of

Estremadura and Granada, and particularly of Andalusia, are most valued. Berenger, whose judgment can be fully depended on, thus enumerates their excellences and their defects: "The neck is long and arched, perhaps somewhat thick, but clothed with a full and flowing mane; the head may be a little too coarse; the ears long, but well placed, the eyes large, bold, and full of fire. Their carriage lofty, proud, and noble. The breast large; the shoulders sometimes thick; the belly frequently too full, and swelling, and the loin a little too low; but the ribs round, and the croup round and full, and the legs well formed and clear of hair, and the sinews at a distance from the bone—active and ready in their paces—of quick apprehension; a memory singularly faithful; obedient to the utmost proof; docile and affectionate to man, yet full of spirit and courage."

GALLOWAYS AND PONIES.

A horse between thirteen and fourteen hands in height is called a GALLOWAY, from a beautiful breed of little horses once found in the south of Scotland, on the shore of the Solway Firth, but now sadly degenerated, and almost lost, through the attempts of the farmer to obtain a larger kind, and better adapted for the purposes of agriculture. There is a tradition in that country that the breed is of Spanish extraction, some horses having escaped from one of the vessels of the Grand Armada, that was wrecked on the neighboring coast. This district, however, so early as the time of Edward I, supplied that monarch with a great number of horses.

The pure Galloway was said to be nearly fourteen hands high, and sometimes more; of a bright bay or brown, with black legs, small head and neck, and peculiarly deep and clean legs. Its qualities were speed, stoutness, and sure-footedness over the very rugged and mountainous country of which it was the native.

THE SHETLAND PONY.

The SHETLAND PONY, or SHELTIE (Plate XXXI), an inhabitant of the extremest northern Scottish Isles, is a very diminutive animal—sometimes not more than seven hands and a half in height, and rarely exceeding nine and a half. He is often exceedingly beautiful, with a small head, good-tempered countenance, a short neck, fine toward the throttle, shoulders low and thick—in so little a creature far from being a blemish—back short, quarters expanded and powerful, legs flat and fine, and pretty, round feet. These ponies possess immense strength for their size, will fatten upon almost anything, and are perfectly docile. One of them, nine hands (or three feet) in height, carried a man of twelve stone the distance of forty miles in one day.

A gentleman was, not long ago, presented with one of these elegant little animals. He was several miles from home, and puzzled how to convey his newly-acquired property. The Shetlander was scarcely more than seven hands high, and as docile as he was beautiful. "Can we not carry him in your chaise?" said his friend. The strange experiment was tried. The sheltie was placed in the bottom of the gig, and covered up as well as could be managed with the apron; a few bits of bread kept him quiet; and thus he was safely conveyed away, and exhibited the curious spectacle of a horse riding in a gig.

The WELSH PONY is one of the most beautiful little animals that can be imagined. He has a small head, high withers, deep yet round barrel, short joints, flat legs, and good round feet. He will live on any fare, and will never tire. About a century and a half ago, pony-hunting was a favorite amusement of the Welsh farmer. The sportsmen ran them down with greyhounds, and then caught them with a lasso.

The EXMOOR PONIES, although generally ugly enough, are hardy and useful. A well-known English sportsman said that he rode one of them half-a-dozen miles, and never felt such power and action in so small a compass before. To show his accomplishments, he was turned over a gate at least eight inches higher than his back; and his owner, who rides fourteen stone, traveled on him from Bristol to South Molton, eighty-six miles, beating the coach which runs the same road.

Among the ponies we may class the Iceland, Lapland, and Finland horses, none of which exceed twelve hands in height. In Iceland thousands run wild in the mountains, and never enter a stable; but instinct or habit has taught them to scrape away the snow, or break the ice, in search of their scanty food. A few are usually kept in the stable; but when the peasant wants more he catches as many as he needs, and shoes them himself, and that sometimes with a sheep's horn. The breed of horses found in the Faroe Islands resembles in most of its points the horses of Iceland. It is, however, rather higher, owing to an infusion of Danish blood.

THE CARRIAGE-HORSE.

In the old coaching days in England the favorite carriage-horse was the Cleveland Bay, which was formed by the progressive mixture of the blood of the race-horse with the original breeds of the country. The points of a good carriage-horse are, substance well-placed, a deep and well-proportioned body, bone under the knee, and sound open feet. At present, it is merely a larger saddle-horse, combining size, strength, and elegance—in other words, a tall, strong, oversized hunter.

THE CART-HORSE.

The CLYDESDALE CART-HORSE (Plate XXXII).—One of the best horses for ordinary heavy work is the CLYDESDALE; it is docile in temper, and possessed of enormous strength and great endurance. It owes its origin to one of the Dukes of Hamilton, who crossed some of the best Lanark mares with stallions that he had brought from Flanders. The Clydesdale is larger than the Suffolk, and has a better head, a longer neck, a lighter carcase, and deeper legs; he is strong, hardy, pulling true, and rarely restive. The southern parts of Scotland are principally supplied from this district; and many Clydesdales, not only for agricultural purposes, but for the coach and the saddle, find their way to the central and even southern counties of England. Dealers from almost every part of the United Kingdom attend the markets of Glasgow and Rutherglen. Mr. Low says that "the Clydesdale horse as it is now bred is usually sixteen hands high. The prevailing color is black, but the brown or bay is common, and is continually gaining upon the other, and the gray is not unfrequently produced. They are longer in the body than the English black horse, and less weighty, compact, and muscular, but they step out more freely, and have a more useful action for ordinary labor. They draw steadily, and are usually free from vice. The long stride, characteristic of the breed, is partly the result of conformation, and partly of habit and training; but, however produced, it adds greatly to the usefulness of the horse."

The PERCHERON breed, justly celebrated for ages, is the model of a light draught-horse. In the days of mail-coaches and diligences, this race was, *par excellence*, the post-horse of France. At the present time it almost exclusively supplies the horses for the omnibuses of Paris and the rapid carriage of merchandise. The brow of these animals is slightly bulging between the orbital arches, which are prominent. The face is long, with a narrow forehead, straight at the top, but slightly bulging out towards the tip of the nose; the nostrils are open and mobile; the lips thick and the mouth large; the ear long and erect; the eye lively, and the countenance animated. Their mane is but poorly provided with hair, but the tail is bushy; the legs are strong and firmly jointed, with rather long shanks devoid of hair. Their coat is generally a dappled gray color, but other colors are not uncommon.

The HEAVY BLACK DRAY-HORSE, which is conspicuous in the brewers' carts in London, is more adapted for show than for utility. They are noble-looking animals, very docile, with round fat carcasses and sleek coats, but they eat a deal of hay and corn, and at long-continued work would be beaten by a team of active horses a good deal lower. Many of these dray-horses stood seventeen hands high at two and a half years old, but all have the fault of slowness, their pace being about three miles an hour; their effectiveness lies in their weight alone.

We have made no mention of the horse as an article of food. M. Figuier writes: "Every one is acquainted with the efforts which, during the last year or two, have been made (and to some extent with success) to introduce horse-flesh for the use of the public. In Paris and some other cities in France, at the present time, it forms no inconsiderable portion of the nutriment of the poor. Prussia and the north of Europe were the first to set the example in this path of economy."

CHAPTER V.

THE WILD AND THE COMMON ASS.

THE WILD ASSES—THE KULAN OR DZIGGETAI—THEIR SPEED—DOMESTICATION—THE WILD ASS OF THE BIBLE—THE AFRICAN WILD ASS—THE COMMON ASS—ITS PATIENCE—ITS INTELLIGENCE—THE EGYPTIAN ASS.

WE follow the usual arrangement, and classify the Ass as a variety of the Horse. Dr. Gray, however, who has been followed by Professor Bell, separates the Ass under the generic name of *Asinus*, leaving the horse alone to fill the genus *Equus*.

THE KULAN.

The KULAN or DZIGGETAI, *Equus hemionus*, is thus described by Pallas: "The Dziggetai cannot be called either Horse or Ass. It is an intermediate form which its first discoverer, Merserschmied, called 'prolific mules.' It is not a mongrel, but a separate species, with a much more beautiful figure than any mule. The body is generally light, the limbs slender, its look wild; the head is rather heavy, but the ears are better proportioned than the mule's are, the hoofs are small, the tail resembles that of a cow. Its color is a light reddish-brown in the summer, but changes in the winter into a grayish-brown. The mane and tail are black, and on the back is a black stripe which becomes broader at the loins." Nothing need be added to this description.

The Kulan is a child of the steppes, and peoples the most dissimilar portions thereof. Although preferring the vicinity of lakes and rivers, it does not avoid the waterless and barren regions, nor even the hills, provided they are steppe-like, that is, without trees. It is found from the eastern slopes of the Ural to the Himalaya, and especially on the Mongolian and Chinese frontier, where it bears the name of "Kiang." It avoids the presence of the pastoral nomads, and only finds itself secure in tracts where even these precursors of civilized man do not linger. Pallas remarked that after the appointment of guards on the Russian frontier, the herds of Kulans became more rare, and today they are driven still further back, but are far from extirpated in the Russian dominions. They may be seen close to the frontier of Europe, and on the east and south of Siberia and Turkestan, but not in such numbers as in the desert steppes of Mongolia and Northwest China, or in the mountains of Thibet. In all this extensive district the vicissitudes of climate compel the Kulan to roam; they form herds at the beginning of winter, and, in bodies of over a thousand, seek in common a neighborhood that can furnish food. In spring, these herds are again dissolved into smaller troops, consisting of a horse and from six to twenty mares, according to the prowess of the leader.

Nothing is more wonderful than the speed and agility of the Kulan; it is as fleet as the antelope, and as sure-footed as the chamois. When pursued, they run in a long and graceful gallop, pausing from time to time to watch the hunters. A herd always puts out one of its number as a sentry; at the sight of any strange object the leading male rushes out to examine it more closely, and will even follow a horseman; but if danger threatens they all turn and fly. The natives of the steppes hunt the Kulan, but unless the hunter is sure of his aim, he does not risk a shot. The sportsman sets out early in the morning, mounted on a light yellowish horse. He stalks his game carefully, always advancing against the wind till he comes to the hill beyond which the Kulans are browsing. He then ties the hairs of his horse's tail closely together, and lets it feed on the summit of the hill. He himself lies on the ground, as much concealed as possible, with his gun ready, at a distance of two hundred yards. The Kulan sees the horse, imagines it to be some strayed mare, and gallops toward her. But as he approaches he stops, and that is the moment to fire. If struck in the breast he falls at the first ball, but sometimes it requires five shots to dispatch him.

The flesh of the Kulan is prized highly by the Kirghises and Tungusians; the hide is used for various purposes.

The Kulan has been brought to Europe, and is as easily trained as a common horse. It will be seen that we regard, with Prycwalski, the Kulan and Kiang as identical with the Dziggetai. This is also the view held by Apollon Rusinsky, who investigated the question, and whose opinion was confirmed by the Kirghis chief whom he consulted.

THE WILD ASS.

The WILD ASS, *Equus onager* (Plate XXXIII), is repeatedly mentioned in the Bible. Xenophon found it near the Euphrates, Pliny in Asia Minor, and it is seen to-day in Palestine. Its home extends from Syria through the Persian dominions to India.

It is considerably smaller than the Kulan, but larger and finer limbed than the common ass. The head is larger than the Kulan's, the lips thick, and closely set with stiff bristles, the ears pretty long, but shorter than in the ass. The prevailing color is of a deep cream tint, which in many specimens varies to a beautiful silver-white, and usually a somewhat lighter stripe runs along the back, in the centre of which is a dark-brown line. The winter coat resembles camel's hair, the summer coat is extremely smooth and soft. The mane stands upright, and is formed of soft woolly hairs; the tuft at the end of the tail is a span long.

In its habits it resembles all other members of the family. An old male conducts the herd of mares and foals, but he is not of so jealous a disposition as the male Kulan, and at times tolerates the presence of a rival. In speed it is not inferior to the Kulan. Porter writes that when he was in the Persian province of Fars, his greyhound gave chase to an animal which his companions affirmed to be an antelope. They followed it at full speed, and, thanks to the hound's skill, got a good sight of it. To his astonishment, it was a wild ass. He resolved to ride it down with an Arab horse of extraordinary speed; but all the efforts of his noble horse were in vain till the wild creature stood still and gave him an opportunity to examine it. But in an instant it was off as quick as thought, curveting, and kicking out in play, as if not in the least wearied by the chase, and was soon beyond the reach of its pursuer.

The senses of the Wild Ass are very acute, and he is thus almost unapproachable; he drinks only every other day, and prefers for food plants like the sow-thistle or dandelion, disliking all fragrant herbs and the true thistle. Like many wild animals, it ascends hills for the purpose of surveying the country, a peculiarity noticed by Jeremiah (ch. xiv. 6): "The wild asses did stand in the high places, they snuffed up the air like dragons." Wood writes: "It is of a very intractable disposition, and even when captured young, can scarcely ever be brought to bear a burden, or draw a vehicle; even foals born in captivity refuse to be domesticated. Many zoologists suggest that this animal is the progenitor of the domestic ass. They are exactly similar in appearance, and are hardly distinguishable by the eye. But while the one is quiet and docile, the other is savage, intractable, and filled with an invincible repugnance to human beings." Some other authorities, however, assert that the best riding asses in Persia and Arabia are Wild Ass foals that have been caught young, and retain all the excellent qualities of endurance possessed by their progenitors.

THE AFRICAN WILD ASS.

The AFRICAN WILD ASS, *Equus tæniopus*, resembles in size and appearance the common Egyptian donkey; but, in his habits, does not differ from the Asiatic species just described. It is found in the regions east of the Nile, being numerous along the course of the Atbara tributary of that river, as far as the Red Sea. It lives in herds exactly like the other wild *Equidæ*. It is very shy and circumspect, and the chase of it is very difficult. Travelers report that herds will often come up to their watch-fires, and remain till some noise or movement sends them bounding back into the night.

All the asses used in Abyssinia seem to be descended from this species. Brehm saw many asses in that country which, he was assured, had been captured when foals and tamed, and he describes them as being docile. A male which he possessed for some time, had a noble carriage and a very affectionate disposition.

In this species the mane is weak and short, the tail-tuft long and strong, the cross on the back very distinct, while the legs have horizontal stripes such as are seen on the legs of the zebra.

THE COMMON ASS.

The COMMON ASS, *Equus Asinus* (Plate XXXIII), is one of the most intelligent of our domestic animals, and is by no means the stupid creature which common parlance makes him.

While the horse is full of pride, impetuosity, and ardor, the ass is mild, humble, and patient, and bears with resignation the most cruel treatment. Most abstemious in its habits, it is content with the coarsest herbage, which other beasts will not touch, even such as thistles and weeds. A small quantity of water is sufficient for it, but this it requires

QUAGGA ASS ZEBRA WILD ASS

pure and clear. It will not, like the horse, wallow in mud or water; and as its master too often forgets to groom it, it performs this duty by rolling itself on the turf or the heather when opportunity offers. It has sharp sight, an excellent sense of smell, and an ear of keen acuteness. If it is laden too heavily, it remonstrates by drooping its head and lowering its ears. "When it is teased," says Buffon, "it opens its mouth and draws back its lips in a disagreeable manner, giving it a mocking and derisive air."

The Ass walks, trots, and gallops like the horse, but all its movements are shorter and slower. Whatever pace it employs, if too hardly pressed, it soon becomes tired; if not hurried, it is most enduring. It sleeps less than the horse, and never lies down for this purpose except when worn out with fatigue. Buffon says that it never utters its long and discordant cry, which passes in inharmonious succession from sharp to flat and from flat to sharp, except when hungry, or desirous of expressing its amorous feelings. Sometimes, however, it brays when frightened.

Attaching itself readily and sincerely, it scents its master from afar, and distinguishes him from all other persons, manifesting joy when he approaches. It recognizes without difficulty the locality which it inhabits, and the roads which it has frequented. When young, it cannot fail to please by its gayety, activity, and gracefulness; but age and ill-treatment soon render it dull, slow, and headstrong.

The Ass carries the heaviest weight in proportion to its size of all beasts of burden; it costs little or nothing to keep, and requires, so to speak, no care; it is a most useful auxiliary to the poor man, more especially in rugged mountainous countries, where its sureness of foot enables it to go where horses could not fail to meet with accidents. It is, therefore, the horse of those of small means; the abstemious and devoted helper of the poor.

The Ass is exceedingly common in all parts of Europe, and many stories of its cunning are told. It can open gates and doors which baffle all the efforts of the horse, and has been known even to fasten them after it passed through. An Ass, attacked by a bull-dog, seized it in his teeth, carried it to a river, and then lay down on it.

Another Ass displayed a singular discrimination of palate, being celebrated for his love of good ale. At one roadside inn the landlady had been very kind in supplying the donkey with a glass of his loved beverage, and the natural consequence was, that the animal could never be induced to pass within a moderate distance of the spot without going for his beer. Neither entreaties nor force sufficed to turn his head in another direction, and his master was in such cases obliged to make the best of the matter, and permit the animal to partake of his desired refreshment. He had a curious knack of taking a tumbler of beer between his lips, and drinking the contents without spilling a drop of the liquid, or breaking the glass. So curious a sight as a donkey drinking beer, was certain to attract many observers, who testified their admiration by treating the animal to more beer. His head, however, was fortunately a strong one, for only once in his life was he ever seen intoxicated, and on that solitary occasion his demeanor was wonderfully decorous, and marked by a comic air of gravity.

But it is in Egypt that the Ass plays the most conspicuous part; in Cairo, donkey-riding is universal, and no one ventures beyond the Frank quarter on foot. Bayard Taylor gives a vivid description of these animals: "The donkeys are so small, that my feet nearly touched the ground, but there is no end to their strength and endurance; their gait is so easy and light, that fatigue is impossible. The passage of the bazaars seems at first quite as hazardous on donkey-back as on foot; but it is the difference between knocking somebody down and being knocked down yourself, and one certainly prefers the former alternative. There is no use in attempting to guide the donkey, for he won't be guided. The driver shouts behind, and you are dashed at full speed into a confusion of other donkeys, camels, horses, carts, water-carriers, and footmen. In vain you cry out '*Bass*' (enough), *Piano*, and other desperate adjurations; the driver's only reply is, 'Let the bridle hang loose!' You dodge your head under a camel-load of planks; your leg brazes the wheel of a dust-cart; you strike a fat Turk plump in the back; you miraculously escape upsetting a fruit-stand; you scatter a company of spectral, white-masked women, and at last reach some more quiet street, with the sensations of a man who has stormed a battery. At first this sort of riding made me very nervous; but presently I let the donkey go his own way, and took a curious interest in seeing how near a chance I ran of striking or being struck. Sometimes there seemed no hope of avoiding a violent collision; but by a series of the most remarkable dodges, he generally carried you through in safety. The cries of the driver running behind, gave me no little amusement. 'The howadji comes! Take care on the right hand! Take care on the left hand! O man, take care! O maiden, take care! O boy, get out of the way! The howadji comes!" Kish had strong lungs, and his donkey would let nothing pass him, and so wherever we went we contributed our full share to the universal noise and confusion."

The color of the Ass is a uniform gray, a dark streak passing along the spine, and another stripe being drawn transversely across the shoulders. In the quagga and zebra these stripes are much more extended. In size, and in other respects, the Ass differs according to its locality. The largest are those of the Spanish Peninsula, which are nearly as large as a horse, and remarkably handsome.

The flesh of the Ass has a disagreeable taste, so that it never can become popular as public food; but that of their foals, on the contrary, is very tender, and differs but little from veal.

As a strengthening agent, or as a mild and light article of food for invalids, the milk of the Ass has long been considered excellent. The Greeks of antiquity made use of it for this purpose. It contains more milk-sugar and less caseous matter than cow's milk; but it should invariably be taken from a young and healthy animal in good condition, which has been fed abundantly on wholesome food. As it is the secretion of a non-ruminant animal, ass's milk is naturally more easy of digestion than the milk of the cud-chewing cow.

CHAPTER VI.

THE ZEBRAS.

THE ZEBRAS OR TIGER-HORSES—THE QUAGGA—THE DAUW, OR BURCHELL'S ZEBRA—HARRIS'S DESCRIPTION OF IT—THE ZEBRA PROPER—HUNTING THE ZEBRA—CROSS BREEDS—THE MULE—THE HINNY—INSTANCES OF THEIR FERTILITY—DARWINISM.

THE three species which we are now about to describe are natives of Africa. They equal the Asiatic Wild Asses in speed and beauty of form, but far surpass them in richness of color and boldness of marking. As they are all distinguished more or less by tiger-like stripes, they have been formed by some writers into a genus named *Hippotigris*, or "Tiger-horse." The Roman emperor Caracalla in the year 211 A.D., is described as slaying with his own hand, in the circus, an elephant, a rhinoceros, and a tiger-horse.

The Zebras, or Tiger-horses, are all characterized by a compact body, powerful neck, a head intermediate between that of the horse and the ass, rather long ears, an upright mane, which is not so thick as that of the horse, and a tail with long hair at the end. They all have the coat marked with bright colored stripes. The south of Africa is their birth-place, although one species probably is found north of the equator.

THE QUAGGA.

The QUAGGA, *Equus Quagga* (Plate XXXII), resembles the horse more than the ass in figure. The body is well made, the head graceful and moderate in size, the ears short, the legs powerful. A short mane rises along the whole neck, the tail is covered with hair from the root, but the hairs are much shorter than in the horse. The ground color of the head, back, and loins is brown, the belly and hairs of the tail white. Grayish-white stripes appear on the head, neck, and shoulders. On the face these marks are close together, and run from the forehead to the muzzle; on the shoulders they run transversely, and are placed wider apart. Ten stripes, visible in the mane also, are on the neck, and four on the shoulders; a few, very faint and at long distances, can be traced on the hind quarters. A broad black stripe runs along the spine, the ears are white within, yellowish-white with brown stripes outside. A full-grown Quagga measures about thirteen hands, or a little under.

It is a native of the plateau of Caffraria, and is tamed without difficulty. The Dutch colonists of the Cape keep them with their herds, and if a hyæna threatens to attack the cattle, the domesticated Quagga will attack and beat down the enemy with its fore-hoofs, ultimately trampling it to death.

THE DAUW.

The DAUW, *Equus Burchellii*, is the noblest of all the tiger-horses, being most like the genuine horse. It is the same size as the Quagga, and possesses a round body with a very arched neck, strong feet, and a mane five or six inches long; its tail resembles that of the Quagga. Its coat is smooth, cream-colored above, and white below. Fifteen narrow black stripes run upward from the nostrils, seven going straight to the forehead, the remainder going obliquely along the cheeks. A black stripe with white borders runs along the spine, ten broad black stripes cross the neck and mane, and contain between them narrow brown bands. Similar stripes surround the whole body, but the legs are free from them, being uniformly white. This species is found in large herds south of the Orange River in South Africa. It avoids the rocky and hilly districts, and is only found on the plains in company with ostriches, antelopes, and gnus. The Dauw is never seen with the quagga or zebra, and avoids their society. It is supposed to seek the company of the ostrich because that gigantic bird is remarkably watchful and circumspect, and this saves the cunning Dauw from the trouble of keeping watch for itself against the approach of an enemy.

Like many other gregarious animals of Southern Africa, the Dauw is found to make periodical migrations for the purpose of supporting itself with the food that has failed in its original district. In times of scarcity, the Dauw, together with several species of antelope, visits the cultivated lands, and makes sad havoc among the growing crops. When rain has fallen, and the forsaken districts have regained their fertility, the Dauw leaves the scene of its plunder, and returns to its ancient pasturage.

The Dauw is capable of a partial domestication, and can be tamed to a considerable extent. It is, however, considered as possessing an uncertain temper, and is of too obstinate a disposition to be of much use to man. By the Kaffirs it is called Peet-sey, and the Dutch colonists have given it the name of Bontequagga.

Respecting its habits in its wild state, Captain Harris writes. "Fierce, strong, fleet, and surpassingly beautiful, there is perhaps no animal in creation, not even excepting the mountain zebra, more splendidly attired, or presenting a picture of more singularly attractive beauty, than this free-born of the desert. A dark pillar of dust rises from the plain, and, undisturbed by any breath in heaven, mounts upward in the clear azure sky like a wreath of smoke—three ill-omened vultures soaring in circles above it. Nearer and more near rolls on the thickening column, until several dark living objects are perceived dancing beneath it. Emerging from the obscurity, their glossy and exquisitely variegated coats glittering in the sun's rays, the head of a column of Burchell's zebras appears, and instantly afterward the serried horde sweep past in gallant array, their hoofs clattering on the hard gravel like a regiment of dragoons. Tearing by at racing speed, straining neck and neck with their shaggy, whimsical-looking bovine allies, the gnus, their own striped and proudly curved necks seem as if they were clothed with thunder, and their snowy tails are streaming behind them. Now the troop has wheeled and halted for an instant to reconnoitre the foe. A powerful stallion advances a few paces with distended nostrils and stately gait, his mane newly hogged, and his ample tail switching his gayly checkered thighs. Hastily surveying the huntsmen, he snorts wildly, and instantly gallops back to his cohort. Away they scour again, neighing and tossing their striped heads abot, switching their mule-like tails in all the pride of fleetness and freedom."

THE ZEBRA.

The ZEBRA, *Equus Zebra* (Plate XXXIII), is the most conspicuous and beautiful of the whole tribe. The general color is a creamy white, marked regularly with velvety black stripes that cover the entire head, neck, body, and limbs, and extend down to the very feet. It is worthy of note, that the stripes are drawn nearly at right angles to the part of the body on which they occur, so that the stripes of the legs are horizontal, while those of the body are vertical. The abdomen and inner portions of the thighs are cream-white, and the end of the tail is nearly black. These stripes are, however, in the male, rather brown and yellow. The skin of the neck is developed into a kind of dewlap, and the tail is sparingly covered with coarse black hair. By the Cape colonists it is called "Wilde Paard," or Wild Horse. It is, however, less like the horse than the Dziggetai or Kulan, described in our previous chapter. It resembles the Dziggetai much more than the Dauw, as it is always found in hilly districts, and inhabits the high craggy mountains in preference to the plains. It is timid, and flies at the sight of any strange object, and in captivity is described as fierce, obstinate, and nearly untamable. A young female, however, that was caught young and sent to the Jardin des Plantes of Paris, was so tractable that it allowed itself to be approached and led almost as readily as a horse. Mr. Rarey, the American horse-tamer, had, he confesses, more difficulty in taming a zebra than any horse which he had seen, but he succeeded at last in subduing it.

The three species just described possess an equal degree of intelligence, and equal acuteness of sense. A certain wildness, cunning, and courage is common to them all, and all have an infinite love of liberty. They defend themselves bravely against all beasts of prey; the hyænas leave them at peace; the lion alone, perhaps, can conquer one of the tiger-horses; the leopard attacks only the weak ones, the adults being able to repel him. As usual, the worst enemy of these harmless creatures is man. The difficulty of the chase and the beautiful skin of the animal attract the European sportsman; the Cape colonists are passionately devoted to it, and the Abyssinians seem also to hunt any zebras that come in their way. Europeans shoot the tiger-horses, the natives spear them, but very commonly the pretty brutes are taken in pitfalls and then slain, for the dead animal alone is of any use to the natives, who regard its flesh as a dainty. The Boers of the Cape never touch it, but leave the animal to be eaten by their Hottentot servants. When wounded, the Zebra gives a kind of groan which Andersson compares to the gasps of a drowning man.

In striking contrast to the ass, the tiger-horses are very silent. The Zebra's voice is rarely heard, and its subdued neighings have a melancholy sound. The Quagga, according to Cuvier, repeats about twenty times the syllables Oa, oa, or Qua, qua, from which the Hottentot name of the species is probably derived. The Dauw repeats about three times the sounds yu, yu, yu.

All these animals bear captivity well, and if well taken care of, will propagate. Very numerous cross-breeds have been procured by various combinations of the Ass and Kulan with the Quagga and Zebra, and the offspring of the Ass and the Zebra has again been crossed with the Pony. In most cases these mongrels were most like their father. A thorough-bred mare in England had, by a Quagga, a filly which resembled its mother in color, being a chestnut with very few stripes. This filly was crossed with a thoroughbred horse, and the foal retained the short mane and some of the stripes of its grandsire. All the future foals of the thorough-bred mare that had been bred to the Quagga were more or less striped, although their sire was a black horse.

THE MULE.

In its size and neck and shoulders, the Mule inherits the fine shape of the Mare. From the Ass it derives the length of its ears, its almost naked tail, its sure-footedness, strong constitution, and its bray. Its hair is short, rough, and generally of a brownish-black color; there are, however, many Mules which have gray or chestnut coats, with a stripe along the back of dark hair, as well as bands of the same shade around the limbs. It is a long-lived animal, even occasionally reaching the age of forty-five to fifty years. Almost omnivorous in reference to herbage, Mules have an advantage that cannot be too highly valued; moreover, a level country or mountainous region equally suits them; provided neither are too damp. Although patient, it will not submit to ill-treatment without bearing malice.

The cross between a stallion and an ass is called a Hinny. It has the small size, long ears, and ungraceful figure of its mother, but derives from its sire the thinner and longer head, the tail covered with hair its entire length, and its neighing voice.

In all mountainous regions the Mule is indispensable; in South America it is to the natives what the camel is to the Arabian. Its strength, endurance, and sure-footedness render it of inestimable value for the transportation of merchandise. In the plantations of our Southern States, and in the farms of the West it is of the utmost service. It is seldom, however, used by us for riding, while the Spaniard and Spanish

American gentleman is as proud of his mule as an American is of his trotter. In older days the mule was the favorite beast of the clergy, and the Pope used to ride about, drawn by a team of white mules.

As a rule, neither the Mule nor the Hinny will produce offspring, but instances where a mule has brought forth are not unknown. In the year 1762, a female mule was crossed with a gray American horse, and gave birth to a beautiful foal of a fox-red color, with a black mane, which was broken in to ride when three years old. She continued to have foals every two years, till she had produced five. Other observations of later date place beyond doubt the occurrence of other similar cases.

The question of the fertility of mules has been often debated. In these days men have more eagerly than ever asserted and denied the doctrine that "all hybrids are sterile," for it is closely involved with the doctrine of the origin of species by natural selection and climatic influences, and therefore opposed to the teaching of those who maintain that every species is derived from one originally created pair which has continued "to bring forth after its kind."

CHAPTER VII.

THE TAPIRS.

THE FAMILY TAPIRIDÆ—THE AMERICAN TAPIR—ITS TRUNK—ITS HABITS—THE TAPIR AS A DOMESTIC ANIMAL—A TAPIR HUNT—PECULIAR MARKS OF THE YOUNG TAPIR—THE MALAY TAPIR—ITS TRUNK—ITS COLOR—DISCOVERY OF THE ANIMAL—CHINESE ACCOUNT—THE PINCHAQUE—BAIRD'S TAPIR.

THE family TAPIRIDÆ consists of a small group of animals which are found in the equatorial forests of South America, and in the Malay Peninsula and Borneo. For a long time only two species were known, but recently Dr. Gray has discovered four species in the Andes of New Granada and Ecuador, and two more in Central America.

The Tapirs are comparatively small, stout-built animals, that appear to stand half-way between the elephants and the swine. They have an elongated, narrow head, slender neck, short rudimentary tail, and legs of moderate length; the ears are short and broad; the eyes oblique and small. The upper lip is prolonged like an elephant's trunk, and hangs down over the lower lip; the fore-feet have four toes, the hind-feet, three; the hair is short, smooth, and thick.

The American Tapir has been known to naturalists for a long time, but the Malay Tapir has only been known since the beginning of this century to Europeans, although it has been described for centuries in Chinese works. As is usually the case when a family has representatives in the old and new world, the old world species seem more highly developed than those that live in the new.

1.—GENUS TAPIRUS.

The AMERICAN TAPIR, *Tapirus terrestris* (Plate XXXIV), was mentioned by travelers a few years after the discovery of America; they regarded it as a kind of hippopotamus, and it appeared in works of

SOUTH AMERICAN TAPIR
MALAYAN TAPIR
BLACK RHINOCEROS

PLATE XXXIV UNGULATA

natural history under the name of *Hippopotamus terrestris.* But, about the middle of the eighteenth century it began to be better known, and at present we possess very accurate accounts of it, and its habits.

The Tapir measures about six feet in length from the nose to the tail, and its height to the withers is about four feet. The body is fat, and terminates in a broad rump. The head, which is pretty large, is compressed on the sides; the eyes are small; the ears elongated, and the animal can contract or enlarge them; the nose is prolonged a few inches, in the shape of a trunk. This addition, which is naked and flesh-colored at the tip, can be diminished to half and elongated to double its quiescent length, but is without that movable finger-like extension which is the characteristic of the elephant's proboscis; so it can be of no use in seizing objects or in sucking up water. The Tapir takes its food directly with its mouth; when it drinks, it raises its contracted trunk in such a way as to prevent its being wetted. The neck is rather long; and the legs are strong and thick. The anterior extremities are terminated in four toes, each of which is provided with a little, short, rounded hoof; the posterior extremities have but three toes. The tail is very short and stumpy. The thick, hard skin of this animal is covered with short hair, very close and smooth, of a more or less dark brown, except under its head, its throat, and the tips of its ears, where it is of a whitish color. The male has on his neck a short mane, composed of stiff bristles, of about an inch and a half in length; this decoration is sometimes seen on the female.

In the densely wooded regions which line the banks of the rivers of tropical America the Tapir lives alone, hidden in the forests and in the most secluded retreats. Following always the same track in its excursions through the woods, it forms well-trodden footpaths, which the sportsman can easily recognize. It sleeps during the day, and wanders at night to seek its food. Sometimes, however, rainy weather brings it from its hiding-place during daylight, when it goes to the swamps, in which it delights to wallow, or to the streams, for it is a great water lover, and can swim and dive with an ease which almost justifies the old naturalists in calling it a "river-horse." Although a large animal, and very strongly made, it falls a victim to many destroyers, the jaguar being the most terrible of its enemies. It is said that when the jaguar leaps upon the Tapir's back, the affrighted animal rushes through the brushwood in hopes of sweeping away its deadly foe, and if it be fortunate enough to gain the river's bank, will plunge into the water, and force the jaguar, who is no diver, to relinquish his hold. The tough, thick hide with which the Tapir is covered is of great service in enabling the animal to pursue its headlong course through the forest without suffering injury from the branches. When it runs, it carries its head very low, as does the wild boar under similar circumstances.

In disposition the Tapir is very gentle, and does not attack human beings except when wounded and driven to bay. It then becomes a fierce and determined opponent, and is capable of inflicting severe wounds with its powerful teeth. The hunter's dogs are often dangerously wounded by the teeth of the despairing Tapir. The voice of the Tapir has a curious, shrill, whistling sound; it is but seldom uttered, and is in ridiculous disproportion to the size of the animal. Schomburgk indeed believed that it was only uttered by the young ones; but the observation of captive Tapirs show that it is the usual voice of the animal. The senses of hearing and scent appear to be equally sensitive, but its sight is weak. During the daytime it is seldom seen, preferring to lie quietly hidden in the deep underwood during the hotter hours of the day, and to emerge at night in order to obtain food, and meet its companions. The nocturnal journeys which the Tapir will make are of considerable extent, and the animal proceeds straight onwards, heedless of bank or river, surmounting the one, and swimming the other, with equal ease. The food of the Tapir is generally of a vegetable nature, and consists of young branches and various wild fruits, such as gourds and melons.

The Tapir is often seen in menageries, and bears captivity well; of course it requires a warm stall and protection from the cold of winter, as, like other importations from the tropics, it is very subject, in our variable climate, to diseases of the lungs. It is reported that in Brazil attempts have been made to tame the Tapir into a really domesticated animal, less for the sake of its food, than for the purpose of making it a draught animal. And indeed it might be worth the trouble of trying the experiment, for they are easily tamed. Frederick Cuvier has given us a few details of the habits of a young Tapir with which he was acquainted. This animal was gentle and confiding, and appeared to have no will of its own. It did not defend its food, but allowed the dogs and goats to partake of it together with itself. When it was let loose into an enclosure, after having been shut up for some time, it showed its joy by running round it several times. It also playfully seized by the back the puppies with which it was brought up. When it was forced to leave a place it liked, it complained by uttering a few plaintive cries. Frederick Cuvier assures us that if the Tapir would be of any use to us, it could be very easily domesticated. Isidore Geoffroy Saint-Hilaire also wished the experiment of domesticating this animal in Europe to be tried; but his idea was never carried out.

"Not less easy to feed than the pig," says Saint-Hilaire, "the Tapir seems to me eminently suited to become one of our domestic animals. When it has no creatures of its own kind to associate with, I have seen it seeking the society of all the animals that were near, with an eagerness without an example in other Mammalia. The Tapir would be useful in two ways to man: its flesh, especially when improved by proper diet, would furnish a wholesome and at the same time an agreeable food; and as it is much larger than the pig, the Tapir might be of great service as a beast of burden to the inhabitants of the south of Europe, and, after a time, to those of colder countries."

The Tapirs are eagerly hunted. The hide is valuable for its thickness and toughness, and whips made from it are exported in large numbers from the Argentine States. The claws are regarded by the Indians as a great medicine, and a sovereign cure for the falling sickness if worn round the neck, or, after having been reduced to powder, taken inwardly. Schomburgk describes a hunt: "As we advanced, we saw a Tapir with its young on a sand-bank, but scarcely had the word 'Maipuri' passed the lips of our Indians, than the animals disappeared in the thick bush that lines the shore. We landed at once, and as soon as we had scrambled through the bush we saw that the fugitives were making for the reeds and prickly grasses that covered the plain. Our dogs were in the boat, and we had no desire to attempt to advance through the reeds, but our Indians glided between the dangerous grasses like serpents. Soon two shots were heard, then shouts of triumph. Our Indians rushed in, and thus we found a road till we came to the prey. While we were standing around it, our dogs came up and took up the scent of the young one. Its whistling cry indicated that it was near the edge of the reed-bed; it soon broke cover, followed by the dogs and thirty Indians whose yells were mingled with the baying of the dogs. The animal grew weaker, and after an obstinate resistance, was secured and carried to our boat."

The young Tapir does not resemble the old ones in color. The back is brown, the upper part of the head is thickly covered with white circular spots, and on each side of the body there run four uninterrupted rows of spots of lighter color which extend over the limbs. As the animal grows older these spots form stripes, and at the end of the second year disappear. The mother is not prolific, as she only bears one young one in a season.

THE MALAY TAPIR.

The MALAY TAPIR, *Tapirus indicus* (Plate XXXIV) is known also by the name of KUDA AGER, two Malay words which signify "River-Horse." It is distinguished from the other members of the genus by its more considerable size, and its more slender body; the head is more pointed, the trunk is more developed, the feet more powerful, the mane is absent. The construction of the trunk-like nose is very remarkable. In the American Tapir this is round; in the Malay Tapir it presents the same section as the trunk of an elephant, that is, round on the upper, flat on the lower surface, and exhibits the finger-like conformation which is so noticeable in the proboscis of the elephant. The color of the Malay species is also very peculiar. The head, the neck, and the body as far as the shoulder-blades, as well as the legs, thighs, and tail are deep black, the rest of the body is white. At some distance, therefore, the animal looks as if it was covered with a sheet. Both the white and black hairs shine in an extraordinary manner. The size of a full-grown male has not been accurately ascertained; an adult female measured nearly eight feet in length, including the tail; she is said to be larger than her mate.

It is astonishing, that with all the intercourse that has existed for centuries between Europe and the East, nothing certain was known respecting this animal until the year 1819. Cuvier had asserted that there was no chance in our days of discovering any large mammal; one of his pupils, named Diard, in reply, sent him a sketch of the Kuda-Ager, or Maiba, as the Hindoos call it. He added: "When I first saw at Barrackpore the Tapir, a sketch of which I send, I was surprised that such a large animal had not previously been noticed, and I was still more surprised when I found in the Asiatic Society the head of a similar animal, with the remark that it was as common in the peninsula as the elephant or rhinoceros." Diard, however, was in error in supposing the creature was quite unknown even to Europeans. A Chinese Natural

History, Pen-tsao-kan-a now describes it as follows: "The ME is like a bear; its head is small, its legs short, its short brilliant hair is black and white. It has the trunk of the elephant, the eyes of the rhinoceros, the tail of a cow, and the feet of a tiger." Pictures of it are very common in Chinese and Japanese Books for Children.

We are still without good accounts of its life in the woods, and even the descriptions of its habits in captivity are far from satisfactory. It may be assumed, however, that it does not differ in its modes of life from the American species already described.

The PINCHAQUE, *Tapirus villosus*, was discovered by M. Raulin during his residence in the Andes. It has a head which resembles strongly that of a fossil Tapir found in the tertiary deposits of the Pampas, the nasal bones being more elongated than in any other species at present existing. This Tapir, from living at a considerable elevation, and therefore in a comparatively cold climate, is entirely covered with long hair which is of a brown color. It is, in consequence, called *Tapirus villosus*, or "The Shaggy Tapir." It does not appear to extend far over the Andes. Gray gives the "Cordilleras" as its habitat, but it seems confined to Ecuador and the United States of Colombia, where it dwells at about eight to twelve thousand feet above the sea.

GENUS ELASMOGNATHUS.

This genus has been formed by Dr. Gray to contain one or two species of Tapirs which inhabit Central America from Panama to Guatemala. He gives the generic characteristics as follows: "The internasal cartilage is ossified nearly the whole length, the bony part produced beyond the end of the nasal." Of the species this learned naturalist enumerates, we need only mention one; the so-called *Elasmognathus Dowii* being a very doubtful form.

BAIRD'S TAPIR, *Elasmognathus Bairdii*, is distinguished by possessing a very short and close fur, the color of which is dark brown, approaching closely to black, but the chest is white, and the cheeks are light brown. The young have the peculiar striped arrangement which is so remarkable in the young of the preceding genus.

Baird's Tapir is a native of the Isthmus of Panama, and has been found as far north as Mexico. It is thought to be the largest of the American Tapirs,

CHAPTER VIII.

THE RHINOCEROS.

THE FAMILY RHINOCEROTIDÆ—GENERAL DESCRIPTION—THE HORN—PECULIAR STRUCTURE OF THE HORN—KNOWN TO THE ANCIENTS—WOODCUT BY ALBERT DÜRER—ARAB SUPERSTITIONS—HAUNTS OF THE RHINOCEROS—A NOCTURNAL ANIMAL—ITS FOOD—ITS HABITS—ITS SENSES—ITS FITS OF RAGE—MATERNAL AFFECTION—ITS FRIENDS THE SMALL BIRDS—CAPTIVE RHINOCEROSES—USES OF ITS HIDE.

THE family RHINOCEROTIDÆ is especially characteristic of Africa and Northern and Malayan India. Four or five species, all two-horned, are found in Africa, where they range over the whole country south of the desert to the Cape of Good Hope. In the east of Asia there are also four or five species which range from the forests at the foot of the Himalayas eastward through Anam and Siam to Sumatra, Borneo, and Java. Three of these Asiatic species are one-horned, the others found in Sumatra and Java are two-horned. All these species are so much alike, that most naturalists do not consider them as forming distinct genera. Gray, however, divides the Asiatic Rhinoceros into two genera, *Rhinoceros* and *Ceratorhinus*, and the African into two, *Rhinaster* and *Ceratotherium*.

The RHINOCEROTIDÆ are all clumsy and unwieldy-looking animals of considerable size, characterized by a remarkably elongated head, the front part of which bears one horn, or two in a line, a short neck, a powerful body, entirely or almost entirely devoid of hair but covered with a thick armor-like hide, a short tail, short stumpy legs, all of which terminate in three toes enveloped in hoofs. The mouth is disproportionately small, the upper lip is developed into a trunk-like process, while the under-lip is square and truncated, the eye is small, the ear rather large. The hide, which is almost impenetrable, is in some species divided by folds of soft and pliant skin into a series of shield-like plates, in others it lies close to the body with only a few slight folds. The few hairs which appear on the animal are confined to the borders of the ears and the tip of the tail, with occasionally a few bristles on the back. The most peculiar feature of the creature is its horn or horns. This is a very curious structure, and worthy of a brief notice. It is in no way connected with the skull, but is simply a growth from the skin, and may take rank with hairs, spines, or quills, being indeed formed after a similar manner. If a Rhinoceros horn be examined—the species of its owner is quite immaterial—it will be seen to be polished and smooth at the tip, but rough and split into numerous filaments at the base. These filaments, which have a very close resemblance to those which terminate the plates of whalebone, can be stripped upward for some length; and if the substance of the horn be cut across, it will be seen to be composed of a vast number of hairy filaments lying side by side, which, when submitted to the microscope and illuminated by polarized light, glow with all the colors of the rainbow, and bear a strong resemblance to transverse sections of actual hair. At the birth of the young animal, the horn is hardly visible, and its full growth is not attained for several years.

As the horn is employed as a weapon of offence, and is subjected to violent concussions, it is set upon the head in such a manner as to save the brain from the injurious effects which might result from its use in attack or combat. In the first place, the horn has no direct connection with the skull, as it is simply set upon the skin, and can be removed by passing a sharp knife round its base, and separating it from the hide on which it grows. In the second place, the bones of the face are curiously developed, so as to form an arch with one end free, the horn being placed upon the crown of the bony arch in order to diminish the force of the concussion in the best imaginable manner. The substance of the horn is very dense, and even when it is quite dry, it possesses very great weight in proportion to its size. In former days it was supposed to bear an antipathy to poison, and to cause effervescence whenever liquid poison was poured upon it. Goblets were therefore cut from this material, and when gorgeously mounted in gold and precious stones, were employed by Eastern monarchs as a ready means for detecting any attempt to administer a deadly drug.

Although the Rhinoceros is at present confined to the Torrid Zone, we have evidence that it once was more widely spread. The traveler Pallas saw in Siberia the feet of a Rhinoceros which had been found in the banks of a river; they were still covered with skin. The head was afterward discovered and sent to St. Petersburg. This fossil species was of a yellow color, with a thick hide destitute of folds, covered with stiff hair projecting from a soft woolly undercoat.

The Rhinoceros has long been known. Its figure occurs in the hieroglyphics of Egypt; many pious persons identify it with the "unicorn" of the Bible, forgetting that "unicorn" is merely a conjectural translation of the Hebrew word designating an animal of the bovine family. Pliny tells us that Pompey exhibited a Rhinoceros B. C. 61, in the arena at Rome. "The Rhinoceros," he writes, "is the born foe of the elephant. It whets its horn on a stone, and, when fighting, aims at the belly." He adds, that at the period of his writing, the animal was found at the island of Meroe in the Nile. Strabo saw one in Alexandria. Martial describes a fight between one and a bear. The Rhinoceros, he says, was very sluggish and slow to engage, but finally tossed the bear as a bull tosses a dog. The Arabian writers are the first to distinguish between the African and Asiatic species, and the Rhinoceros plays a conspicuous part in the Arabic tales of magicians. Marco Polo is the first European in modern times to mention the animal, and in 1513 King Emmanuel of Portugal received a living specimen from the East Indies. Albert Durer made a wood-cut representing it; unfortunately he was guided by a very imperfect drawing sent from Lisbon to Germany. For about two hundred years this wood-cut was the only representation of the Rhinoceros. Chardin, at the beginning of the last century, sent a better sketch from Ispahan, and at present accurate representations of most of the species have been published.

Although all the *Rhinocerotidæ* have the same general traits, yet each species has its own peculiarities, some being exceedingly irritable and bad-tempered, others harmless and gentle. They are everywhere more dreaded than the elephant. The Arabs of the Soudan believe that magicians assume the form of the monster, which is accursed from the beginning:—"Not the Lord, the all-creating, made them, but the Devil, the all-destroying, and the Faithful must have nothing to do with them. The Mussulman must quietly get out of their road, in order that he may not defile his soul, and be rejected at the day of judgment."

The favorite haunts of the Rhinoceros are well-watered regions, swampy lands, rivers that overflow their banks, and lakes with sedgy and muddy shores, near which grassy plains extend. Before a creature of such weight, and protected with such an armor, the densest jungle opens its most impenetrable thickets, and the most terrible thorns are powerless. Hence we find them in great numbers in the forest lands, from the shore up to the height of ten thousand feet above the sea. Some species seem to prefer the elevations. The Java Rhinoceros is more numerous in the hills where many grassy pools and swamps are scattered, the African Rhinoceros, which lives in the prickly mimosa thickets of Central Africa, is not rare in West Abyssinia at seven thousand feet above the sea. Water is indispensable; every day the huge animal rolls himself in the mud; for in spite of its thick hide, it is very sensitive to the attacks of insects, against whose stings nothing but a good coating of mud can protect it. Plunging into the soft mud, they lie and grunt for pleasure. When the coating of mud dries and falls off, the Rhinoceros seeks to get temporary relief from his insect pests by rubbing himself against the trunks of the trees.

The Rhinoceroses are more active by night than by day; they dislike great heat, and sleep during the noontide in some shady spot. They sleep very deeply, and are easy to approach when they are thus buried in repose. Gordon Cumming reports that even the little birds, which always accompany the Rhinoceros, and warn him of danger when sleeping, in vain endeavored to awaken one which he was preparing to shoot. Some Hottentots, led by the creature's loud snores to its sleeping-place, put their guns close to his head and fired; the sleeper never stirred; they loaded again, and killed it at the next discharge. About midnight the Rhinoceros takes a mud-bath, and goes to his feeding-grounds, where he lingers for hours. Afterward, he roams wherever he pleases; he passes through the bush and jungle, never changing his course except to avoid the larger trunks, and in India he forms long, straight paths, where all the shrubs at the sides are broken down, and the ground trodden hard; the elephant, on the contrary, pulls up by the roots the brush that stands in his way. These paths always lead to water, and would be very useful to the traveler if he could be sure of not meeting their constructor.

As regards food, the Rhinoceros is to the Elephant what the Ass is to the Horse. He loves hard fodder, thistles, reeds, prairie grass, and the like. In Africa it eats the prickly mimosa, especially a low, bushy variety which, on account of its crooked thorns, has been called the "Wait-a-bit thorn." During the rainy season, the Rhinoceros approaches nearer to the cultivated land, and does incalculable damage to the farmers, for his appetite is enormous, and even in captivity the animal requires nearly sixty pounds weight of food a day. The Indian species uses its finger-like lip to seize tufts of grass, and to convey them to its mouth. As the throat is enormous, it swallows its food half-masticated. Small trees or shrubs are often dug up by the aid of its horn, which it inserts beneath their roots. But in their food, also, the various species differ. The Indian Rhinoceros prefers twigs; the Sumatran, grass; the African, the branches of the mimosa, which it cuts off "as if with a pair of shears," while it is poisoned by the Euphorbium which the White Rhinoceros eats with impunity.

The Rhinoceroses pass their lives eating or sleeping. Unlike the elephants, they do not live in herds; their troops never exceed ten members. In this society there is little harmony, each lives for itself, and does what it chooses. Between the sexes, however, intimate and permanent friendships are formed, and pairs may be often seen which seem to act in common; and in captivity both sexes display a tender affection for their mates. Usually the Rhinoceros walks slowly, and is awkward in lying down or rolling, but he is not so unwieldy as he looks. He can not turn actively, nor can he climb; but when once in motion on level ground he runs pretty quick at a brisk *trot*, for he does not *pace* like the camel or elephant. He holds his head very low, and when enraged shakes it from side to side, and raises his stumpy tail. He can keep up this trot for a considerable time, and is dangerous even to well-mounted hunters, especially where the jungle is thick, while elephants with a howdah have no chance with him. He swims well, and has been seen to dive to the bottom of pools for the purpose of digging up aquatic plants with his horn.

The Rhinoceros possesses a very acute sense of hearing and of smell. He often pursues his enemy by scent alone, and when enraged, regards neither the number nor quality of his opponents. He charges straight on a line of armed men, or on the most inconsiderable object. Red colors provoke his anger, even when the wearer or bearer of the offensive garment is quite peaceable. His anger knows no limits; he cuts at the bushes near him, or plows long furrows in the ground. A writer, describing the single-horned species, says that "it is a mistake to suppose that the horn is their most formidable weapon. I thought so myself at one time," he adds, "but have long been satisfied that it is merely used in defence, and not as an instrument of offence. It is with their cutting-teeth" (lower canines) "that they wound so desperately. I killed a large male," this writer asserts, "which was cut and slashed all over its body with fighting; the wounds were all fresh, and as cleanly made as if they had been done with a razor—the horn could not have been used here." Another one he had wounded halted and, out of pure rage, cut at the jungle right and left, exactly as a boar uses his tusks. "A medical friend's servant, who was sauntering through the forest, was actually disemboweled by a Rhinoceros. He examined the wound immediately, and I heard him say afterward that if it had been done with the sharpest instrument, it could not have been cleaner cut. This could not have been done with the horn."

The extraordinary irritability of the Rhinoceros conceals in a great degree its real intelligence. While far inferior to the elephant, it yet is superior to most of the ruminants; and in captivity soon learns to know its keeper, and submit to his wishes. We still, however, know too little of the animal in its wild life. The female defends its young at all hazards. A hunter in India discovered a female with its young one; the mother rose up and slowly retired, pushing the little one before her with her snout. The hunter rode up and made a cut at her with his sabre, but the weapon only left a few white marks on the solid hide. The mother endured patiently all his blows till her offspring was safe in the bush. Then she turned, gnashing her teeth, on the assailant. His horse had the sense to run away, the Rhinoceros pursued, smashing and crashing through all obstacles. When she reached the hunter's attendants, she charged them too, and when they climbed some trees, she attacked them; the huge trunks quivered under the blows of the enraged beast, which was finally shot in the head. A young Rhinoceros was born on board a ship in London in 1872, but, to the great loss of science, it died in three or four days, deeply lamented by Mr. Bartlett.

Pliny's remark about the hostility of the Rhinoceros and the Elephant is a mere fable, of course. But if the Rhinoceros has no hereditary enemy, it has an hereditary friend. All African travelers relate that the Borélé is attended by a small bird which lives on the parasitic insects abounding on its patron's ample hide, and which wakens him from slumber when danger draws nigh. "They are his best friends," writes Gordon Cumming, "and seldom fail to rouse him. I have often chased a Rhinoceros for miles, and during the chase the birds sat upon his back. When I put a ball into his shoulder, they fluttered about six feet above him, screaming, and then resumed their station. Sometimes boughs of trees swept them away, but they soon returned. I have shot a Rhinoceros by night, and the birds remained with it till morning, thinking it was asleep, and tried to awaken it when I came up." Other enemies than man the Rhinoceros need not fear. Lions and tigers know that their claws are powerless, and that the paw which can prostrate an ox would not be felt by the armor-clad pachyderm. Man, however, in all regions, pursues the Rhinoceros, and modern improvements in musketry have rendered him an easy victim to the European sportsman. When unaccustomed to the sight of intruders, the animal takes to flight; but when it has been repeatedly disturbed, it does not wait to be attacked, but commences hostilities, and fights to the death.

The Rhinoceros is constantly found in our menageries, where it shows itself, as a rule, good-tempered, ready to take food from the spectators' hands, and tame enough to be allowed to walk about in a paddock. It ought to be bathed in or sprinkled with lukewarm water every day. How long they will live is not known, but captive ones in India have been said to have attained the age of forty-five. The use which can be made of the Rhinoceros does not compensate for the damage he does. He is intolerable where regular cultivation exists. All parts of the animal, however, have their uses. The blood, as well as the horn, are highly esteemed in the East for their medicinal qualities. Cups and drinking-vessels of rhinoceros-horn are deemed a necessary in the house of a wealthy man, for such cups reveal the presence of poison in any fluid poured into them. The horn is used also for the handles of sabres, and when polished, has a beautiful, soft, yellowish tint. Shields, breast-plates, dishes, and other utensils are made from the hide; the flesh is eaten, the fat and marrow employed as salves of magic efficacy. The Chinese say that after swallows' nests, lizards' eggs, and little dogs, nothing is such a dainty as the tail of a rhinoceros.

CHAPTER IX.

THE ASIATIC RHINOCEROSES.

THE ONE-HORNED RHINOCEROS—THE INDIAN RHINOCEROS—ITS THICK HIDE—MODE OF HUNTING—
THE WARA OR JAVANESE RHINOCEROS—THE EMPEROR BABER—THE TWO-HORNED RHINOCEROS
OF BADAK OF SUMATRA—THE HILL-EATING RHINOCEROS—THE ROUGH-EARED RHINOCEROS.

DOCTOR GRAY, in his last Catalogue of Mammalia, groups the Asiatic Rhinoceroses into two genera, *Rhinoceros* and *Ceratorhinus*. The former comprehends those species which have the armor-like hide divided into distinct shield-like plates, the latter those in which the hide has divisions only at the shoulders and the loins; the former have one horn, the latter two; there are also differences in the dentition of the genera.

THE ONE-HORNED RHINOCEROS.

The INDIAN RHINOCEROS, *Rhinoceros Indicus* (Plate XXXV), attains a length of twelve feet, including its tail, and stands about five feet high; its weight is about forty-five hundred pounds. Its head is short and triangular; its mouth, of a moderate size, has an upper lip, which is longer than the lower, pointed and movable. It has in each jaw two strong incisor teeth. Its eyes are small; its ears are rather long and movable. The horn upon its nose is pointed, conical, not compressed, sometimes two feet in length, and slightly curved backwards. This singular weapon is composed of a cluster of hairs closely adherent; for when the point is blunted, it is often seen divided into fibres resembling the hairs of a brush. This horn is, however, very firm and hard, of a brownish-red on the outside, of a golden-yellow inside, and black in the centre.

The thick, horny hide lies upon a thick layer of loose cellular tissue, and is thus capable of being easily moved to and fro. This integument is divided into plates by regular deep folds which exist even in the newborn young ones. The skin in these folds is very soft and thin, while the hard plates, which can be lifted up by hand, are as thick and solid as boards. The first fold runs just behind the head around the neck, forming below it a kind of dewlap; from this fold two oblique folds, one on each side, run backwards and upwards, deep at first, but vanishing at the withers; from the centres of these folds a pair of creases runs forward to the back of the neck. Behind the withers is a very conspicuous deep fold which descends behind the shoulder-blade, and then turns horizontally over the fore-arm and passes in front of it. Another deep fold commences at the loins, and descends obliquely in front of the thigh, sending out other folds toward the tail. The hide appears divided into three grand pieces, one on the head and shoulders, another one on the body, the third on the hind-quarters. The animal is thus defended with a shield on its back, one on each shoulder, one on the rump, and one on each thigh. This hide has been compared to a suit of armor of well-adjusted pieces. It is, however, so thick and hard that, without these creases or folds, the animal, imprisoned as it were in its armor, could scarcely move. It is of a dark color, nearly bare, generally provided with only a few coarse and stiff hairs on the tail and ears, occasionally with curly woolly hairs on certain parts of the body. But each of these shields is everywhere covered with irregular, round, more or less smooth, tubercles of horn, which lie so thick together on the outside of the legs that these limbs look as if they were covered with scale-armor. The skin in the folds is of a dark flesh-color.

The tubercular prominences on the hide sometimes assume an extraordinary degree of development. In the old wood-cut by Albert Durer, the Rhinoceros is represented with horns on the shoulder-plates as well as on the nose. In the Zoological Gardens of Antwerp, a Rhinoceros eighteen years old was remarkable for the size of these tubercles; some were as large as hazel-nuts, others on the collar-bone and on the skin before the ears attained a length varying from two to five inches. In the centre of the neck there was a group of five perpendicular horns, one of which was over three inches in height. These little horns fell off from time to time, and were quite different in structure from the wart-like knobs which occur on the sides of the animal. A Rhinoceros in the Zoological Gardens of Moscow did actually shed a horn, and another grew in its place. There can be no doubt this happens with wild animals.

The Indian Rhinoceros is at present restricted to the Terai, an unhealthy, marshy tract at the foot of the Himalayas, skirting Nepaul,

WHITE RHINOCEROS
INDIAN RHINOCEROS

Sikkim, and Bhotan, being more common in the eastern portion of the district, and may perhaps occur in the hill ranges east and south of the Brahmaputra River.

In India, in former times, the Rhinoceros was hunted on light, quick horses. The huntsmen followed it from afar off, and without any noise, till the animal became tired and was obliged to lie down and sleep. Then the sportsmen approached it, taking care to keep to leeward, for it has a very acute sense of smell. When they were within shot, they dismounted, aimed at the head, fired, and galloped away; for if the Rhinoceros is only wounded, it rushes furiously upon its aggressors. When struck by a bullet, it abandons itself wholly to rage. It rushes straight forward, smashing, overturning, trampling under foot, and crushing to atoms everything which comes in its way.

THE JAVANESE RHINOCEROS.

The JAVANESE RHINOCEROS, *Rhinoceros sondaicus*, obtains its English epithet from the belief that it was a species peculiar to Java. This is not the case, for it is found in Tenasserim and the Sunderbunds, and is often confounded with the Indian Rhinoceros. The difference, indeed, between the two species is not sufficiently striking to be noticed by ordinary beholders unless the two animals were close together. The *Rhinoceros sondaicus* is about a third less in size than the Indian Rhinoceros; its coat of mail is much the same except that the tubercles on the hide are smaller and of uniform size, while the polygonal facets of the skin have a few small bristles growing upon a depression in the centre of each. The strong fold or plait at the setting-on of the neck is continued across the shoulders; the neck-folds are less heavy and pendulous, and the posterior plate crossing the buttock from the tail is less extended.

The Javanese Rhinoceros, or Wara as it is called by the natives, is reported to be a timid animal, but an instance is related of one attacking a sailors' watering-party in Java. It is diffused, more or less abundantly, over the whole Indo-Chinese region and the Malayan peninsula, but is not found in Sumatra. In Java it is found in the most elevated regions, ascending with astonishing swiftness even to the tops of the mountains. Its retreats are discovered by deeply-excavated passages which it forms along the declivities of the hills. In Bengal it is found not only in the Sunderbunds, but in the Rajmahal hills, near the Ganges, and several have been killed within a few miles of Calcutta; it lives in these cultivated districts on growing rice and vegetable roots.

In the time of Baber, the founder of the Great Mogul dynasty, the Rhinoceros occurred near Benares, in Central India. In his notice of the animals peculiar to Hindustan, after describing the elephant, the imperial author remarks: "The Rhinoceros is another. This also is a huge animal. The opinion prevalent in our countries that a Rhinoceros can lift an elephant on its horn is probably a mistake. It has a single horn over its nose upwards of a span in length; but I never saw one of two spans." From this it would seem that the particular species referred to is *Rhinoceros sondaicus*, inasmuch as Baber would probably have been able to obtain larger examples of the horn of *Rhinoceros indicus*. "Out of one of the largest of these horns," he adds, "I had a drinking-vessel made and a dice-box, and about three or four fingers' bulk of it might be left. Its hide is very thick. If it be shot at with a powerful bow drawn up to the arm-pit with much force, the arrow enters three or four fingers' breadth. They say, however, that there are parts of its skin that may be pierced, and the arrows enter deep. On the sides of its two shoulder-blades, and of its thighs, are folds that hang loose and appear at a distance like cloth housings. It is more furious than the elephant, and cannot be rendered so tame. It strikes powerfully with its horn, with which, in the course of these hunts, many men and many horses were gored." This description of a One-horned Rhinoceros is unmistakable, but it is strange to read that these animals could be killed with arrows.

The Wara is hunted for the sake of its horn, which is in great demand in China. Pitfalls are dug in the paths it is known to traverse, and the bottom of the pits planted with sharp stakes; the whole is then covered with branches and twigs. The Rhinoceros comes along, unsuspecting, treads on the treacherous boughs, and is either impaled at once, or at all events rendered helpless. The adults are slain because they cannot be extricated from the pit; the young ones are led away captive.

THE TWO-HORNED ASIATIC RHINOCEROS.

The SUMATRAN RHINOCEROS or BADAK, *Rhinoceros sumatranus*, is the basis of Gray's genus *Ceratorhinus*. It is a comparatively small animal, which certainly never much exceeds four feet in height; but its horns sometimes attain a beautiful development, more especially the anterior one, which is much longer than the other, slender except at base, and has a graceful curvature backward, more or less decided in different individuals; the other, or posterior horn, is not placed close behind the first, as in the different two-horned African species, but at a considerable distance from it, and it has a corresponding backward curvature. An anterior horn of this small Rhinoceros in the British Museum measures thirty-two inches along its front, and is seventeen inches in span from base to tip. The posterior horn is very thick, short, and pyramidal. A pair of horns of this species is a beautiful object when carved and polished, and set with the bases upward and on a parallel in a carved black wooden stand, similar to those upon which Chinese metallic mirrors are mounted. The wealthy Chinese give such extravagant prices for fine specimens that they are exceedingly difficult to be got hold of by any one else. A pair upon the head was estimated to be worth twenty-five dollars; and the price, as usual, increases with the size and length to a sum much higher. The natives of Sumatra assert that sometimes a third horn is seen, but this is doubtless such a development as that already mentioned in our description of the Indian Rhinoceros.

This Rhinoceros has a smooth hide, thinly but conspicuously covered with short, coarse, black hair; there are folds about the neck, a distinct fold behind the fore-quarters, a slight crease before the hind-limbs, but nothing comparable to the coat of mail of the one-horned varieties. The hide is rough, but easily cut with a knife. Both lips are broad and non-prehensile, and the animal therefore grazes rather than browses. The tail terminates in a thin tuft. The head is peculiarly long, the neck short and heavy, and the limbs very clumsy. It is a very quiet creature, and an adult male has been seen to fly in terror from a native wild dog.

The Asiatic Two-horned Rhinoceros has been supposed, until recently, to be peculiar to the island of Sumatra, as the smaller one-horned Rhinoceros is to that of Java; but both of them are widely diffused over the Indo-Chinese countries, and throughout the Malayan peninsula, the smaller one-horned being likewise found in Java, and the Asiatic two-horned also in Borneo as well as in Sumatra. The two-horned species has been killed in one of the hill ranges immediately to the southward of the Brahmaputra River, so that its range may be said to extend northward into Assam. It is worthy of notice that the full-grown female of this species becomes very speedily tame and tractable. It is probable that Rhinoceroses described as existing in the southeast of China belong to this species. The Burmans speak of the Fire-eating Rhinoceros, and Professor Oldham's camp-fire was attacked by one which proved to be of the two-horned species.

All the three species above described occur in the southern provinces of British Burmah: the *Rhinoceros indicus* in the high range called the Elephant-tail mountain, the *Rhinoceros sondaicus* in the extreme south, and the *Rhinoceros sumatranus* between the tenth and seventeenth degrees of north latitude. The first is the shyest, the last is the wildest, the middle one the mildest in character.

The ROUGH-EARED RHINOCEROS, *Rhinoceros lasiotis*, is the name given by Sclater to the small one-horned elephant found in the peninsula of Malacca and Farther India. It is usually considered to be identical with the Badak. The first specimen to which a separate species was given was captured by a strange accident. Some English officers on duty in the northern part of the Bay of Bengal, collecting elephants for the army, heard that a rhinoceros had got into some quicksand and could not get out by its own efforts, but two hundred of the inhabitants of the vicinity had succeeded in rescuing it by casting ropes over it, and had fastened it to two trees, where it was in the best condition, and so fierce that they durst not let it loose. Captain Hood and Mr. Weekes proceeded to the spot with eleven elephants; they found a female rhinoceros eight feet long and four feet high, with horns only slightly developed. They placed it between two elephants and took it to Chittagong, where it became tame very soon. It was a difficult matter to induce the elephants to assist in the capture, and when the savage beast was fastened to them, a scream from it threw them repeatedly into alarm. At last the march began till a river was reached, which had to be crossed by swimming. The elephants entered it boldly, but the rhinoceros refused and was dragged through the stream by her two companions, for she refused to swim a stroke. Finally, the rhinoceros was brought safe to Calcutta, and thence shipped to Europe.

CHAPTER X.

THE AFRICAN RHINOCEROS.

THE BORELE OR LITTLE BLACK RHINOCEROS—THE SWORD-HUNTERS OF ABYSSINIA—THE KEITLOA
—ITS FIERCENESS—THE MOHOHOO OR WHITE RHINOCEROS—HUNTING ADVENTURE OF MR.
OSWELL—THE KOBAOBA—PROBABILITY OF ITS EXTINCTION.

THE African Rhinoceroses are all two-horned, but are formed by Gray into two genera: *Rhinaster* with a prehensile upper-lip, and *Ceratotherium* with a non-prehensile upper-lip, which are known to travelers as the Black and White Rhinoceros respectively. Each genus has two species.

THE BLACK RHINOCEROSES.

The BORELE or LITTLE BLACK RHINOCEROS, *Rhinoceros bicornis* (Plate XXXIV), is the commonest of the African species, and is easily distinguished from the Asiatic two-horned species by its upper-lip and the shape of its horn. The foremost horn is of considerable length and inclines backward, while the second is short and conical.

The Borele is a very fierce and dangerous animal, and is more feared by the natives than even the lion. Although so clumsy in shape and aspect, it is really a quick and active creature, darting about with lightning speed, and testing the powers of a good horse to escape from its charge. Like many other wild animals, it becomes furiously savage when wounded, but it will sometimes attack a passenger without the least provocation. On one occasion an angry rhinoceros came charging down upon a wagon, and struck his horn into the bottom plank with such force as to send the wagon forward for several paces, although it was sticking in deep sand. He then left the wagon and directed his attack upon the fire, knocking the burning wood in every direction, and upsetting the pot which had been placed on the fire. He then continued his wild career in spite of the attempts of a native who flung his spear at him, but without the least effect, as the iron point bent against the strong impenetrable hide, which covered its huge carcass.

The skin of this animal does not fall in heavy folds, like that of the Asiatic species, but is nevertheless extremely thick and hard, and will resist an ordinary leaden bullet, unless it be fired from a small distance. The skin is employed largely in the manufacture of whips, or jamboks, and is prepared in a rather curious manner. When the hide is removed from the animal it is cut into strips of suitable breadth and laid on the ground. These strips are then hammered for some time in order to condense the substance of the skin, and when they are dry are carefully rounded with a knife and polished with sandpaper. One of these whips will continue serviceable for several years. The horn of the Borele, from its comparatively small dimensions, is not so valuable as that of the other species, but is still employed in the manufacture of drinking cups and sword-handles. Its value is about half that of ivory.

When wounded, the Black Rhinoceros is a truly fearful opponent, and it is generally considered very unsafe to fire at the animal unless the hunter is mounted on a good horse or provided with an accessible place of refuge. An old experienced hunter said that he would rather face fifty lions than one wounded Borele; but Mr. Oswell, the well-known African sportsman, always preferred to shoot the Rhinoceros on foot. The best place to aim is just behind the shoulder, for, if the lungs are wounded, the animal very soon dies. There is but little blood externally, as the thick loose skin covers the bullet-hole, and prevents any outward effusion. When mortally wounded the Black Rhinoceros generally drops on its knees.

One of a party of Namaquas shot a Borele, and, approaching to what he thought the carcass, stood astride of it and stabbed it. At the touch of the cold steel the beast rose up and made off at full speed with its dismayed rider. The Borele stopped when it had run forty or fifty paces and was killed by a lucky shot.

"The Borele," says Mr. Chapman, "is a dumpy, plump-looking animal of a very dark color, very lively in his actions, always on the trot, very nervous, wary, and fidgety, often flying round in a fury whether he has observed danger or not, making the hunter sometimes believe that he has been discovered. When he fancies he sees or hears anything, he lifts one foot, tosses up his horn and nose and sinister little eyes, and presents altogether a picture of the most intense and earnest scrutiny and attention, wheeling round with great rapidity, and by his active gestures and startling snorts often rendering the nerves of the inexperienced hunter very unsteady. On the whole his actions are those of a lively pig." This Rhinoceros extends as far north as Abyssinia, where it was seen by the traveler Bruce. In that country it is pursued and killed by the *agageer* or Sword-hunters. Two men ride on the same horse. One dressed, and armed with javelins; the other naked, with nothing but a long sword in his hand. The first sits on the saddle, the second rides behind him on the horse's rump. Directly they have got on the track of the quarry, they start off in pursuit of it, taking care to keep at a great distance from the Rhinoceros when it plunges into the thickets, but the moment it arrives in an open spot they pass it, and place themselves opposite to it. The animal, in a rage, hesitates for a moment, then rushes furiously upon the horse and its riders. These avoid the assault by a quick movement to the right or the left, and the man who carries the long sword lets himself slide off on to the ground without being perceived by the Rhinoceros, which takes notice only of the horse. Then the courageous hunter, with one blow of his formidable sabre, cuts through the tendon of the ham or hock of one of the monster's hind legs, which causes it to fall to the ground, when it is dispatched with arrows and the sword.

THE KEITLOA.

The KEITLOA, *Rhinoceros Keitloa*, is distinguished from the Borele by having horns of nearly equal length. The hind horn, which is straight, grows to two feet and a half or more in length, being often as long as the anterior horn, although as a rule the latter is the longer. The upper-lip is very pendulous, the neck is somewhat long, the head is not thickly covered with wrinkles. At its birth the horns are only indicated by a prominence on the nose, and at six years of age are only nine inches long. The Keitloa is a terribly dangerous opponent, and its charge is so wonderfully swift that it can hardly be avoided. One of these animals that had been wounded by Mr. Andersson, charged suddenly upon him, knocked him down, fortunately missing her stroke with her horns, and went fairly over him, leaving him to struggle out from between her hind legs. Scarcely had she passed than she turned and made a second charge, cutting his leg from the knee to the hip with her horn, and knocking him over with a blow on the shoulder from her fore-feet. She might easily have completed her revenge by killing him on the spot, but she then left him, and plunging into a neighboring thicket, began to plunge about and snort, permitting her victim to make his escape. In the course of the day the same beast attacked a half-caste boy who was in attendance on Mr. Andersson, and would probably have killed him had she not been intercepted by the hunter, who came to the rescue with his gun. After receiving several bullets, the rhinoceros fell to the ground, and Mr. Andersson walked up to her, put the muzzle of the rifle to her ear, and was just about to pull the trigger, when she again leaped to her feet. He hastily fired and rushed away, pursued by the infuriated animal, which, however, fell dead just as he threw himself into a bush for safety. The race was such a close one, that as he lay in the bush he could touch the dead rhinoceros with his rifle, so that another moment would probably have been fatal to him.

The Keitloa is of a dark neutral gray color, as seen from a distance. This animal droops behind, and has a stiff, clumsy, and awkward walk. He feeds on bushes and roots, is nervous and fidgety when discovered, but confines his movements generally only to the head and horns, moving them about in an undecided manner, first one way, then the other. He is not so excitable as the Borele. But both are fierce and energetic animals, and so active and swift that they cannot be overtaken on horseback. The Keitloa, it may be added, is more an inhabitant of rocky hills, while the Borele loves the thorny jungle. The Keitloa, as well as the Borele, extends as far north as Abyssinia. It exceeds, however, the latter in height, sometimes measuring six feet at the shoulder.

"Both species," writes W. C. J. Andersson, "are extremely fierce, and, excepting the buffalo, are, perhaps, the most dangerous of all the beasts of Southern Africa. Seen in its native wilds, either when browsing at its leisure or listlessly sauntering about, a person would take this beast to be the most stupid and inoffensive of creatures; yet, when his re is roused, he becomes the reverse, and is then the most agile and terrible of animals. The Black Rhinoceroses are, moreover, subject to sudden paroxysms of unprovoked fury, rushing and charging with inconceivable fierceness animals, stones, and bushes; in short, every object that comes in their way." "The Black Rhinoceros," writes Gordon Cumming also, "is subject to paroxysms of sudden fury, often ploughing up the ground for several yards with its horns, and assaulting large bushes in the most violent manner. On these bushes they work for hours with their horns, at the same time snorting and blowing loudly, nor do they leave them in general until they have broken them to pieces. During the day they will be found lying asleep, or standing indolently in some retired part of the forest, or under the base of the mountains, sheltered from the power of the sun by some friendly grove of umbrella-topped mimosas. In the evening they commence their nightly ramble, and wander over a great extent of country. They usually visit the fountains between the hours of nine and twelve o'clock at night; and it is on these occasions that they may be most successfully hunted, and with the least danger."

The food of the Black Rhinoceros, whether the Borele or the Keitloa, is composed of roots, which the animal ploughs out of the ground with its horn, and of the young branches and shoots of the wait-a-bit thorn. It is rather remarkable that the black species is poisoned by one of the Euphorbiaceæ, which is eaten with impunity by the two white species.

THE WHITE RHINOCEROSES.

The WHITE RHINOCEROS or MOHOGOO, *Rhinoceros simus* (Plate XXXV), so called from its pale color, is a very different animal from those of which we have been treating. It grows to more than six feet and a half high at the withers, where there is a sort of square hump, and the head is a foot longer than in the Keitloa, being nearly one-third of the entire length of the body, with an exceedingly long anterior horn, attaining to more than four feet in length, while the hind horn is very short, not exceeding seven or eight inches. "Its color," remarks Mr. Chapman, "is of such a light neutral-gray, as to look nearly as white as the canvas tilt of a wagon." His fellow-traveler, Mr. Baines, describing a freshly-killed one, tells us that "the skin was of a light pinky-gray, deepening into a bluish neutral tint on parts of the head, neck, and legs. The limbs, shoulders, cheeks, and neck were marked with deep wrinkles, crossing each other so as to have a lozenge-shaped reticulated appearance; but the only approach to a fold was a slight collar-like mark across the throat. The mouth was very small, and the limbs were dwarfish compared with the bulk of the carcass. The eyes were small and set flat on the side of the head, with no prominence of brow, and in such a position that I should doubt very much the assertion that the Rhinoceros can see only what is straight before it. I should think, on the contrary," continues Mr. Baines, "that anything exactly in front would be absolutely hidden from its view." Mr. Chapman estimated the weight of one of these White Rhinoceroses as being probably not less than five thousand pounds avoirdupois.

"The male," he says, "measures six feet eight inches at the withers, carries his head so low that the chin nearly sweeps the ground, and is constantly swaying his head to the right and left when suspicious. The calf, instead of going behind or at the side, always precedes the dam, and when fleeing is helped on by her horn or snout. The back of this animal is tolerably straight, the croup being as high, or even higher, than the withers. It moves each ear alternately backwards and forwards when excited, and the ears, when thrown forward, turn as if on a pivot, so as to bring the orifice innermost. In the other African Rhinoceroses the two ears are moved together, and not alternately. The ears are pointed or tufted."

This animal is of a comparatively mild and gentle disposition; and, unless in defence of its young, or when hotly pursued, or wounded, will very rarely attack a man. "It is gregarious in families," remarks Mr. Chapman, "the individuals comprising which are greatly attached to each other; and it utters a long sound, and not such a startling, whistling snort as the Borele does. It is an indolent creature, and becomes exceedingly fat by eating grass only." Elsewhere, he remarks of a herd of eight which he observed at a drinking-place: "The Rhinoceroses, all of which were of the white kind, occupied each twelve minutes to drink their fill, after which they wallow in the mud, or else go to their regular sleeping-places. At these their dung is found accumulated sometimes to the amount of a ton or more. They like the warmth of the manure to lie in. The sounds emitted by these animals is something like the coughing of a horse, and when in distress, a stifled asthmatic cry; when in pain they squeal like a storm-whistle." According to Gordon Cumming and others, their flesh is excellent, and even preferable to beef. The speed of this species is very inferior to that of the others, so that a person well mounted can easily overtake and shoot them.

But in spite of its usual gentleness the White Rhinoceros is sometimes pugnacious, as the following anecdote related by Mr. Oswell displays: "Once as I was returning from an elephant chase, I observed a huge white rhinoceros a short distance ahead. I was riding a most excellent hunter—the best and fleetest steed that I ever possessed during my shooting excursions in Africa—at the time; but it was a rule with me never to pursue a rhinoceros on horseback, and simply because this animal is so much more easily approached and killed on foot. On this occasion, however, it seemed as if fate had interfered. Turning to my after-rider, I called out: 'By heaven! that fellow has got a fine horn.' I will have a shot at him.' With that I clapped spurs to my horse, who soon brought me alongside the huge beast, and the next instant I lodged a ball in his body, but, as it turned out, not with deadly effect. On receiving my shot, the Rhinoceros, to my great surprise, instead of seeking safety in flight, as is the habit of this generally inoffensive animal, suddenly stopped short, then turned sharply round, and having eyed me most curiously for a second or two, walked slowly toward me. I never dreamed of danger. Nevertheless, I instinctively turned my horse's head away; but strange to say, this creature, usually so docile and gentle—which the slightest touch of the reins would be sufficient to guide—now absolutely refused to give me his head. When at last he did so, it was too late; for, notwithstanding the rhinoceros had only been walking, the distance between us was so inconsiderable, that by this time I clearly saw contact was unavoidable. Indeed, in another moment I observed the brute bend low his head, and with a thrust upward, strike his horn into the ribs of the horse with such force as to penetrate to the very saddle on the opposite side, where I felt its sharp point against my leg. The violence of the blow was so tremendous as to cause the horse to make a complete somersault in the air, coming heavily down on its back. With regard to myself, I was, as a matter of course, violently precipitated to the ground. While thus prostrated, I actually saw the horn of the infuriated beast alongside of me; but seemingly satisfied with his revenge, without attempting to do farther mischief, he started off at a canter from the scene of action. My after-rider having by this time come up, I rushed upon him, and almost pulling him off his horse, leaped into the saddle; and without a hat, and my face streaming with blood, was quickly in pursuit of the retreating beast, which I soon had the satisfaction to see stretched at my feet."

The flesh of the White Rhinoceros is apt to be rather tough, but is of good flavor. The best portions are those which are cut from the upper part of the shoulder and from the ribs, where the fat and the lean parts are regularly striped to the depth of two inches. If a large portion of the meat is to be cooked at one time, the flesh is generally baked in the cavity of a forsaken ant-hill, which is converted into an extempore oven for the occasion; but if a single hunter should need only to assuage his own hunger, he cuts a series of slices from the ribs, and dresses them at his fire. The hide of the Mohogoo is enormously thick, and gives a novice no little trouble to get it from the body, as it is as hard as a board, and nearly as stiff. An adept, however, will skin the animal as quickly and easily as if it were a sheep.

THE KOBAOBA, *Rhinoceros Oswellii*, is much rarer than either of the preceding species, and is found far in the interior, mostly to the east of the Limpopo River. The peculiar manner in which this species carries its horns makes it a very conspicuous animal. In all the other species the horns are curved, and incline rather backward; but in the Kobaoba the foremost horn is nearly straight, and projects forward, so that when the animal is running the tip of the horn nearly touches the ground. Indeed, the extremity of an adult Kobaoba's horn is generally rubbed down on one side, owing to the frequency with which it has come in contact with the earth. The head of this and the preceding species is always carried very low, forming a singular contrast to the saucy and independent manner in which the Borele carries his head.

The long horn of this species sometimes exceeds four feet in length, and is almost straight. The best ramrods are manufactured from it, and a ramrod four feet long has been seen. Mr. Chapman, however, believes that the Kobaoba is only an old Mohogoo. He writes: "I believe that wherever guns are to be found at present the White Rhinoceros is not allowed to reach its prime, and will soon be extinct. In newly-opened countries we always find long-horned Rhinoceroses at first. These are selected and shot by every new-comer for their long horns. I have never found a person yet who could conscientiously say that he had seen a young or middle-aged Kobaoba that was distinguished from a Mohogoo — not even a Bechuana or Bushman." That traveler, however, nevertheless believes in the existence of a second species of flat-lipped and grass-eating African Rhinoceros, but he gives no description of it.

CHAPTER XI.

THE HIPPOPOTAMUS.

THE HIPPOPOTAMUS OR RIVER-HORSE—DESCRIPTION—HABITS—FAVORITE HAUNTS—FOOD—VIOLENCE WHEN PROVOKED—MATERNAL AFFECTION—MODES OF HUNTING—PITFALLS AND DOWNFALLS—HARPOONING—THE HIPPOPOTAMUS IN CAPTIVITY—THE SMALL OR LIBERIAN HIPPOPOTAMUS.

THE family HIPPOPOTAMIDÆ contains only one genus, the *Hippopotamus* or "River-horse," as we call it by a literal translation of the Greek name of this monstrous animal which haunts the rivers of tropical Africa. The ancient Egyptians more appropriately named it the "River-swine," and it seems to have been in old times common in the Nile River; delineations of it occur in the ancient wall-paintings of the Old Kingdom of Memphis. To hunt the River-horse seems to have been one of the favorite sports of the Egyptian nobles, who are depicted in the act of harpooning it, or dragging it ashore by means of iron barbs attached to ropes. The Bible describes it under the name of Behemoth: "He eateth grass like an ox, his bones are as strong pieces of iron, he lieth under the shady trees, in the covert of the reeds and fens, he drinketh up a river." The Greek and Roman writers make repeated mention of it, and give pretty accurate descriptions of it. At present, however, the Hippopotamus has receded as civilization has advanced, and is only found high up the course of the Nile. In the year 1600 two River-horses were captured at the Damietta mouth of the Nile, but none are now to be seen in Egypt, nor even in Nubia, where, at the beginning of this century, Ruppell saw them in numbers. In the Soudan, however, and in all the larger rivers and lakes of Africa, they still abound. Wherever men have firearms the river-horse has vanished; where men have only lances it holds its ground. It is not till we are in the interior of the continent that we see the beasts represented in the hieroglyphic paintings; there, together with the pavian and the crocodile, the elephant and the rhinoceros, we meet the Hippopotamus.

GENUS HIPPOPOTAMUS.

The HIPPOPOTAMUS, *Hippopotamus amphibius* (Plate XXXVI), is, after the elephant and the rhinoceros, the largest of terrestrial mammalia; its absolute height is about five feet, as its legs are extremely short, but the actual bulk of its body is very great indeed. The head is nearly square and very bulky; the small ears and the oblique eyes, as well as the immense nostrils, are the highest points of a plane formed by the face and brow; the muzzle is large and swelling; the upper-lip, descending in front and at the sides, partly hides the under-lip; the nostrils are surrounded by a muscular apparatus which closes them under water; the upper part of the head is devoid of hair, and is pinkish in color. The rest of the huge body, which is equally hairless, is of a brown color, curiously marked with lines like the cracks in old varnish, and dappled with some very black spots which can only be seen in certain lights and on close examination. The tail is very short and has a few bristle-like hairs. Each foot has four toes, each enclosed in its hoof. The mouth is enormously large, extending nearly from eye to eye, and is armed with teeth which differ from those of the SUIDÆ, less by their number than by their shape; the incisors of the lower jaw lie almost horizontally with their points turned forward, while those of the upper-jaw are placed vertically; the canines are large and curved, forming almost a semicircle, and have their outer surface deeply channeled. The teeth are very solid and close-grained, and are extremely white in color; a large tooth will weigh five to eight pounds. With their teeth the Hippopotamus can cut grass as neatly as if mown with a scythe, and can sever a pretty thick stem of the bushes by the river.

The Hippopotamus is, as its name implies, most aquatic in its habits. It generally prefers fresh water, but it is not at all averse to the sea, and will sometimes prefer salt water to fresh. It is an admirable swimmer and diver, and is able to remain below the surface for a very considerable length of time. In common with the elephant, it possesses the power of sinking at will, which is the more extraordinary when its size is considered. It usually prefers the stillest reaches of the river, where it finds an abundance of aquatic plants. Only when this supply of food is scanty does it quit the stream. Possessed of an enormous appetite, having a stomach that is capable of containing five or six bushels of nutriment, and furnished with such powerful instruments, the Hippopotamus is a

HIPPOPOTAMUS
PLATE XXXVI UNGULATA

terrible nuisance to the owners of cultivated lands that happen to be near the river in which the animal has taken up his abode. During the day it is comfortably asleep in its chosen hiding-place, but as soon as the shades of night deepen it issues from its den, and treading its way into the cultivated lands makes sad devastation among the growing crops. Were the mischief to be confined to the amount which is eaten by the voracious brute, it would still be bad enough, but the worst of the matter is that the Hippopotamus damages more than it eats by the clumsy manner of its progress. The body is so large and heavy, and the legs are so short, that the animal is forced to make a double track as he walks, and in the grass-grown plain can be readily traced by the peculiar character of the track. It may therefore be easily imagined that when a number of these hungry, awkward, waddling, splay-footed beasts come blundering among the standing crops, trampling and devouring indiscriminately, they will do no slight damage before they think fit to retire.

The Hippopotamus is a gregarious animal, collecting in herds of twenty or thirty wherever food is plentiful, and never straying far from its feeding-place. In favorable spots where woods clothe the banks of the stream and aquatic vegetation is abundant in the water, the monsters of the river are soon discovered. At intervals of three or four minutes the traveler observes a vapory column rising about a yard above the surface, and hears a peculiar snort or bellowing, and beholds the shapeless head of the creature appear, a reddish or reddish-brown mass with two points, the ears, and four protuberances, the eyes and nostrils; often indeed only the latter are visible. The rest of the huge carcass is seldom exhibited. The approach of even a large boat does not disturb or alarm the animal, which stares with stupid astonishment at the intruders without interrupting its ascent and descent in the waters. In the narrower and shallower streams where the dry season leaves much of the channel dry, the hippopotami form deep troughs in the bed of the river, in which they can dive and hide themselves. Sometimes several of these troughs, each of which can contain four or five, are united by channels. They never leave the water except in utterly unfrequented spots, where they come to land and lie, half asleep, in the reed-beds, as happy and complacent as so many swine wallowing in the mud. Small birds, such as the *Ilyas Egyptiacus*, walk over the huge bodies and pick off sundry parasites, and, according to the Arabs, act as guardians to the slumbering monsters. It is a fact that at the slightest cry of these birds they retire into the water, but, except where they have learned to dread man, they are careless and sluggish in their movements. The day thus passes in alternate dozing and waking, but at night the herd becomes livelier. The grunting of the males rises to a roar, and all the troop begin to sport in the stream; they regularly follow boats on a night voyage, and cause with their snorting and grunting, their bellowing and splashing, an endless tumult. They swim with remarkable activity, rising, sinking, wheeling or advancing with consummate skill, and when progressing peacefully, scarcely disturb the water. When angered, however, they spring forward with a violence that sets the water in a turmoil, and sometimes destroy or upset boats. When disturbed while they are sleeping or basking on the bank, they show that they are more active than they seem; they will plunge from a height of six yards and send behind them a wake like a small steamboat.

When feeding, on landing, the Hippopotamus is not only destructive to the fields of the husbandman, but dangerous to man and beast. With blind fury it attacks everything which comes in its way, and those who describe it as a peaceable, good-natured animal, can never have seen it in anger. Even when in the water the monster is not to be trusted implicitly. Lieutenant Vidal, in sailing up the Tembi River, in southwestern Africa, suddenly felt his boat raised up and the steersman flung overboard. The next instant a giant hippopotamus appeared, and rushed with open jaws at the boat; it seized it with its terrible teeth and tore out seven of the planks. The boat had probably grazed the monster's back with its keel. On land the River-horse is still more dangerous. Here they cannot be relied on to take flight; they rush on like a savage boar and seize the object of their fury, tearing it with their teeth and trampling it down with their feet. A single bite has been known to kill a man. A female with her young one is most to be dreaded, for every object seems to provoke her fury. She never leaves the little one for a moment, and watches every movement; at times she plays awkwardly with her calf; at times she carries it on her back. The young sit astride of her short neck, and the mother seems to rise to the surface more often than she herself requires in order to let her offspring breathe. It is not advisable to approach a female when thus engaged. One whose calf had been speared on the previous day made at the boat in which Dr. Livingstone was sitting, and drove her head against it with such force that she lifted the bows completely out of the water.

There are various modes of hunting this mischievous but valuable animal, each of which is in vogue in its own particular region. Pitfalls are universal throughout the whole hippopotamus country, and lure many an animal to its destruction without needing any care or superintendence on the part of the men who set the snare. They are simply pits dug across the path of the Hippopotamus, and provided with a sharp stake in the centre. There is also the " down-fall," a trap which consists of a log of wood, weighted heavily at one end, to which extremity is loosely fixed a spear-head well treated with poison. This terrible log is suspended over some hippopotamus path, and is kept in its place by a slight cord which crosses the path and is connected with a catch or trigger. As soon as the animal presses the cord the catch is liberated, and down comes the armed log, striking the poisoned spear deep into the poor beast's back, and speedily killing it by the poison, if not from the immediate effects of the wound.

The white hunter of course employs his rifle and finds that the huge animal affords no easy mark, as, unless it is hit in a mortal spot, it dives below the surface and makes good its escape. Mortal spots, moreover, are not easy to find, or when found, to hit; for the animal soon gets cunning after it has been alarmed, and remains deeply immersed in the water as long as it is able, and when it at last comes to the surface to breathe, it only just pushes its nostrils above the surface, takes in the required amount of air, and sinks back again to the river bed. Moreover, it will often be so extremely wary that it will not protrude even its mouth in the open water, and looks out for some reeds or floating substances which may cover its movements while breathing. As a general rule it is found that the most deadly wound that can be given to a hippopotamus is on the nose, for the animal is then unable to remain below the surface, and consequently presents an easy mark to the hunter. A heavy ball just below the shoulder always gives a mortal wound, and in default of such a mark being presented, the eye or the ear is a good place to aim at.

The most exciting manner of hunting the Hippopotamus is by fairly chasing and harpooning it, as if it were a whale or a walrus. This mode of sport is described very vividly by Mr. Andersson.

The harpoon is a very ingenious instrument, being composed of two portions, a shaft measuring three or four inches in thickness and ten or twelve feet in length, and a barbed iron point, which fits loosely into a socket in the head of the shaft, and is connected with it by means of a rope composed of a number of separate strands. This peculiar rope is employed to prevent the animal from severing it, which he would soon manage were it to be composed of a single strand. To the other end of the shaft a strong line is fastened, and to the other end of the line a float or buoy is attached. As this composite harpoon is very weighty it is not thrown at the animal, but is urged by the force of the harpooner's arm. The manner of employing it shall be told in Mr. Andersson's own words: " As soon as the position of the hippopotami is ascertained, one or more of the most skillful and intrepid of the hunters stand prepared with the harpoons, while the rest make ready to launch the canoes should the attack prove successful. The bustle and noise caused by these preparations gradually subside. Conversation is carried on in a whisper, and every one is on the *qui-vive*. The snorting and plunging become every moment more distinct; but a bend in the stream still hides the animals from view. The angle being passed, several dark objects are seen floating listlessly on the water, looking more like the crests of sunken rocks than living creatures. Ever and anon one or other of the shapeless masses is submerged, but soon again makes its appearance on the surface. On, on glides the raft with its sable crew, who are now worked up to the highest state of excitement. At last the raft is in the midst of the herd, who appear quite unconscious of danger. Presently one of the animals is in immediate contact with the raft. Now is the critical moment. The foremost harpooner raises himself to his full height, to give the greater force to the blow, and the next instant the fatal iron descends with unerring accuracy in the body of the hippopotamus. The wounded animal plunges violently, and dives to the bottom; but all his efforts to escape are unavailing. The line or the shaft of the harpoon may break; but the cruel barb once imbedded in the flesh, the weapon (owing to the toughness and thickness of the beast's hide) cannot be withdrawn. As soon as the hippopotamus is struck one or more of the men launch a canoe from off the raft, and hasten to the shore with the harpoon-line, and take a round-turn with it about a tree or bunch of reeds, so that the animal may either be 'brought-up' at once, or, should there be too great a strain on the line, 'played' (to liken small things to great) in the same manner as the salmon by the fishermen. But if time should not admit of the line being passed round a tree or the like, both line and 'buoy' are thrown into the water, and the animal goes wherever he chooses. The rest of the canoes are now all launched from off the raft, and chase is given to the poor brute, who, so soon as he comes to the surface to breathe, is saluted with a shower of light javelins. Again he descends, his track deeply crimsoned with gore. Presently and perhaps at some little distance he once more appears on the surface, when, as before, missiles of all kinds are hurled at his devoted head. When thus beset the infuriated beast not unfrequently turns upon his assailants, and either with his formidable tusks, or with a blow from his enormous head, staves in or capsizes the canoes. At times, indeed, not satisfied with wreaking his vengeance on the craft, he will attack one or other of the crew, and with a single grasp of his horrid jaws either terribly mutilates the poor fellow, or it may be, cuts his body fairly in two. The chase often lasts a considerable time. So long as the line and the harpoon hold the animal cannot escape, because the 'buoy' always marks his whereabout. At length, from loss of blood or exhaustion, Behemoth succumbs to his pursuers."

When an animal is killed the rejoicings are great, for not only is the ivory of great commercial value, but the flesh is very good eating, and the hide is useful for the manufacture of whips and other instruments. The fat of the hippopotamus, called by the Cape colonists "zee-koe speck," or sea-cow bacon, is held in very high estimation, as is the tongue and the jelly which is extracted from the feet. The hide is so thick that it must be dragged from the creature's body in slips, like so many planks, and is an inch and a half in thickness on the back, and three-quarters of an inch on the other portions of the body.

The Hippopotamus may be seen in many zoological gardens. These specimens have all been captured when young after the death of their mother. The Roman emperors had some specimens, but from the time of Heliogabalus down to the year 1850 only one hippopotamus reached Europe. In the year 1849 the British consul in Egypt expressed to Abbas Pasha a desire to have a hippopotamus. Orders were sent to the governor of Nubia to forward one to Cairo. He sent some troops up the river to procure the desired animal. The troops did not fall in with their prize till they had reached a distance of fifteen hundred miles above Cairo. A large female hippopotamus being wounded, was in full flight up the river; but presently a ball or two reached a mortal part, and then the maternal instinct made the animal pause. She fled no more, but turned aside, and made toward a heap of brushwood and water-bushes that grew on the banks of the river, in order (as the event showed) to die beside her young one. She was unable to proceed so far, and sank dying beneath the water. The action, however, had been so evidently caused by some strong impulse and attraction in that direction, that the party instantly proceeded to the clump of water-bushes. Nothing moved—not a green flag stirred; not a sprig trembled; but directly they entered, out burst a burly young hippopotamus-calf, and plunged head foremost down the river-banks. He had all but escaped, when amidst the excitement and confusion of the picked men, one of them who had "more character" than the rest, made a blow at the slippery prize with his boat-hook, and literally brought him up by burying the hook in his fat, black flank. Two other hunters—next to him in presence of mind and energy—threw their arms round the great barrel-bellied infant, and hoisted him into the boat, which nearly capsized with the weight and struggle. The hunting-division of the army, headed by the commander-in-chief, arrived at Cairo with their prize on the 14th of November, 1849. The journey down the Nile, from the place where he was captured, viz., the White Nile, had occupied between five and six months. This, therefore, with a few additional days, may be regarded as the age of this hippopotamus on reaching Cairo. The color of his skin at that time was for the most part of a dull red-dish tone.

When these animals have been allowed to lead their natural mode of life they have bred regularly. In captivity the mother displays the same watchfulness, the same resolution to defend her offspring, which is so conspicuous in their wild state. Her jealousy is so great that she renders the rearing of the young one a very perilous task, and is apt to injure it by her awkward attempts to aid it. This difficulty has been successfully overcome in many instances, and a young Hippopotamus is frequently the attraction of traveling menageries.

The SMALL HIPPOPOTAMUS, *Hippopotamus liberiensis*, is found in certain portions of West Africa. Its most distinguishing feature is the presence of only two incisor teeth in the lower jaw. It is much smaller than the River-horse of Egypt or South Africa.

CHAPTER XII.

THE PECCARIES.

THE SWINE FAMILY—GENERAL CHARACTERISTICS—THE PECCARIES—THE COLLARED PECCARY—ITS COURAGE AND FIERCENESS—THE WHITE-LIPPED PECCARY—ITS HABITS—METHODS OF HUNTING THE PECCARY—FLESH OF THE PECCARY.

THE family SUIDÆ comprises *five* genera, the best known of which is that embracing our common domestic Hog. Gray, after a careful study of all the specimens in the British Museum, has thought himself justified in forming the family into *three* sub-families,—SUINA or Swine, DICOTYLINA or Peccaries, and the PHACOCHERINA or Wart-hogs; but these groups approach so closely to each other that they cannot be regarded as really forming sub-families.

The members of this family, when compared with the heavy, unwieldly creatures we have just described, may be regarded as gracefully formed. The head is conical with a truncated snout, which, although possessed of considerable mobility, is used only for the purpose of rooting in the ground, and distinguishing by its tactile powers the substances which are suitable for food. They have on all their feet two large middle toes, armed with strong hoofs, while the lateral toes are too short to touch the ground. The canine teeth bend upward so as to form projecting tusks, which in some species attain an extraordinary development.

The SUIDÆ are found in all parts of the globe except New Holland. Large woodland swamps, where the bush is thick and the grass high, are their favorite haunts; they love the neighborhood of water, and are fond of rolling in the mud. Most of them are gregarious animals, nocturnal in their habits, and much more active than they seem; their walk is pretty quick, their gallop a succession of leaps, each accompanied by a grunt. They all are excellent swimmers, and will cross the sea from one island to another. Their senses of smell and hearing are excellent, but the eye is small and the sight is far from sharp. As a rule they fly from danger, but when hard pressed fight gallantly, using their tusks with great force and skill. The males defend the females, and the females their offspring, with great devotion. They are all omnivorous in the full sense of the word; they eat whatever is eatable; many, however, are from circumstances restricted to a vegetable diet.

In general, all the species are very prolific, and the young ones are remarkably pretty and active creatures, without any of that appearance of imperfection which usually characterizes new-born animals. They grow quickly and attain maturity soon, and hence they are always found in large numbers.

I. GENUS DICOTYLES.

The two species of this genus are peculiar to South America, extending from Mexico to Paraguay. They spread also northward into Texas, and as far as the Red River of Arkansas. The tusks in this genus are not turned outward, and the exterior lateral toe of the hind-foot is wanting. The generic name *Dicotyles* means "double cupped," and is given to the animal on account of a peculiar open gland on the back.

THE COLLARED PECCARY.

The COLLARED PECCARY or TAJACA, *Dicotyles torquatus* (Plate XXXVII), is a small pig about three feet long; the head is short, the muzzle slender, and the tail short. The thick bristles which cover its body are dark-brown with yellow and black rings, and are somewhat longer on the back of the neck. The prevailing color is dark-brown, but on the shoulders and round the neck is a broad band of a yellowish-white color, from which this species has obtained its name. The open gland on the back always discharges a fluid of a most fetid odor, which however seems to be grateful to the possessors of it, as they are often seen mutually rubbing each other's backs with their snouts.

Although the Peccary is a very harmless animal to outward view, it is a very dangerous enemy, in spite of its light weight (fifty to sixty pounds) and its short tusks; for these tusks are shaped like a lancet, double-edged and acutely pointed, and inflict terrible wounds. No animal can withstand the united attacks of the Peccary; fear is a feeling to which it is an utter stranger, and even the jaguar is forced to abandon the contest and to shrink from encountering the circular mass of Peccaries. Schomburgk, whose accounts are perfectly trustworthy, writes: "As we were passing through a woody oasis I heard a peculiar noise like the galloping of horses. With the cry '*Poinka!*' the Indians cocked their guns and drew their bows, and soon an innumerable herd of peccaries came in sight. When they saw us they stopped in their charge for an instant, then grunting loudly rushed past us. I was so surprised by the sudden appearance of the creatures that I forgot to shoot at first, and raised my gun to make up for lost time, but my arm was seized by an Indian. When the main herd was past and some stragglers came in sight, the Indians began to use bow and gun. They affirmed that it was most dangerous to fire into the middle of a herd, for the peccaries dispersed in all directions, and tore with their tusks every living thing that came in their way; while if the stragglers only are attacked, the main body pursues its course." In Webber's *Romance of Natural History* there is a very amusing account, too long to be quoted, of the consternation caused during a bear hunt by a charge of peccaries which scattered men, dogs, and bear in a common confusion. Another traveler writes as follows: "While pushing my way through a wood my dog started a peccary; suddenly eight or ten burst through the underwood, and before I could realize the scene had finished my unlucky companion with their sharp teeth. I suddenly found myself surrounded; I killed several, but it was no use; my ammunition was soon expended, and it was only by clubbing my gun that I fought my way to a tree, and with more than one wound from their incisors, reached a secure position. Here I remained besieged till they dispersed."

THE WHITE-LIPPED PECCARY.

The TAGNICATE, or WHITE-LIPPED PECCARY, *Dicotyles labiatus*, is larger than the preceding animal, assembles in larger herds, is fiercer in its disposition, and works more woe to the farmer.

The White-lipped Peccary derives its name from a band of white hairs that crosses the upper jaw, and covers nearly the whole of the lower. The color of the adult animal is black-brown, flecked with a gray grizzle, but when young it is striped. A slight mane runs along its neck, and its ears are fringed with long and stiff hairs. It is a most mischievous animal, as it makes long marches over the country, ravaging the crops in its progress, and always choosing, with a perversely excellent taste, the best maize and grass. Its cry is a sharp, shrill grunt, and when angry, it clashes its teeth smartly together, producing a sound which is recognizable at some distance, and is very useful to the hunters, as it serves to give timely notice of the animal's approach.

In all the woody lowlands of South America, and even as high as three thousand feet above the sea, both species of Peccary are abundant. The Tagnicates roam the forests in herds of hundreds, under the guidance of the strongest male, changing their abode every day, and always on the move. In their journeys, neither the open country, which otherwise they seldom visit, nor the rivers delay them. If they come to open ground, they cross it at full speed; if they encounter a stream, they plunge in without hesitation. The herd advances in close order, the males leading, each female followed by its young. Their approach is heard from some distance, not so much by reason of their grunting as by the crashing of the brush through which they are dashing. Their migrations are performed in search of food, for they live chiefly on fruits and roots, varying this sometimes by devouring serpents, lizards, and the larvæ of insects. In their movements they resemble the common swine, but have neither its voracity nor its filthiness; they only eat what they need, and only wallow during the extreme heats.

The usual resting-place of the Peccary is in the hollow of a fallen tree, or in some burrow that has been dug by an armadillo and forsaken by the original inhabitant. The hollow tree, however, is the favorite resort, and into one of these curious habitations a party of Peccaries will retreat, each backing into the aperture as far as he can penetrate the trunk, until the entire hollow is filled with the odd little creatures. The one who last enters becomes the sentinel, and keeps a sharp watch on the neighborhood. The native hunters take advantage of this curious habit to immolate great numbers of these animals. There are two methods of Peccary killing—by the gun, and by the sword and pitchfork.

In the former method the hunter takes up his temporary abode in some concealed spot that commands the entrance of the tree or hole in which the Peccaries are known to sleep. As soon as the sentinel has assumed its post, the hunter takes a careful aim at the forehead, and kills it with a single ball. Its place is at once taken by its successor, which, in its turn is shot, and so on till the whole family is exterminated. Schomburgk describes the chase of the Peccary, and states that when dogs are employed they are trained to cut off stragglers and keep them at bay till the hunter comes up. If the sportsman has no dog, he mounts a tree and imitates the bay of a hound. The Peccaries, with every bristle set on end, surround the tree from which the voice of their hereditary foe is heard, and gnash their teeth. If the hunter is armed with bow and arrows he can kill several of the animals, but the report of a gun scatters them. At times this method has an unfortuate termination. An Indian who had thus attracted a herd to a tree, was unlucky enough to find the branch upon which he was seated give way beneath him; he caught hold of another as he fell, but his feet were so near the ground that the Peccaries began to bite them. He managed to struggle out of the way, when the herd attacked the musket which he had dropped, and tore the stock to pieces. The Indians are always glad to drive the Tagnicates into a river, for although they swim well, they are comparatively helpless. The Indians leap into the stream and hit them on the snout with a stick; a few blows dispatches a Peccary. The Tagnicates are hunted for their flesh as well as on account of the damage they do. In the plantations a large pit is often dug, into which dogs and men drive the intruders, which are then killed by lance-thrusts.

The young Peccary can be easily tamed, and becomes as tame as our pigs; it is soon attached to its new home and new companions, especially to the men about it, whose voices it knows and whom it loves to accompany in their walks. It will announce the approach of strangers by its grunts, and attacks fiercely all strange dogs. The Tagnicate especially shows itself susceptible of domestication.

The flesh of both the species of Peccary is eaten by the Indians; it is said to have a pleasant taste, not at all like pork. If the animal has been hunted, the dorsal gland must be at once cut out, otherwise the whole flesh contracts its disgusting odor.

CHAPTER XIII.

THE TRUE SWINE.

THE GENUS SUS—RELIGIOUS PROHIBITIONS—THE BOAR OF VALHALLA—THE BOAR'S HEAD—THE WILD BOAR OF EUROPE—HUNTING THE WILD BOAR—THE WILD HOG OF INDIA—THE DOMESTIC HOG—ENEMIES OF THE HOG—BREEDS OF HOGS—THE BERKSHIRE—TRICHINOSIS.

THE true swine are divided into *three* genera, one of which, *Babirusa*, is confined to the islands of Celebes and Boura; another, *Potamochœrus*, is wholly African; while the third, *Sus*, is found throughout Asia and Europe. It is a remarkable fact that these hardy, omnivorous animals should have entirely died out in North America, except a few peccaries in the southern part, to which they are comparatively recent immigrants.

GENUS SUS.

This genus comprises *fourteen* species, and its generic characteristics are three incisors, one canine tooth, triangular in shape and curved upwards, with seven molars in each jaw, forming a total of forty-four teeth; a thin curling tail, and in the female eight to ten teats.

All the species of swine are unclean to the Jew and the Mohammedan, and it is almost impossible for us to form an idea of the horror and loathing with which the ancient Hebrews regarded the hog. We are told in the Book of Maccabees that the Jews allowed the temple to be dedicated to Olympian Jupiter, they submitted to the abolition of the Sabbath, and they consented to walk in the procession of the feast of Bacchus, but when swine's flesh was put before them, they refused to touch it. The reason for the Mosaic prohibition is not easy to discover. It is by no means certain that the flesh of the hog is harmful in hot climates, and in its wild state the animal is very cleanly. In the West, on the other hand, the hog has always furnished a staple article of food; the Roman and Greek physicians considered pork the most nutritious meat, and the ancient trainers gave it exclusively to the athletes and gladiators of the arena. Our Teutonic ancestors believed that the flesh of the boar Sœhrimnir was the daily food of the gods and heroes who dwelt in Valhalla. King Ragnar Lodbrog, when dying, sang "I am filled with joy when I think of the feast preparing for me in Odin's palace." Reminiscences of this mythical repast gave rise to the introduction of the boar's head at solemn festivals. At the Christmas dinner at New College, Oxford, a boar's head is brought to table in a long procession, while the choristers of the chapel march before it singing a Latin hymn. We need not mention the enormous proportions to which the American trade in pork has attained. It is not too much to say that pork is the chief article of diet in the world, and supplies the bone and muscle of the Chinese porter, as well as of the Irish peasant, or our own farmers.

THE WILD BOAR.

The WILD BOAR, *Sus aper* (Plate XXXVII), is still common in the forests of the continent of Europe. Its whole body is covered with blackish-brown bristles which form a sort of mane on the neck when the animal is excited. The lower canine teeth curve outward and upward, and give a severe wound; the toes are close together, and in walking the hind-foot is brought forward as far as the heel, and a little outside of the front-foot. Till the age of six months old, the young of the Wild Boar is, like the young peccary, striped in alternate bands of light and brown fallow color. The old males live alone, but the females continue with their young ones for at least two years. In deserted forests troops of females are found, which live on good terms, and combine for mutual defence, forming a circle when attacked.

In summer, Wild Boars are principally to be met on the outskirts of forests, in the approaches to fields or vineyards, and near swamps, where they retire during the heat of the day to refresh themselves by wallowing in the muddy water. In autumn, they permanently reside in the forests, in the heart of which they establish their winter retreat.

Dark, damp localities are generally chosen for their lairs; here they lie hid during the whole day, and only leave in the evening or at night to seek their food. They dig up the ground in search of worms and the larvæ or grubs of cockchafers; and they also devour reptiles, birds' eggs, and all the young animals they can surprise. Field-mice, moles, and even young rabbits are likewise favorite food. Acorns, chestnuts, and beech-nuts constitute a large portion of their vegetable diet. They often lay waste fields of potatoes, maize, and other grain. A whole crop is sometimes destroyed by these animals in a single night. When they root up the soil in search of their food, they invariably proceed in a straight line; and as the furrows which they make are as broad as their heads, experienced sportsmen can thus tell the size of the animal whose track they are following.

In Germany and France the chase of the Wild Boar occupies a prominent position in the list of field sports, but it is occasionally dangerous sport. This savage animal is not alarmed by the pursuit and the barking of dogs; but the sound of horns, the cries of the sportsmen, and the report of guns terrify it. It runs with a rapidity and a lightness which surprise us when we consider its heavy, thickset figure. Its route is invariably straight, and if any imprudent hunter does not get out of its way, he is certain to be upset; but it will not turn from its course to attack any one. If it is wounded, it changes its tactics, and rushes on all within reach. When fatigue or loss of blood prevents its flight, it places its back against a bush or tree, and makes a most vigorous resistance. Those hounds which approach too closely are frequently ripped up. But there is always found, in a well trained pack, some intelligent and knowing member, which keeps baying the game at a safe distance, and confuses the boar with its ferocious barking until a favorable moment occurs, when, with a bound, it seizes the game at its weak point—the ear. The furious animal is then powerless, and is easily slain by the hunter. It is considered, however, more sportsmanlike to await the charge of the boar with the boar-spear in hand. The chase is then a most exciting one, for the boar is a terrible antagonist, his charge is made with lightning swiftness, and, together with his furious eyes and lips dripping with foam, he is a sufficiently formidable foe to disconcert any one who is not possessed of good nerves and a steady hand. The animal has an awkward habit of swerving suddenly from his course, snapping at the spear-head and breaking it from the shaft. He also, when the hunter is on horseback, will charge at the horse instead of the rider, and rising on his hind legs, in order to give the blow greater force, will lay open the horse's flank and instantly disable it.

The flesh of the Wild Boar is esteemed a great delicacy; it is more tender and savory, and less fat than that of the domestic animal.

The Wild Boar, such as we have described it, is found in France, Germany, Belgium, Poland, Austria, the Danubian Principalities, Southern Russia, and Spain. In Asia it extends from the Caucasus to the Amoor, and is probably identical with the Wild Boar of Syria and Asia Minor, which is called scientifically *Sus libycus*. Beyond the limits just mentioned occur other species which still require investigation. Such are the Crested Hog, *Sus cristatus* of India, the *Sus andamanensis*, the *Sus barbatus* of Borneo, *Sus vittatus* of Java, *Sus leucomystax* of Japan, and *Sus senaarensis* of North-eastern Africa, and others. Of these we need only mention the first.

The WILD HOG OF INDIA, *Sus cristatus*, is superior in size, strength, and swiftness to the European species.

This animal is a sad plague to the agricultural population of India, as it makes terrible havoc among the crops, and is especially fond of frequenting the sugar-canes, eating them and chopping them into short lengths, which it forms into hut-like receptacles for its young. The boar is a most fierce and savage animal, and if driven from the cane-brake, will rush at any man or animal that may be within his reach, and cut them terribly with his sharp tusks. Even the sow can do considerable damage with her teeth, but instead of ripping like her mate, she bites sharply and rapidly. When the animal is fairly roused, and takes to his heels, he puts the mettle of the swiftest and stanchest horse fairly to the test, and even on ground where the horse has all the advantage, he will frequently distance his pursuers, and regain his domicile in the cane-brake. Among the plantations are numbers of old disused wells, the sides of which have fallen in and were never properly filled up. In these wells the wild hog loves to lie, for the mouth of the well is so overgrown with thick verdure that the aperture is scarcely visible even to a person that stands on its brink, while from those who are not aware of its precise locality it is entirely hidden.

The spear is generally employed in boar-hunting, or "pig-sticking," as the sport is familiarly termed, and is either thrown from the horse's back, or is held like a lance and directed so as to receive the animal's charge. When driven to bay, the Indian boar is as savage an animal as can be imagined, as with flashing eyes and foaming mouth he dashes first at one and then another of the horsemen, sometimes fairly driving them from the spot, and remaining master of the field. The unevenness of the ground over which the chase is pursued, adds to the difficulty of the sport.

THE DOMESTIC HOG.

The Domestic Hog is a wild boar which has, by long servitude, been modified physically and morally. Captivity, however, has by no means deprived it of its courage. When Baron de Rutzen settled in England, he stocked his park with some wild boars from Germany, which became the terror of the neighborhood. One day, however, a most domestic boar belonging to a farmer escaped from his stye, immediately gave battle, and slew and nearly devoured the champion of the intruders.

The Hog is not more voracious than the cow, the dog, or the sheep, for each of these animals will eat to repletion if furnished with a large amount of food, and will become inordinately fat in consequence of such high feeding. In its wild state it is never found overloaded with fat, and, as has already been seen, is so active an animal that it can surpass a horse in speed, and is so little burdened with flesh that it can endure throughout a lengthened chase. Neither is it naturally a dirty creature, for in its native woods it is as clean as any other wild animal. But when it is confined in a narrow stye, without any possibility of leaving its curtailed premises, it has no choice, but is perforce obliged to live in a constant state of filth. We may remark that it can eat poisonous herbs, such as hemlock, and is not injured by poisonous snakes. Common experience proves that it can exterminate the rattlesnake.

Leigh Hunt characterizes a pig as an animal "having a peculiar turn of mind; a fellow that would not move faster than he could help irritable, retrospective, picking objections, and prone to boggle; a chap with a tendency to take every path but the proper one, and with a sidelong tact for the alleys." Man takes advantage of this peculiar obstinacy, and induces the pig to go one way by pulling his tail in an opposite direction. The moral and mental philosophy of a pig's existence is thus ingeniously set forth by Sir Francis Head: "With pigs, as with mankind, idleness is the root of all evil. The poor animal, finding that he has absolutely nothing to do, having no enjoyment, nothing to look forward to but the pail which feeds him, must eagerly (or, as we accuse him, greedily) greet its arrival. Having no business or diversion—nothing to occupy his hours—the whole powers of his system are directed to the digestion of a superabundance of food. To encourage this, nature assists him with sleep, which, lulling his better faculties, leads his stomach to become the ruling power of his system—a tyrant that can bear no one's presence but his own. The poor pig thus treated, gorges himself, sleeps, eats again, sleeps, awakes in a fright, screams, struggles against the blue apron, screams fainter and fainter, turns up the whites of his little eyes, and dies!"

But though the progress of modern civilization may not have advanced far into pigdom, yet do we occasionally hear of shrewd knacks and habits acquired by these animals. The jungle-hog of India, we are told, makes his bed of meadow-grass; this he cuts down with his teeth, as if it were done with a scythe, and piles it up into oblong heaps, as regularly as thatch on houses. When he has thus collected a large heap, he creeps under it to rest; when he leaves it he creeps out at the other end without disturbing it. He remains quite invisible when in his house, but leaves a loop-hole through which to have an eye upon his enemies. In China it is not rare to see harnessed together a woman, a horse, a pig, and an ass. In Minorca an ass and a hog are occasionally yoked together to a plough; and Pennant speaks of a Morayshire farmer who used a cow, a sow, and two horses, to form his team. In Hertfordshire a farmer once went to St. Alban's market in a small cart drawn by four hogs; and a hog has been known to submit to the saddle and bridle. In some parts of Italy and France pigs are employed to hunt for truffles. A string is tied to the animal's leg, and he is led into the fields where truffles grow; wherever he stops, smells the soil, and roots up the ground, there truffles will be found. A good pig of this kind is worth about forty dollars.

When King Louis XI of France was sick, all means were tried in vain to amuse him, till one of his courtiers discovered a peasant who was the happy possessor of some young pigs that danced to the bag-pipe. The creatures were dressed up with a coat and hat, scarf and sword. They could jump about to the music pretty well, but could only with great difficulty stand on their hind legs. Other dancing pigs have been exhibited in modern times, and the "learned pig" could play euchre, exhibiting a capacity hardly to be expected from so maligned a creature. Among other calumnies, there is prevalent an idea that, whenever the Hog takes to the water, he cuts his own throat with the sharp hoofs of his fore-feet. This, however, is by no means the case, for the animal is an admirable swimmer, and will often take to the water intuitively. In one of the Moray Islands, three domestic pigs belonging to the same litter swam a distance of five miles; and it is said that if they had belonged to a wild family, they would have swum to a much greater distance.

The Domestic Pig may be said to be a manufactured production, a veritable monster when compared with the primitive pig. The Hog is either kept in a stye continually, or allowed, as in districts where acorns, chestnuts, and the like abound, to run about during some portion of the year. When the animals are confined, the styes ought to be kept clean and warm. The fecundity of this animal is remarkable. A sow usually has two litters a year, each of from twelve to fifteen; a Leicestershire sow is reported to have given birth to three hundred and fifty-five young ones in twenty litters, and the famous military engineer Vauban, when discussing the provisioning of towns, calculated that one sow would produce in ten generations 6,434,838 pigs. Yet the sow is often a very bad mother, and constant watching is required to prevent her eating her young as soon as they are born.

The BERKSHIRE HOG (Plate XXXVI) is one of the modern favorite varieties. It is hardy, rapid of growth, and furnishes excellent pork, and firm bacon. The Windsor and Leicester or Harrison-breed become in ten or twelve months so fat, that the neck, face, and eyes almost disappear.

Every part of the Hog, down to the bristles, is turned to some useful purpose. The head-quarters of the pork-trade are Cincinnati and Chicago, from which cities an enormous export goes on to all parts of Europe, in addition to the supplies sent to our eastern cities.

In a work intended for the popular eye, we may be allowed to diverge from the natural history of the Hog and say a few words on a disease respecting which alarm occasionally arises. This disease is caused by the consumption of uncooked pork, and is called TRICHINIASIS. The trichine is a minute worm, with difficulty visible to the naked eye, for it has scarcely as large a diameter as a very fine hair, and in length is rarely over two millimetres. It is found in the intestines, where it lives and produces its young, which are at first in the grub or worm state. When pork containing the trichine grubs is eaten by man, these pass into his intestines. But this abode not suiting them, they cut their way out, and get into the veins, when they are carried along with the blood in the circulating torrent, and finally lodge in the muscles. This is the part of the human form which is preferred by the trichine. It gnaws, separates, and dissects the muscular and tendinous fibres, producing intolerable pain. This disease has made the greatest ravages in the North of Ger-

BABYROUSSA WILD BOAR
WART HOG COLLARED PECCARY
DOMESTIC HOG

many, where raw ham is much eaten. France, however, seems to have till lately enjoyed complete immunity from it.

Although this epidemic only rarely appears, we will state the best means for preventing its development. They are as follows:

1st. Watch carefully over the food of the pigs, and never give them animal substances about which there is the least suspicion; 2d. Inspect carefully the pork, if possible, with a microscope; 3d. Cook most thoroughly every piece of pork, bacon, ham, etc., before use.

The experiments which have been made to determine the amount of cooking that is necessary to destroy the trichines give the following results:

1st. The trichines are killed in hams by a protracted salting, or in sausages by subjecting them to strong smoking, continued for twenty-four hours. 2d. They resist ordinary smoking for three days; if prolonged, however, it appears to destroy them. 3d. Cooking pork by boiling is not certain to kill them, *unless performed most thoroughly*.

CHAPTER XIV.

THE RIVER-HOGS, BABYROUSSA, AND WART-HOGS.

THE RIVER-HOGS—THE PENCILLED HOG—THE BUSH-HOG, OR BOSCH-VARK—EDWARDS' RIVER-HOG—THE BABYROUSSA—ITS PECULIAR TUSKS—THE WART-HOGS HIDEOUS APPEARANCE—THE AFRICAN WART-HOG, OR VLACKE-VARK—THE WART-HOG OF ÆLIAN OR ENSALUS.

WE will conclude our account of the *Suidæ* with *three* genera belonging to Africa and the islands of the Eastern Ocean. They are all very remarkable in appearance, but as yet we have almost no information respecting their modes of life while they enjoy their savage freedom. They are all gregarious, all omnivorous, all fond of low-lying swampy lands for their dwelling-place.

THE RIVER-HOG.

The River-Hogs are the handsomest specimens of the whole hog-family; they have a long face, a moderately long snout, large but narrow and sharp-pointed ears, which are ornamented with a tuft of hair, and a tail with a bushy tip. All the *three* species are African. A marked peculiarity in these creatures is a long protuberance between the eye and nose.

GENUS POTAMOCHŒRUS.

The PENCILLED HOG, *Potamochærus pictus*, has been known since the beginning of the sixteenth century. It is much smaller than the European boar, but attains a length of four feet, and stands nearly two feet high. The hide is covered with short, soft, thickly-placed, smooth-lying bristles which grow long on the cheeks and under-jaw, forming a short mane on the back of the neck, a bushy brush under the eye, and a pair of whiskers on the cheeks. The color is a reddish-yellow, or a brownish-red shading off into yellow. This beautiful bright color extends over the back, neck, and sides; but the forehead, cranium, and ears, as well as the legs, are black. The long hairs which we have described on the face and around the eyes and the mane are white, or yellowish-white. The young ones, like all young wild swine, are striped, and are very lively and prettily marked.

Although this hog was known and had been seen in captivity by Markgrave in the sixteenth century, we still know little respecting its wild life on the coast of Guinea and near the Camaroons River. The first living pair which was brought to Europe arrived at the London Zoological Gardens in 1852, and specimens are still rare in collections. They do not, in captivity, display any great divergence in their habits from the common wild boar. They may be described as good-natured, on the whole, not permitting their keeper to take liberties with them, but not displaying the vicious temper of the much smaller peccaries. They required to be well-sheltered from the English climate, and seemed to enjoy rolling in their deep bed of straw in which they half-buried themselves. Like the domestic sow, these hogs ate their offspring, not merely when newly-born, but gradually during the first week.

The BUSH HOG, *Potamochœrus Africanus*, is another species, inhabiting South Africa; it is somewhat larger than the one just described, and has a more savage expression, which does not belie its temper. It is called by the Dutch Boers the BOSCH-VARK, and chiefly frequents hollows or excavations in the forests. It is very variable in color, some specimens being dark brown, others brown and white, others bright chestnut. An English traveler writes: "Where the locality is sufficiently retired and wooded to afford shelter to the bush-bucks which I have mentioned, we may generally expect to find traces of the Bush Pig. The Bush Pig is about two feet six inches in height and five feet in length; his canine teeth are very large and strong, those in the upper-jaw projecting horizontally; those in the lower, upwards. He is covered with long bristles, and taking him all in all, he is about as formidable-looking an animal, for his size, as can be seen. The Bosch-Varks traverse the forests in herds, and subsist on roots and young shrubs. A large hard-shelled sort of orange, with an interior filled with seeds, grows in great quantities on the flats near the Natal forests; this is a favorite fruit of the wild pigs, and they will come out of the bush of an evening and roam about in search of windfalls from these fruit-trees.

The Kaffir tribes, although they refuse to eat the flesh of the domestic pig, will still feast without compunction on that of its bush brother. In the bush I always found the Kaffirs disinclined to encounter a herd of these wild swine, stating as their reason for doing so that the animals were very dangerous; they also said that the wounds given by the tusks of this wild pig would not readily heal. The Kaffirs are much annoyed by these wild pigs, which destroy their gardens, and leave narrow openings in the fences through which the Busch Vark walks only to fall into a deep pit in which is a pointed stake, where he is killed by the assegais of the Kaffirs. The tusks are considered great ornaments, and are worn suspended round the neck by a string.

EDWARDS' RIVER-HOG, *Potamochœrus Edwardsii*, is an allied species, found only in Madagascar, where it indicates a later migration from the mainland than in the case of the other mammalia of the island.

THE BABYROUSSA.

The Babyroussa is an exceedingly formidable-looking animal. Its extraordinary tusks, with their long upward curve, have given rise to the native name by which it is commonly known. Babi-rusa means "stag-hog," and characterizes the creature very imperfectly. This appellation, when literally translated, led many old travelers to suppose that it was derived from the speed and agility of the animal, but in these respects it bears no comparison to the fleet and bounding stag.

GENUS BABYROUSSA.

This third division of the *Suidæ* contains only *one* species.

The BABYROUSSA, *Babyrusa Alfurus* (Plate XXXVII), is found only in the Celebes islands. It reaches a considerable size, measuring nearly four feet in length, and standing two feet and a half high. The back is arched, the feet have four toes, the hide is thick, rough, and wrinkled in deep folds near the ears and neck. The general color is a dirty gray; the tail is thin and pendulous.

This strange creature is notable for the curious manner in which the tusks are arranged, four of these weapons being seen to project above the snout. The tusks of the lower jaw project upward on each side of the upper, as is the case with the ordinary boar of Europe, but those of the upper jaw are directed in a very strange manner. Their sockets, instead of pointing downward, are curved upward, so that the tooth, in filling the curvatures of the socket, passes through a hole in the upper-lip, and curls boldly over the face. The curve, as well as the comparative size of these weapons, is extremely variable, and is seldom precisely the same in any two individuals. The upper tusks do not seem to be employed as offensive weapons; indeed, in many instances they would be quite useless for such a purpose, as they are so strongly curved, that their points nearly reach the skin of the forehead. The female is devoid of these curious appendages.

From all accounts, the Babyroussa seems to be a very fierce and dangerous animal, being possessed of great strength, and able to inflict terrible wounds with the tusks of the lower jaw. A naval officer who had experienced several encounters with this creature, spoke of it with great respect, and seemed to hold its warlike abilities in some awe. The adult male Babyroussa is considerably larger than the boar of England, and the officer above mentioned affirmed that he had seen them as large as donkeys. It is a very good swimmer, and will take to the water for its own gratification, swimming long distances over arms of the sea.

Except in the island of Celebes and the adjacent islands, the Baby-roussa has not been discovered either in the Moluccas or in New Guinea, although its tusks have been seen in the hands of the Papuans. Its habits are like those of other swine; its gait is a quick trot, its gallop is lighter than the wild boar's. The first pair brought to Europe was given by Marcus, the Dutch governor of the Moluccas, to the French naturalists Quoy and Gaimard. They seemed to feel the cold weather exceedingly, shivering continually, and hiding themselves, even in summer, in their straw; they both died of the change of climate. Others, however, have been brought to London.

THE WART-HOGS.

The animals of this genus are grouped by Gray into a sub-family. They are the most repulsive looking of all the *Suidæ*; they are strongly built, and remarkable for their hideous head, and the peculiar dentition. The sides of the face bear three extraordinary protuberances, one pointed and movable, at one time turned upward, at another time hanging beneath the eye; a second projects out laterally, while the third runs along the under jaw up to the corner of the mouth. The eyes are small and project like those of the hippopotamus, and are surrounded by a large semicircular fold; the ears are pointed, the snout broad and oval.

The short limbs have four hoofs; the fore-legs have a broad callosity on the pastern joint. The hide, with the exception of some long hairs on the cheeks and neck, is covered with short, sparse bristles. The canine teeth are directed upward and outward; the last grinder is remarkably elongated. The color of the animal is reddish-brown, and it sometimes reaches the length of five feet and the height of two feet six inches. There are *two* species of the genus.

GENUS PHACOCHŒRUS.

The AFRICAN WART-HOG, *Phacochærus Æthiopicus* (Plate XXXVII), has very large protuberances on the face, and very strong, curved tusks. It is called VLACKE-VARK by the Dutch settlers at the Cape, and INGOR-LOOK by the Betshuani. They live in holes in the ground, are gregarious, and very dangerous to the unwary traveler.

This animal is not devoid of sagacity, as was proved by Gordon Cumming: "I selected the old boar for my prey, and immediately separated him from his comrades. After ten miles of sharp galloping, we commenced ascending a considerable acclivity, where I managed to close with him, and succeeded in turning his head toward my camp. He now reduced his pace to a trot, and regarded me with a most malicious eye, his mouth a mass of foam. He was entirely in my power, as I had only to spring from my horse and bowl him over. I felt certain of him, but resolved not to shoot as long as his course lay in the direction of my wagon. At length, surprised at the resolute manner in which he held for my camp, I headed him; when, to my astonishment, he did not in the slightest swerve from his course, but trotted along behind my horse like a dog following me. This at once aroused my suspicions, and I felt certain that the cunning old fellow was making for some retreat; so I resolved to dismount and finish him. Just, however, as I had come to this resolution, I suddenly found myself in a labyrinth of enormous holes, the haunt of the ant-bear. In front of one of them the wild boar pulled up, and charging stern foremost into it, disappeared from my disappointed eyes, and I saw him no more. I rode home for my men, and returning, we collected grass and bushes, and tried to smoke him out, but in vain." Smith asserts that the Wart-Hog seldom declines a combat; none but the most expert huntsmen dare attack it; it rushes out suddenly, strikes right and left, and fights to the death.

The first living Wart-Hog from the Cape of Good Hope came to the Hague in 1775. At first it seemed quiet, but it soon showed its savage temper by killing its keeper.

ÆLIAN'S WART-HOG, *Phacochærus Æliani*, is more widely extended than its congener, being found in the Cape de Verde Islands, New Guinea, Abyssinia, and Mozambique. It is called "Haroja" by the Abyssinians, "Dosar" by the Somali, and "Halaf" by the Arabs. Buffon gives it the appellation of "Engallo." It is easily distinguished from the foregoing by the presence of incisor teeth, and by a slight depression of the bones of the forehead. Its color is an earthy-brown, thinly clothed with bristles, except along the spine, where a well-developed mane, six or eight inches long, appears. The tail is nearly naked, but tufted at the tip.

Both species live chiefly on roots and bulbs, which they dig up with their powerful tusks, aided by a kneeling position to facilitate the lever-like action; they vary their diet, however, with the larvæ of insects, worms, and lizards, and even devour any carrion that may come in their way. Specimens of both species have been brought to Europe for various Zoological Gardens, and no difference in their habits can be seen; they are both unsusceptible to kindness, and never display any attachment to their keepers. The females are not so savage as the males, but are equally devoid of all affection towards man. The flesh of the Wart-Hog is, according to the accounts of Schweinfurth, less palatable than that of the wild boar; when fresh, it produces severe dysentery. The Abyssinian Christians, as well as their Mohammedan neighbors, deem it unclean, and refrain from tasting it, differing in this respect from the Kaffir tribes in the south of the African Continent.

CHAPTER XV.

THE CAMEL.

THE RUMINANTS—THE CAMELIDÆ—THE CAMELS OF THE OLD WORLD—THE ARABIAN CAMEL, OR DROMEDARY—THE CAMEL IN THE EAST—THE CAMEL IN EUROPE—THE CAMEL IN AFRICA—ITS FOOD—ITS POWERS OF RESISTING THIRST—ITS SPEED—MODE OF RIDING—ITS BEHAVIOR WHEN LOADING—ITS VICES—ANECDOTE OF LATIF PASHA—ITS VALUE—THE TWO-HUMPED CAMEL OF BACTRIA.

THE remaining families of the order with which we are now engaged are sometimes placed together as an order *Ruminantia*. The division thus made is superfluous, yet all the animals that "chew the cud," although different in many respects from each other, have a certain general resemblance. They all live on vegetable food, they devour grass, leaves, young shoots, and roots, as well as grain. Most of them are, even when wild, of greater benefit than injury to man, although where the cultivation of the soil has been carried to a great height, some species cannot be tolerated by the husbandman. Both the wild and tame ruminants minister in many ways to the service of man. Their flesh and their hide, their horns and their hair, are all of high value. In a state of domesticity the Ruminants are docile, patient, and enduring, and are, in fact, indispensable. Three families alone have, up to this time, remained undomesticated, the Muskdeer, the Giraffes, and the Antelopes. All the others contain some members which man has made his slave and assistant, while all the wild species are objects of the chase; and on this account are treated with almost royal honors by lovers of the "noble art of Venery."

THE CAMELS.

The family CAMELIDÆ consists of a very restricted group of *two* genera, comprising *six* species; the majority of the species now existing only in a state of domestication. The leading characteristics of the

CAMEL

family are cushions developed more or less on the soles of the feet, the absence of horns, and a cleft upper lip. North Africa, Central Asia, and Southwestern America are the original homes of these animals. The two species found in the Old World are entirely domestic animals, no longer occurring in a state of nature; in the New World two of the four species are domesticated, but two still retain their independence.

GENUS CAMELUS.

The *two* species of this genus are both natives of the Old World; they are remarkable for their size, and for the possession of one or two humps. They have a small, strongly arched head; their ears are slightly developed, but their sense of hearing is excellent. The eyes have oblong horizontal pupils, are gentle in expression, and protected by a double eyelid; their power of sight is very great. The nostrils are situated at some distance from the extremity of the upper lip, and externally appear like two small slits which the animal can open or close at will. The lips constitute a very delicate organ of touch, and the camel has an extremely acute sense of smell. The head, which is as a whole exceedingly repulsive in appearance, is carried on a long neck, which, when the animal moves slowly, describes a graceful arched curve. The Camel possesses callosities on the breast, knees, and insteps, as well as on the *patella* and heels. The feet are bifurcated. The two toes on each foot are not enveloped in horn, and have only on the last joint a somewhat short and hooked nail. A hard and callous sole covers the bottom of the toes—a characteristic which enables them to walk with ease on loose sand, where the elephant would be useless, and the horse soon exhausts its strength. The hump of the camel is a very curious part of its structure. The Arabs say that it feeds upon its hump, for when it suffers from privation and fatigue the protuberance diminishes, and will often nearly vanish at the end of a long and painful journey. As the animal is intended to traverse the parched sand-plains, and to pass several consecutive days without the possibility of obtaining liquid nourishment, there is an internal structure which permits it to store up a considerable amount of water for future use. For this purpose, the honeycomb cells of the "*reticulum*" are largely developed, and are enabled to receive and to retain the water which is received into the stomach after the natural thirst of the animal has been supplied. After a Camel has been accustomed to journeying across the hot and arid sand-wastes, it learns wisdom by experience, and contrives to lay by a much greater supply than would be accumulated by a young, inexperienced animal.

Buffon has said that the Camel is the real treasure of the East. It feeds the inhabitants with its milk and flesh, and its hair furnishes them with clothes. Without its use as a beast of burden, nations separated from each other by vast stretches of desert could not carry on any trade. Without it, the Arab could not inhabit those arid countries in which he dwells. With it, this "ship of the desert," as the Eastern nations have called it in their figurative and symbolical language, life is possible even in such places as Buffon has called " the blank spots in nature."

THE CAMEL OF ARABIA.

The CAMEL, *Camelus Arabicus*, or DROMEDARY, *Camelus Dromedarius*, (Plate XXXVIII), has only one hump. It attains a height of about six to seven feet, and a length of from nine to ten feet. Many varieties are found, arising from difference of location. The Camel of the desert is a thin, tall, long-legged creature; that of the cultivated regions of North Africa is a heavy, stout animal. Between a *Bisharen* or pure-bred camel of the Nomad tribes, and a common pack-camel of Egypt there is as great a difference as between the pure Arab steed and a Flemish dray-horse. The term Dromedary is ordinarily restricted to the high-bred animals used for riding.

The Camel is found at present only as a domestic animal, in the regions of Africa lying north of the twelfth degree of latitude, and of the extreme west of Asia. Its distribution coincides with that of the Arabian race. From Arabia it has spread eastward through Syria to Persia, where the two-humped, or Bactrian camel appears, and westward through Africa to the shores of the Atlantic Ocean. There is no representation of this very remarkable beast in the Egyptian monuments, but it is mentioned in Egyptian documents of the fourteenth century before Christ. In the Bible, the earliest mention of the Camel is in the Book of Genesis, where we read that Abram took on his journey "sheep and oxen and camels" (Gen. xii. 16). Many years after, we are told how Rebekah offered water to Jacob's camels at the well, and when Jacob was about to leave Laban camels are mentioned (Gen. xxiv. 10-19) as forming a considerable portion of his wealth. One of the plagues that fell upon Egypt (Exod. ix, 3) destroyed the camels with the other cattle. Job had three thousand, and afterward six thousand camels; and the Amalekites possessed "camels without number, as the sand by the seashore for multitude" (Judges vii, 12), and the Reubenites took from the Hagarites fifty thousand camels. By the Mosaic law the Camel may not be eaten; but its milk was probably used by the Jews, for the Hebrews had no such aversion to this one of the "prohibited animals," as they always evince toward the hog. They used it as a beast of draught, for Isaiah (ch. xxi, v. 7) speaks of a "chariot of camels," but naturally its chief use was as a beast of burden.

The Camel is essentially a desert animal, and attains his highest development in the driest and hottest regions. In Egypt, where food is plentiful, he loses his most precious qualities; near the Equator, where the vegetable world approximates in character to that of South America or Southern Asia, he no longer can live. Up to the twelfth degree of latitude he flourishes; a couple of degrees farther, he dies without any explicable cause. The reason seems to be, that the Camel can thrive in dry heat, but dies in moist heat. Some attempts to domesticate the Camel in Europe have been made. In 1622, Ferdinand de Medici introduced some into Tuscany, and at San Rossore, near Pisa, the camels live on a wide sandy plain as happy as in their original home; they numbered in 1840 one hundred and seventy-one. In Southern Spain the Camel has been found to succeed admirably. In 1858 camels were employed in Texas and Arizona. But the greatest success has attended the introduction of them into Australia.

In the North and East of Africa camels are kept in vast numbers. Many Arab tribes have them by thousands and hundreds of thousands. Before the railroad from Cairo to Suez was open, at least six hundred camels a day were employed in conducting the service between those places. Still greater is the number of camels employed in the great caravans between the North of Africa and the negro countries. The tribe of Tibbo have two hundred thousand of them, the Berbers possess more than a million. In Arabia, the province of Nedjed is the most famous for them, and it supplies Syria, Hedjaz, and Yemen with them, beside sending thousands to Asia Minor. The number of camels that perish yearly in the passage of the desert cannot be counted, but travelers tell that the tracks of the caravans are for miles lined with their bones so closely that the road seems to pass through a heap of bleaching skeletons.

The Camel lives entirely on vegetables, and is not particular what he eats. The driest and most withered herbage of the desert, sharp reed-grass, half-dead twigs enable him to support life for weeks. Under certain circumstances, an old mat or an old date-basket is a delicacy for him. In Soudan, the houses which are thatched with grass required to be protected from his attacks. No thorns or prickles seem to affect his palate; he will eat twigs of the mimosa, the spikes of which will even penetrate the sole of a boot. When at evening the caravan rests, the camels are turned out, and they rove from tree to tree, eating off every accessible branch. But more juicy food is always acceptable to them; in the fields of millet, beans, peas, and corn of all kinds, they work fearful havoc. During long journeys through the desert, when it is necessary to keep the burden as light as possible, the Arab takes a little barley for his camel, and gives him about two hands-full. But, in general, the leaves of various trees are their favorite food. If the Camel is living on juicy plants, he can go without water for weeks. Most of the accounts respecting the animal's power of abstaining from water are, however, fables. Stories that they can travel fourteen or twenty days without water are laughed at by the Arabs. In the dry season, the Camel on a journey must not only have sufficient food and water, but a rest of nearly one day in four. Only in rare cases, usually when one of the expected wells is dry, is the Camel permitted to go so long without water. The opinion that the large cells in the first division of the stomach serve as reservoirs for water, and that in extreme need the traveler can, by killing his camel, obtain water, is declared by Brehm to be unfounded. That distinguished traveler, who knows the camel as well as a farmer knows the horse, says that he had put inquiries to all camel-drivers, and they all affirmed that they had never heard such a monstrous lie. He tells us that he has seen camels killed which had been drinking the day before, and convinced himself that it was utterly impossible to drink any fluid which had been for twenty-four hours in the stomach mixed with the Camel's food and gastric juices; the very smell of a Camel's stomach when cut open being intolerable. But we must remember that men dying of thirst are not very particular as to the quality of the liquid which may save their lives. The water obtained from the stomach of the Camel is light-green in color, very unpleasant, but hardly more so than that which is carried in leathern bags on the Camel's back, exposed to the heat of the sun.

It is a pleasant sight to see the thirsty camels approach a well. Stupid as they are, they never forget where they have once drank. They raise up their heads, snuff the air with half-shut eyes, lay their ears back, and begin to run at a pace which compels the rider to hold on firmly to his saddle. When the well is reached, they crush and crowd each other in their haste to drink, uttering horrible bellowings. They drink for three minutes without interruption, their bodies swell visibly, and when the journey is resumed, the fluid is heard awash in the stomach like water in a half-filled cask.

When one sees a camel standing in repose, one can scarcely believe that it can vie in speed with a horse. But it can do so. The camels of the desert can accomplish distances and speeds such as no other tame creature can attempt. The ordinary pace, indeed, is not very rapid. "In crossing the Nubian desert," says Captain Peel, "I paid constant attention to the march of the camels, hoping it might be of some service hereafter in determining our position. The number of strides in a minute with the same foot varied very little, only from thirty-seven to thirty-nine, and thirty-eight was the average; but the length of the stride was more uncertain, varying from six feet six inches to seven feet six inches. As we were always urging the camels, who seemed, like ourselves, to know the necessity of pushing on across that fearful tract, I took seven feet as the average. These figures give a speed of 2.62 geographical miles per hour, or exactly three English miles, which may be considered as the highest speed that camels, lightly loaded, can keep up on a journey. In general, it will not be more than two and a half English miles. My dromedary was one of the tallest, and the seat of the saddle was six feet six inches above the ground." But the post-riders, mounted on camels of good blood, perform journeys of incredible rapidity. On such a camel Mohammed Ali rode from Cairo to Alexandria, a distance of seventy-five miles, in twelve hours. Another camel did the same distance in nine hours, including the loss of an hour in twice crossing the Nile. Such are the camels which the Towaregs use for war and in their predatory excursions. Mounted on his fleet dromedary, the warrior (Plate XXXVIII) crosses with inconceivable speed the desert that separates him from the tribe he is going to attack, or the home to which he is returning.

The Arabs desire three things in a good Camel; he must have a soft back, must not require the whip, and must not scream when he gets up or lies down. The first requisite is a necessary one, even for an Arab. Nothing is so disagreeable as the gait of the Camel. An Englishman declares that any one who wishes to practise camel-riding at home can do so by taking a music-stool, screwing it up as high as possible, putting it into a cart without springs, sitting on the top of it cross-legged, and having the cart driven diagonally over a newly ploughed field. The Arab mode of riding is to pass one leg over the upright pommel, which is merely a wooden peg or stake, and to hitch the other leg over the dangling foot. When the Camel increases its speed to a gallop, the rider grasps with his right-hand the cantle of the saddle, for the movement of the animal when galloping throws the rider forward with a violent jerk. The second requisite of a good camel, that is, not to require the whip, we can all understand. The third, however, can only be appreciated by those who have been present when a pack-camel has been loaded. The voice of the Camel cannot be described. Groans, moans, growls, bellowings, and screams follow in a most extraordinary succession. A traveler gives the following account of the Camel's behavior.

"The camel, when he is not eating or drinking or being loaded, is invariably chewing the cud. His long, crooked jaw is in perpetual motion, and when he is told to lie down to receive his burden, he does so without varying this incessant masticatory process. He awkwardly bends his fore-knees, drags his hind-legs under him, and comes to the ground with a curious kind of flop. All this time his long, melancholy face shows not the slightest indication that he knows what he is lying down for; and this unmistakable hypocrisy, I think, stamps the Camel as an animal of a very high order of intellect. But in a few seconds the expression on the Camel's face undergoes a striking alteration. As he sees the driver approaching him with a box on his shoulder he seems at last to understand the indignity and torture to which he is about to be submitted, and the astonishment, virtuous indignation, and dismay on the ill-used animal's countenance ought certainly to make some impression on the stony heart of the driver. They never have the slightest effect. The man binds the first box on the wretched animal's back, and goes away to get another. Then the Camel, wisely abandoning his efforts to move man to compassion, points his hairy nose upward, and howls his wrongs to the skies.

"Never in circus, pantomime, or show, have I seen anything half so ludicrous as the Camel's appearance at that moment. His upper lip is curled back from the teeth, his under lip doubles up and drops down as though he had no further use for it, his great mouth opens so wide that one can see about half a yard down his throat; and out of the cavern thus revealed come a series of the most astonishing howls that ever startled the air—howls of such abject misery that it is difficult to avoid the conclusion that the Camel's heart is breaking; and this impression is strengthened by the tears that flow copiously down the wailing animal's elongated cheeks, and drip from the end of his nose. In the utterance of each note of woe, the Camel seems to be exerting the utmost power of his lungs, but he is all the time holding a large force in reserve, and as the driver adds box after box to the pile on his back, a howl more resonant and heart-rending than the last testifies to each addition to the creature's misery; and never, except when he is absolutely engaged in trumpeting his agonies into space, are the great watery eyes of the Camel removed from the person of his persecutor; they follow him wherever he goes, and express through their tears contempt, indignation, astonishment, and dismay. I think it must have been this extraordinary habit on the part of the 'ship of the desert' that gave rise to the well-known expression: 'It's the last straw that breaks the Camel's back.' But the Eastern driver has no fear of any such catastrophe, and piles up the load until it reaches almost the proportions of an elephant's burden. Then, the cases being bound fast with ropes, the Camel is told to rise, and the animal, feeling that he has conscientiously done his whole duty by entering his protest at every stage of the work, contentedly accepts the unavoidable result, stops his tears, suppresses his cries, gets up on his feet, and, resuming his occupation of chewing the cud, is ready for the week's march that usually lies before him."

Brehm describes the Camel as the most unamiable, stupid, obstinate, and bad-tempered creature that can be imagined. Compared with it an ox is sensible, a mule is docile, an ass a loveable animal. Maliciousness and stupidity are the bases of its character; cowardice, wrong-headedness, irrational obstinacy, ingratitude, and a hundred other vices must be added to fill up the picture. The Camel exhales an odor nearly as offensive as the skunk's, his cries torture the ear, and its unspeakably dull-looking head on its long ostrich-like neck do not conciliate one's regard. When we know that no Camel ever shows the slightest affection for his master, but on every opportunity deliberately resists every wish, we may find some excuse for the hard hearts of the Arabian camel-drivers. The cowardice of the Camel is extraordinary. The roar of a lion scatters a caravan, every beast throws off its burden and rushes away. The howl of a hyæna, the sight of a monkey or of a dog terrifies it. It lives with no animal on terms of friendship. The horse cannot endure it, and on its part it seems to regard all animals with the dislike which it displays toward man. Its intelligence is small; it has only one great talent, namely, that of making a man mad.

The young Camel, when just born, is a strange-looking creature; it is covered with long, thick, soft woolly hair, the hump is small, the callosities on the joints scarcely visible. As it grows the wool increases in thickness and length, and the creature is strikingly like the South American species. It is weaned at the beginning of the second year, and then its training commences. It is accustomed to do without sleep, and to suffer the extremes of hunger, thirst, and heat. In its third year it is usually set to work, and by the end of the fourth year is considered capable of being employed in lengthened journeys. The camels for riding are trained to pace, not to gallop. The saddle is a dish-like frame perched on the hump, and held by three strong girths. The bridle consists of a leathern thong surrounding the creature's muzzle. No bit is used. Around the saddle hang weapons, holsters, cartridges, and bags of dates and water-skins, and the saddle itself is covered with sheep-skins dyed red or blue. The saddle of the pack-camel is not fastened by girths, but kept in position by the burdens which are skilfully arranged to balance. It is often stated that if a camel has been overloaded, it will, even when the load is removed, refuse to rise; this is not true. It remains lying when overladen, because it cannot rise, but when the weight is lessened, it gets up at once.

In desert-journeys the load of a camel is about three hundred and fifty pounds. In Egypt, however, the drivers piled up such loads that the government interfered, and fixed the weight to be placed on the unfortunate animal at five hundred and fifty pounds. One day Latif Pasha was sitting in his courtyard holding judgment in true Oriental fashion. A huge, heavily laden camel stalked in through the open gateway. "What does the beast want?" cried the Pasha. "See, its load is illegal! Weigh it." The camel had been carrying one thousand Arab pounds. The driver next appeared, in some astonishment. "Do you not know," thundered the Pasha, "that your camel ought to carry only seven hundred pounds? By the beard of the Prophet, and by the Almighty who has created man and beast, I will show you what it is to torture an animal! Give him five hundred blows." The driver was duly bastinadoed. "Now go!" said the judge; "if your camel ever brings a complaint again, it will be worse for you."

The Camel is subject to many diseases, especially in the low latitudes. In the Soudan, a fly is said to be very destructive. Most camels, however, die in their vocation. The simoom, or hot wind of the desert, is their deadliest foe. They scent the poisonous blast hours before the storm bursts. They become alarmed and anxious long before the sweltering heat which heralds the tempest is felt by man, and hurry forward at full speed. When the tempest breaks, no inducements can urge them onward; they lie down with their heads stretched out on the ground to leeward. They undoubtedly suffer as much as their riders, who, after the simoom has passed, feel as if every limb was broken, and who experience that feeling of weariness which usually is felt only after long sickness. When the camels are loaded, after the hot blast has passed, they show in every step what torments they feel. Their thirst increases, their weakness becomes more pronounced as they advance. A Camel suddenly falls; no words, no lashes move him to rise. The Arab, with tears in his eyes, leaves him to his fate; next morning the Camel is dead; before noon comes, the vultures are circling round the carcass, and before evening, the hyaena cannot find enough to satisfy its hunger.

Independently of its value as a beast of burden, the Camel is most precious to its owners, as it supplies them with food and clothing. The milk mixed with meal is a favorite dish among the children of the desert, and is sometimes purposely kept until it is sour, in which state it is very grateful to the Arab palate, but especially nauseous to that of a European. The Arabs think that any man is sadly devoid of taste who prefers the sweet new milk to that which has been mellowed by time. A kind of very rancid butter is churned from the cream by a remarkably simple process, consisting of pouring the cream into a goat-skin sack, and shaking it constantly until the butter is formed. The flesh of the Camel is seldom eaten, probably because the animal is too valuable to be killed merely for the sake of being eaten. Sometimes, however, in a season of great festivity, a rich Arab will slay one of his camels, and calling all his friends and relations to the banquet, they hold high festival upon the unaccustomed dainty. The long hair of the Camel is spun into a coarse thread, and is employed in the manufacture of broad-cloths and similar articles. At certain times of the year, the Camel sheds its hair, in order to replace its old coat by a new one, and the Arabs avail themselves of the looseness with which the hair is at these times adherent to the skin to pluck it away without injuring the animal.

The name of caravan has been given to companies formed in the desert by the assemblage of travelers, who thus, through numbers, avoid the insults and robberies of the brigands scattered around and over its immense confines. These caravans use Camels and Dromedaries for their beasts of transport; the former are loaded with the baggage and provisions, the latter are reserved to carry the travelers. When all are loaded and ready to start, an Arab, who acts as guide, precedes them, the Camels and Dromedaries following in line. This guide sings a monotonous and modulated plaintive song, indicating to his attendants by the quickness or slowness of its measure, when they are to increase or slacken their pace. When the guide's voice ceases, the whole troop halts and kneels to be unloaded, and they are turned loose to gather the scanty herbage found in the vicinity.

THE CAMEL OF BACTRIA.

The BACTRIAN or TWO-HUMPED CAMEL. *Camelus Bactrianus* (Plate XXXIX), is distinguished by two humps, one at the withers, another at the rump.

The general formation of this animal; its lofty neck, raising its head high above the solar radiations from the heated ground; its valve-like nostrils, that close involuntarily if a grain of drifting sand should invade their precincts; its wide cushion-like feet, and its powers of abstinence, prove that, like its Arabian relative, it is intended for the purpose of traversing vast deserts without needing refreshment on the way. This species is spread through Central Asia, Thibet, and China, and is domesticated through a large portion of the East.

It is the beast of burden for the traffic between China and South Siberia or Turkestan. The Kirghis Tartars set less store by it than the horse, but the Mongols prize it as the Arabs prize the Dromedary. The Bactrian Camel is much better tempered than the Arabian species; it is much more docile; its complaints are only murmurs, not the ear-splitting cries of the Dromedary. It is, however, equally stupid and cowardly. A hare springing up, says Przewalski, throws it into abject terror; a stone by the roadside frightens it out of its senses. It screams at the sight of a wolf. It feeds on the vegetation of the steppes which other animals avoid; salt is indispensable for it; it drinks the salt-water of the brackish lakes and licks up the crystals deposited on their margins.

The shape of the animal necessitates a peculiar saddle. Thick folds of felt are laid on the two humps, and on them is placed the framework to which its load is fastened. A strong animal will carry four hundred and fifty pounds a distance of about twenty miles a day. In summer it can go for three, in winter for five to eight days without food or drink, and it requires to rest one day in eight. The journey from Pekin to Kiatka is accomplished in about a month, after which it has a rest of about a fortnight; but in the West it is never worked so hard. When the hair begins to fall off it is treated more kindly; it is covered with felt coverings, and is allowed a respite from labor. The Bactrian Camel receives unceasing care. The Mongols regard it as the most fragile of animals, for although it can stand the icy storms of the winter and the toils of journey, it falls an easy victim to the climate of their summers. In winter it is never unsaddled; in summer the saddle is removed as often as possible, but never before the animal has become cool. It keeps up with the caravan as long as it can, but when it falls, it refuses to rise, and is left to the care of the inhabitants of the nearest village.

The height of the Bactrian Camel is rather more than that of the Arabian. Its color is brown, varying into sooty-black, or dirty-white.

CHAPTER XVI.

THE LLAMAS.

THE AMERICAN CAMELIDÆ—THE GENUS AUCHENIA—THE GUANACO—ITS HABITS—THE LLAMA—ITS USE AS A BEAST OF BURDEN—THE ALPACA OR PACO—ITS WOOL—THE VICUÑA—INDIAN HUNTS.

THE American representatives of the family CAMELIDÆ are, like all American animals, smaller than their Old World kindred. They are to the Camel, as the Puma is to the Lion. The former, indeed, being dwellers in the mountain ranges, cannot be expected to attain the size of the "ships of the desert." They are distinguished from the camels not only by their less size, but by the absence of a hump, by their large ears and eyes, and their long woolly coats. The feet are very different from those of the true camels, the toes being completely divided, and furnished with a small rough cushion beneath, and a strong claw-like hoof above, so that the foot can take a firm hold of the rocky and uneven ground on which they live. Altogether, they have a very ovine look, suggesting to the spectator a long-legged, long-necked sheep.

GENUS AUCHENIA.

Naturalists are not agreed as to the number of species into which the South American camels ought to be divided. Some attribute the differences between the various kinds to the effects of domestication, but the natives—with whom the investigator, Tschudi, agrees—always assert that there are *four* distinct species. Two of these, the Guanaco and Vicuña, are still wild; the Llama and Alpaca, on the other hand, have been tamed since time immemorial. They were the beasts of burden in Peru when the Spanish conquerors first entered the kingdom of the Incas, and the Peruvians themselves have a tradition that the taming of these creatures took place in those far back days when their gods used to walk the earth. The first European invaders everywhere found herds of Llamas in the possession of the Andean mountaineers, and describe the animals in unmistakable terms. Xerez, who wrote of Pizarro's conquest of Peru, says: "Six leagues from Caxamarca, Indians live who have flocks of sheep; some of them are so large that they are used as beasts of burden." Don Pedro de Cieza, in 1541, writes: "In no part of the world are there so many strange sheep as in Peru, Chili, and some provinces of La Plata. They are the most useful creatures that God has created, for the natives could not live without them. They are about the size of an ass, with the neck of a camel, and the look of a sheep. They are very tame and docile. The natives call them Llamas. Another kind is called Guanaco; these are somewhat larger, and run with great agility. Another variety of these sheep is named Vicuñas; they roam about in the desert; their wool is finer than that of the Merino sheep. There is also another sort called Paco, with still longer wool. Without these animals, intercourse with Potosi would be impossible." This passage is quoted to show that three hundred years have made no change whatever in the characteristics of these animals, and that therefore there is good reason for making *four* species of these useful creatures.

All the *Auchenia* dwell on the lofty plateaux of the Cordilleras. They prefer the colder regions, and are only found in the low country at the extreme south of the chain of the Andes in Patagonia. Near the equator they live at a height of four thousand yards above the sea. During the rainy season the wild varieties retire to the highest ridges of the mountains, and in the dry season descend into the valleys. They live in herds, very often of several hundreds, and are hunted with great ardor by the various Indian tribes.

THE GUANACO.

The GUANACO, *Auchenia guanaco*, is about the size of the Virginia deer, standing a little over six feet high. The body is relatively short and compact, the neck thin and curved, the upper lip prominent and deeply cleft; the nostrils are capable of being closed; the ears are about half the length of the head, and very mobile; the legs are slender and long, the toes cleft to the middle, the soles have rough cushions, but the fore-legs have not the callosities which the other varieties possess, like the camel. A coat of long, rich, but sparse hair covers the body; it consists of a short, fine layer of wool, through which grows longer and thinner hair. This coat is short on the face, but on all other parts of the animal except the legs, forms a wool-like fleece, which, however, is never as fine as that of the Llama. The general color is reddish-brown.

The Guanaco extends along the Cordilleras from the island of Tierra del Fuego to Northern Peru, and is especially abundant in the southern part of the chain of the Andes. It prefers mountain plateaux, but Darwin found it in great herds in the low plains of Southern Patagonia. It ascends, during spring, when the herbage is coming up at high elevations, as far as the line of eternal snow, but in the dry season it returns to the lower valleys. It avoids the snow-fields; its feet are not adapted for walking on the slippery surface. At times the Guanacos make long journeyings, and Darwin saw herds of them on the coast at Bahia Blanca. They are not afraid of the sea, and swim fearlessly.

The herds usually consist of from twelve to thirty members, although one of five hundred has been seen. Generally the herd consists of females under the guidance of an old male. It wanders by day from one valley to another, browsing with pauses; it never feeds by night. It seeks a drinking-place morning and evening, and the Guanacos seem to prefer brackish water to fresh. The movements of these animals are quick and lively, although they have not the speed often attributed to them; a good horse can soon overtake them. Their gait is a short gallop, and is, like the camel's, a *pace*. They climb admirably, and run like the chamois on the verge of the steepest precipices, where a man cannot find standing-room. When reposing, they lie down like the camel on the breast and legs, and chew their cud in a dreamy state of complacency.

The Guanacos are usually shy and wild, flying away when anything suspicious comes in sight. The leader of the herd precedes it by some paces, and keeps a good look-out while the females are feeding. When danger is approaching, he sets up a neighing scream, and all the flock takes flight, the leader bringing up the rear. Like many animals, however, they are curious, and Darwin says that if the hunter lies on the ground and kicks his legs in the air, they will draw near to see what strange creature it may be; even the report of his rifle does not frighten them, they think it part of the performance.

The Guanaco, in common with the other species, is rather short-tempered, and has a very unpleasant habit of displaying its anger by

LLAMA ALPACA VICUGNA
BACTRIAN CAMEL
PLATE XXXIX UNGULATA

discharging a shower of half-digested food and saliva over the offender. Formerly, this salival discharge was thought to be acrid, and capable of raising blisters upon the human skin. This, however, is fortunately not the case, although the assault is eminently disagreeable, on account of the ill scent of the ejected liquid. In its wild state the Guanaco seems to have little or no idea of resistance, being easily held by a single dog until the hunter can come up and make sure of his prize. But in domesticated life, it seems to imbibe a spirit of combativeness, for it will kick with both hind legs, and deliver severe blows with the knees of those limbs. With his rivals the male fights desperately, with teeth and feet.

The Guanaco is hunted for its flesh and hide; it is pursued by dogs, driven to bay, and then lassoed. In the lowlands they are often killed by the gun. A wounded Guanaco at once hastens to a river, in order to die on its banks; indeed they all, when they feel death approaching, seem to seek certain definite places. "On the banks of the Santa Cruz," Darwin writes, "the ground beneath the bushes was white with bones. These bones were not broken or gnawed, and could not therefore have been dragged there by beasts of prey. The Guanacos had crept there to die." The Guanaco cannot be domesticated. When quite young it can be taught to follow its master like a dog, but when it grows old it uses every effort to escape. It can be easily kept in captivity.

THE LLAMA.

The LLAMA, *Auchenia llama* (Plate XXXIX), is found most abundantly in Peru. It is larger than the Guanaco, and is distinguished by possessing callosities on the breast and fore-knee. Its color varies; white, black, piebald, skewbald, dark-brown, fox-red, and even yellowish animals have been seen. The Llama is of the greatest service; it not only supplies food, but transports all merchandise. Its usual load is one hundred and fifty pounds, but some will carry a hundred pounds more. It can march at the rate of ten leagues a day for five days in succession. Its gait is so steady, that little fastening is requisite to keep the load on its back. About three hundred thousand are in constant use for bringing silver bars from the mines to the smelting works, and carrying back provisions. Acosta writes: "I have often wondered at the sight of these herds of sheep transporting two or three thousand bars of silver, worth three hundred thousand ducats, without any other attendance than a few Indians. They love the cold air of the mountains, and retain their health even when covered with ice and snow." "Nothing can be prettier," another traveler says, "than a train of these animals marching one behind the other in the greatest order, following a leader who is adorned with a tastefully decorated halter, a bell, and a little flag on his head. Thus they march along the snowy summits of the Cordilleras, or the sides of the precipices where not even a mule can go. The conductor needs no goad or whip to urge them on or guide them. They advance quietly and without a rest." Like the Guanacos, they are curious and keep constantly looking around them, and are equally timid, dispersing at any alarm, when the conductor has an arduous task in collecting them. The use of the Llama as a beast of burden is gradually, however, being superseded by that of the ass.

The flesh of the Llama is a favorite article of diet, and that of the lamb of one year is a great dainty; the older animals are killed to obtain "jerked" or dried meat.

The Llama is seen in nearly every Zoological Garden, even in the collection in the Central Park, New York. It appears to be more amiable when kept in the society of its congeners, and soon knows its keeper, but towards strangers it shows itself easily irritated, and if provoked, discharges its filthy spittle in his face. It lives and breeds in captivity and requires very little care.

THE ALPACA.

The ALPACA or PACO, *Auchenia paco* (Plate XXXIX), is smaller than the Llama, very like a sheep, with a long neck, and well-shaped head. The wool is long and very soft, and white or black in color.

The Alpacas live in large herds, which pasture all the year on the plateaux, and which are only collected to be shorn. There is, perhaps, no more obstinate animal than the Alpaca. When one is separated from the herd, it flings itself on the ground, and neither coaxing nor flogging can make it get up; it will rather die than stir; the only way to induce it to exert itself is to bring up another herd, and then it condescends to join itself to them. The wool attains the length of nearly four inches, and has from time beyond the memory of man been spun into garments. The coarser wool is called by the Indians *Hanaska*, the finer *Cumbi*; they make from it table-cloths and other things which are remarkable for wearing well and having a smooth surface. The Incas of Peru had great masters in the textile art. The most skilful lived near the Lake of Titicaca. They dyed the wool with various herbs. The present Indians have lost the art, except in its ruder manufactures of coverlets and cloaks. The best wool is sent to Europe, where, as we all know, it is spun and woven into a variety of articles.

All attempts to acclimatize Alpacas in Europe have failed. Nor do they succeed any better in Australia. An Englishman named Leeds was sent out by the government of New South Wales to procure the animals from Bolivia. But the Bolivian government forbade the exportation, and only after great difficulties did he manage to ship three hundred Alpacas. Five years later, after the government had spent fifteen thousand pounds, scarcely a dozen of the animals were alive, while the young ones born from the imported ancestors were in a very poor condition. There are, however, many places where they might be domesticated, but it is not worth the while to do so, such places being already occupied by more profitable animals. The Alpacas are enduring, require little attention, and breed fast, and in addition to their wool supply good flesh. They are never used for carrying burdens, but kept solely for the hair and flesh. To obtain the former, the herds are annually driven in and shorn, which is no light task with an animal so full of natural obstinacy. The shearing over, they are again turned loose.

THE VICUÑA.

The VICUÑA, *Auchenia vicuña* (Plate XXXIX), is intermediate in height between the Llama and the Alpaca. It is distinguished by its short, kinky wool, of exceptional fineness. The top of the head, the back of the neck, the back and loins, are covered with hair of a peculiar reddish-yellow color. The lower side of the neck and of the limbs are of a bright ochre-yellow; the hair on the breast and stomach, which measures nearly five inches, is white. The Vicuñas in the rainy season seek the crest of the Cordilleras; but they avoid stony districts, because their hoofs are tender. In summer they revisit the lowlands, following the supply of food. They browse all day long. The herds consist of a male and five or six females, the former always keeping watch. He gives the alarm at the sight of danger, and when the herd retreats, covers the rear.

In the hills where they dwell, it is impossible to ride them down. The sportsman has to shoot them. By a strange contrast, a herd of guanacos disperse when the male leader is killed, but the Vicuñas, *even be on dead led*, gather around the wounded leader, and can be shot down to the very last one. If the first shot kills a female, then the whole flock scurries away.

The Vicuñas, according to Acosta, are very timid, and flee at the approach of the hunter, driving their young before them. They do not increase in numbers rapidly, and during the reign of the Incas, the subjects of the kingdom were forbidden to hunt them, the chase being reserved for the monarch himself. Since the Spanish conquest they have diminished considerably, for the conquerors hunt the males and females alike. The Indians do not use firearms in the chase. They take stakes and immense coils of rope. The former they place about twelve to fifteen feet from each other, and unite by the rope which is raised about two feet from the ground. A large circle is thus enclosed, leaving an entrance two hundred feet broad. The women who accompany the hunters hang rags of all kinds on the rope. The men, many of whom are mounted, then make a wide circuit and drive the Vicuñas toward the enclosure. When enough have been collected, the entrance is closed. The timid animals dare not pass the ropes with the fluttering rags, and are easily lassoed, or caught with the *bolas*. These latter consist of three balls, two heavy, one light, of lead or stone, united by a long thong. The lighter ball is taken in the hand, and the others whirled around the head. All are then let go, and fly toward the object aimed at, which they at once encircle in the thong. If the hind-legs of the animal are aimed at, the bolas envelop them so tightly that every motion is impossible. The Vicuña is then killed, and its flesh distributed among those who participate in the hunt; the fells are given to the church. Bolivar, the founder of the Republic of Bolivia, forbade the slaughter of Vicuñas; he ordered them to be shorn when captured. But this operation has proved almost impossible. The hunts in the time of the Incas were grand affairs; thirty thousand Indians were assembled; the circuit they formed would be fifty miles, and all wild animals were driven into the toils. The bears, cougars and foxes were slain, as well as a certain number of stags, roebucks, and vicuñas. At present, it is said that if guanacos are thus enclosed they break through the ropes and are followed by the vicuñas. Hence, skilful huntsmen do not drive the former,

Tschudi was present at one of these hunts, which lasted for a week, at which one hundred and twenty vicuñas were killed.

The young Vicuña is easily tamed, but, like all its kindred, becomes mutinous as it grows older. A clergyman had a pair which lived with him quietly for four years. At the end of that time the female ran away, and sought to join a herd of her wild kindred. They drove her away with bites and pushes, and she remained a solitary wanderer on the hills. For months, sportsmen and travelers encountered her, for she was conspicuous by the bright collar which had been placed around her neck, but she fled from their approach.

It may be added, that since the establishment of the Alpaca manufacture by Sir Titus Salt at Bradford, England, the wool of all the *Auchenia* has become an article of commerce, and is worked up, either alone, or mixed with other textile materials, into numerous and beautiful fabrics. Dress-goods, umbrellas, gloves, and other articles too numerous to mention are woven from the wool of the various species. The hair of the guanaco is the most valuable, being longer in the staple, and of a silky, soft texture. The natives of Peru prefer, however, that of the vicuña, from which they weave robes and mantles which have quite the appearance of silk, and which, as the hair does not need to be dyed, last for a very long time.

CHAPTER XVII.

THE MOUSE DEER.

THE TRAGULIDÆ OR HORNLESS DEER—DISPUTES OF NATURALISTS—THE KANCHIL—ITS APPEARANCE AND HABITS—ATTEMPTS TO INTRODUCE IT TO EUROPE.

THE family TRAGULIDÆ is a group of small, deer-like animals with tusks in the upper-jaw, and having some structural affinities with the camel. The musk deer, *Moschus moschiferus*, was formerly classed in this family, but a minute examination of its structure by the celebrated French naturalist M. Milne-Edwards has shown it to be more closely allied to the true deer. The Chevrotains or Mouse-deer, as the TRAGULIDÆ are called, differ from the Musk-deer by having only three divisions in the stomach, and by the absence of a gland to secrete musk. They are usually divided into *two* genera: TRAGULUS comprising *five* species, which range over all India to the foot of the Himalayas and Ceylon, and through Assam, Malacca, and Cambodia to Sumatra, Borneo, and Java; and HYOMOSCHUS, forming *one* species, found in Western Africa.

Naturalists, however, are by no means agreed respecting the classification of the *Tragulidæ*, and many deny that the differences that exist are such as to justify the creation of *two* genera, or even the number of species assigned. As a matter of fact, all classification depends upon anatomical peculiarities, rather than on anything which the ordinary observer would detect. In a work intended for the general public, we have no hesitation in neglecting the refined distinctions of the professional naturalists. Fancy, therefore, a delicate deer-like creature with a slender, graceful head, no horns, beautiful bright eyes, and legs as thick as a lead-pencil, with the tiniest hoofs, a short, little tail, and soft, smooth hair,—and you have the Mouse-deer.

The KANCHIL, *Tragulus pygmæus* or *Javanicus* (Plate XLII.), is, including the tail, about a foot and a half long, and stands about eight inches high. The hair on the head is of a reddish fawn color, almost black on the top of the skull; on the back it is brown with a reddish-yellow shade, strongly marked with black along the spine; the abdomen is white. A white streak runs from the under-jaw to the shoulder, then, lower down, comes a dark streak which divides in the throat, and encloses another band of white. The limbs are fawn-yellow, the upper arm reddish, the feet dull fawn color. The difference of color is produced by the peculiar coloring of the hairs. On the back they have the lower half white, then bright yellow occurs, while the tips are black. In proportion, therefore, to the degree in which the black points fall off, or as the light ring is more or less prominent, does the general color of the hide differ. The white spots consist of pure white hair. The older males have projecting canine teeth. The hoofs are of light-brown horn.

Java, Singapore, Penang, and other neighboring islands with the Malay Peninsula are the homes of this charming little creature. Allied species represent it in Sumatra, Borneo, and Ceylon. In Java it lives in the hills, rather than in the lowlands, preferring the lower edge of the primæval forests which cover the hills. There it lurks in the fringe of brushwood, from which it can sally forth and in a few minutes reach the grassy slopes below. It is never seen in herds. During the day it lives retired, reposing and ruminating in the densest thickets. When twilight comes, it sallies forth to browse on leaves, herbs, and berries. Water is indispensable for its existence.

All the movements of the little animal are extremely graceful and light, and marked with great vivacity. It can take long leaps, and skilfully avoids all difficulties in its way. But its delicate limbs soon refuse their office, and it would be easily the prey of its enemies, if it did not possess a peculiar method of defence, which in this country is usually considered to be the especial prerogative of the opossum. When it cannot conceal itself in the underwood, and plainly sees that escape is impossible, it lies down and pretends to be dead. The hunter stoops to grasp his victim, but before his hand can reach it, the Mouse-deer has made a spring and disappeared like lightning. The natives of Java assert that the male Chevrotain has another way of escaping the attacks of beasts of prey. They say it leaps up in the air and hangs itself by its long canine teeth to a projecting branch! Sir Stamford Raffles tells us that the Javanese can find no better means to describe a consummate trickster, than the words "He is as cunning as a kanchil."

Various attempts have been made to introduce the Mouse-deer into Europe. A pair presented to the Queen of England was forwarded by her Majesty to the Zoological Gardens in the Regent's Park, London, but they died in two months. Brehm had several in the Thiergarten of Berlin; they were very cleanly, always dressing and licking their fur. Their large and beautiful eyes lead one to expect a great amount of intelligence, but this is not the case; the tiny creature is placid and quiet, passing the day in eating, chewing the cud, and sleeping. Its delicate soft voice is a weak piping.

"By the kindness of the government," writes Bodinus, "we received a pair of chevrotains. In spite of great attention, fresh grass, clover, bread, milk, and oats, they could not be brought into good condition. Under these circumstances I changed their diet to berries. They devoured them eagerly, and good results followed. The large eyes grew brighter, the hair became sleeker and more brilliant, the body plumper, and I was convinced that the tiny creatures could be kept in perfect health on milk and bread, and some green stuff.

The Javanese, who call the animal Poetyang, hunt it zealously, and esteem highly its tender and sweet flesh. The slender legs are often mounted in gold and silver, and used for tobacco-stoppers.

CHAPTER XVIII.

THE DEER.

THE CERVIDÆ—THEIR ANTLERS—THE PROCESS OF GROWTH OF THE ANTLER—THE SHEDDING OF THE VELVET—HABITS OF THE CERVIDÆ—THE VARIOUS GENERA—THE ELK OF THE OLD WORLD OR THE MOOSE OF THE NEW WORLD—THE ELK OF SWEDEN—THE MOOSE OF CANADA—HABITS—MODES OF HUNTING.

NO group in the whole order can be more easily described than the family CERVIDÆ. They are ruminant animals, with antlers or horns that are shed. Compared with this striking characteristic, all other distinguishing features are superfluous. The Deer form an extensive group of animals equally adapted for inhabiting forests or open plains, the Arctic regions, or the Tropics. They range, in fact, over the whole of the great continents of the globe, with the one striking exception of Africa, where they are only found on the shores of the Mediterranean.

The distinctive characteristic, then, of the animals of this group consists in the texture, shape, and manner of growth of their frontal protuberances. These projections, which are called *antlers*, and not horns, are bony, solid, and more or less branching; they are also devoid of the horny casing which exists in all hollow-horned Ruminants. They fall off, and are renewed periodically every year up to a certain age.

In the adult individual the antler is composed of a cylindrical or flattened stem, according to the genus, which is called the *brow-antler*, from which branch out at intervals slighter and shorter additions, called *tines* or *branches*. The base of the brow-antler is surrounded with a circle of small bony excrescences, which afford a passage to the blood-vessels intended to provide for the growth of the antler; these are called *burrs*.

The whole process of the growth, shedding, and renewal of the antlers is so peculiar, and so entirely different to anything which we see in any other animal, that a somewhat lengthy account of the process is necessary. Soon after birth, the place destined by nature to bear the antlers is indicated by a considerable development of bone on the forehead. During the sixth or eighth month of the animal's existence, two small elevations or *knobs* make their appearance; these remain permanently during its whole life. From each of these *knobs* grows a cartilaginous prolongation, which soon assumes a bony texture.

Until they become perfectly hard, these two early sprouts are protected against any external friction by a kind of velvety skin, which serves as a vehicle for the calcareous matter, and dries up as soon as ossification is accomplished, the beast freeing them from the velvet by rubbing its head against a tree. The technical name for these bony developments during the first year of the deer's life is that of *dags*. They are simply straight horns. They are shed in the second year, and are then replaced with *antlers*, which bear one *branch* or *tine*. In the spring of the third year the same process is repeated, but the new *antlers* bear an additional *branch* or *tine*, and this augmentation of the number of the *tines* at each annual reproduction of the *antlers* goes on till the animal reaches maturity. The junction of the *antler* with the bony frontal development or *knob* is very close. The root of the antlers throws out processes of various size which fit into corresponding depressions in the *knob*, while processes running from the *knob* in turn insert themselves into depressions in the root of the *antlers*. A close dove-tail joint is thus formed, and the union is so perfect, that in a newly-grown antler it cannot be detected, even by a vertical section through the *antler* and the *knob*; but when the antler is full grown and hardened, it can be seen in the section as a very fine zigzag line. If violence is applied to an *antler* before it is ready to be shed, it does not break off at this joint, but the whole *knob* is torn away from the frontal bone. The shedding is effected by the weight of the antlers. Some days previous to this event the skin around the *knobs* becomes loose; the deer avoids striking or pushing with his antlers, and betrays that there is an unaccustomed sensation in the place. There is usually an interval, sometimes of only a few minutes, sometimes of several days, between the dropping off of the two antlers. The deer hangs his head and ears, and shows clearly that if the process is not painful, it is all events uncomfortable. When one of the antlers is fallen, he carries his head on one side, and keeps shaking it, so as to get rid of the other. Immediately after the shedding, the reproduction of the bony ornaments of the head begins. This renewal is a very curious phenomenon, and we owe to the patience of Dr. Sœmmering the best account of the whole interesting process: "Immediately after the shedding of an antler, the lower surface of it is dry, or rather not bloody, the blood-vessels being quite dead and empty. At the edge of the *knob*, between the *burrs*, the openings of numerous canals are visible, containing arterial vessels which spring from the external carotid, and are developed in an extraordinary manner while the antler is growing. When the antlers have both been shed, the deer seems worn out and spiritless; he feels that he is defenceless, and seeks solitary places. The round surfaces on which the antlers rest are covered with a coat of blood and lymph, and surrounded by a ring containing vessels which contribute to the displacement of the antlers. The rush of blood to the *knobs* is checked by the old antler, the vessels become congested and form this thick ring, cutting off the antler from the brow, and undermining it. It is from this vascular ring that vessels proceed which contribute a secretion of calcareous matter for the formation of the new growth. The *dags*, or first year's horns, rest on a protuberance of the frontal bone which diminishes in height each year as each successive shedding of the antlers removes a layer from it. By the fourteenth day a vascular tumor has filled up the place left by the shed antler, and spread over the hairy skin of the *knob*, especially in the anterior region, to form the lowest or eye-tine. On the twentieth day the protuberance, which is burning hot to the touch, begins to be covered with white hair, and increases rapidly. By the thirtieth day all the points of the future antler are indicated by more or less prominent divisions and folds, and over the edge of the hairy knob is a bluish vascular ring, the beginning of the *burrs*. On the forty-fifth day the last division has not taken place, but by the fiftieth day all the tines are pretty long. The upper portion of the antler is completely formed by the eightieth day, but is still covered with very sensitive velvet. By the hundred and twentieth day the antler is fully grown and the tines ossified to the end."

Such a process is without parallel in natural history. As soon as the growth is complete, and the ossification finished, the *burrs* increase in size; they strangulate the blood-vessels, and the velvet loses its vitality, and is soon rubbed off against the branches or trunks of trees. The horns annually fall off early in the year. As a rule, antlers are found only on males, but have been seen on barren females. It is probable that the various functions devolving on the female, such as gestation, parturition, and giving suck, divert the nourishing fluids from the head and concentrate them in other organs, and this is the physiological cause which deprives the weaker sex of antlers.

The modes of life of the *Cervidæ* are as different as their dwelling-places. Some live, like chamois, on the rocks, some hide themselves in the thick forests, some frequent arid steppes, others, swamps and morasses. They are all social animals, many kinds forming large herds. During the summer the males usually leave the herds, and may often be seen wandering about together. At other times, they are deadly foes, and terrible combats ensue. The female has one or two young ones annually; no prettier group of animals can be seen than a hind and her fawns, with their large, soft, innocent eyes, and graceful movements. But neither the majestic appearance of the male, with his branching antlers, nor the tenderness of the female, have saved the deer from the fate which attends all animals that man cannot enslave! Their presence cannot be reconciled with modern agriculture. In America they are rapidly becoming scarcer as the country is filled up, and in Europe they are merely preserved, some species for sport, some for ornament, otherwise they would long ago have disappeared.

If we include the Musk-deer, and an allied genus lately discovered in China, the family *Cervidæ* must be distributed in *eight* genera, most of which contain only one species.

Nearly all the members of this family are remarkable for the elegance of their shape, the dignity of their attitudes, the grace and vivacity of their movements, the slenderness of their limbs, and the sustained rapidity of their flight. They have a very short tail; moderately sized and pointed ears; their nostrils are generally situated in a muzzle, and their eye is clear and full of gentleness. In most of the species there is, below the internal angle of the eye, a small depression, called a tear-pit, which is nothing but a sort of gland, secreting a peculiar fluid. This gland is not, as might be supposed from the name, the place from which the tears proceed.

GENUS ALCES.

The ELK or MOOSE ranges all over Northern Europe and Asia as far north as East Prussia, the Caucasus, and North China, and over Arctic America to Maine on the East, and British Columbia on the West. Wallace assigns to the genus only *one* species; adding that the American

MOOSE DEER   WAPITI or ELK

species may, however, be distinct, although very closely allied to that of Europe. Much confusion exists in the use of both the names Elk and Moose. The animal commonly known in America as the Elk is the Wapiti, *Cervus Canadensis*. In Ceylon, Elk designates a large species of deer, while in Africa, it is applied to an antelope. The true Elk is the animal known in our Northeastern States as Moose.

THE MOOSE OR ELK.

The ELK, *Alces palmatus* or *Alces malchis* (Plate XLI), has been long known in Europe. Cæsar found it in the Black Forest, and several of the later Roman emperors exhibited specimens in their triumphs. In the great German epic, the Nibelungen Lied, the name Elk occurs; it is applied to an animal found in all parts of Germany, and antlers of elks are often discovered in the woods of Brunswick, Hanover, and Pomerania. In this last province large herds existed in the sixteenth century. But the decrease has been rapid. In the seventeenth and eighteenth centuries only a few could be found in Saxony and Siberia. A herd has been preserved to the present day in the Royal Forest near Tilsit, which in 1874 numbered seventy-six members. The Elk is now confined in Europe to Norway, Sweden, and Russia. In Asia it extends to the Amoor River, abounding near the Lena and Lake Baikal, ranging northward as far as any tree will grow.

The Moose or Elk is the largest of all the deer tribe, attaining the extraordinary height of seven feet at the shoulders, thus nearly equaling an ordinary elephant in dimensions. The horns of this animal are very large, and widely palmated at their extremities, their united weight being so great as to excite a feeling of wonder at the ability of the animal to carry so heavy a burden. To support such a load a short and very thick neck is necessary. It does not reach its full development until its fourteenth year. The muzzle is very large, and is much lengthened in front, so as to impart a most unique expression to the Elk's countenance. The color of the animal is a dark brown, the legs being washed with a yellow hue. Its coat, which is composed of coarse, rough, and brittle hair, rises into a small mane on the nape of the neck, and on the spine. The long black hair under the throat forms a kind of beard, and in the male animal covers a considerable protuberance. Its speed is very great, and its endurance wonderful, but the pace is usually a trot.

not a gallop. It has been known to trot uninterruptedly over a number of fallen tree trunks five feet in thickness.

The Elk swims with great facility. During the summer it submerges its whole body, except the head, and in this way preserves itself from the stings of the horse-fly; thus it passes the greater portion of the day, while it principally subsists upon aquatic herbage. It is also partial to damp forests and marshy localities. This animal feeds off the ground with difficulty, on account of the shortness of its neck; in order the better to reach the grass, it kneels or straddles its fore-legs. It prefers, however, to browse off the young shoots, buds, and bark of trees, thus furnishing hunters with sure proofs of its vicinity. When the ground is hard and will bear the weight of so large an animal, the hunters are led a very long and severe chase before they come up with their prey; but when the snow lies soft and thick on the ground, the creature soon succumbs to its lighter antagonists, who invest themselves in snow-shoes and scud over the soft snow with a speed that speedily overcomes that of the poor Elk, which sinks floundering into the deep snow-drifts at every step, and is soon worn out by its useless efforts.

The skin of the Elk is extremely thick, and has been manufactured into clothing that would resist a sword-blow, and repel an ordinary pistol-ball. The mad emperor of Russia, Paul, carried on a regular war of annihilation against the Elks, as he considered elk-skin the only material fit to be made into breeches for his cavalry. The flesh is sometimes dressed fresh, but is generally smoked like hams, and is much esteemed. The large muzzle or upper lip is, however, the principal object of admiration to the lovers of Elk-flesh, and is said to be rich and gelatinous when boiled, resembling the celebrated green fat of the turtle.

Among the Carnivora, the chief enemies of the Elk are the same as those of the reindeer, namely, the bear, the wolf, and the glutton.

The Elk, when captured young, may be completely tamed without difficulty. It recognizes the person who takes care of it, and will follow him like a dog, manifesting considerable joy on seeing him after a separation. It goes in harness as well as the Reindeer, and can thus perform long journeys. For two or three centuries it was used for this purpose in Sweden, but the custom is now given up. It is impossible to understand why hardly any attempts have been made to domesticate such a useful animal in those climates suited to it, and thus prevent the destruction which threatens to entirely extirpate the race.

The Elk lives in a family composed of a male, a female, and the young of two generations.

THE MOOSE.

The MOOSE, *Alces Americanus*, which is also called the BLACK ELK and the FLAT-HORNED ELK, is said by the naturalists who form it into a separate species to be distinguished from the European variety by deeper indentations in the palmated antlers, by the slightness of the beard, and the darker color of its coat. Its antlers are larger than those of the European Elk, sometimes weighing seventy-five pounds and measuring thirty-two inches long and thirty-one broad. "The Moose deer," Hamilton Smith writes, "is higher than a horse, and when seen fully grown and in all its spread of antlers, makes a striking impression."

The Moose is confined to the northern regions of the continent, Canada, and New Brunswick, sometimes appearing in Maine and Northern New York. Franklin saw it at the mouth of the Mackenzie River in 65° north latitude. Its habits and food, as well as its general configuration, are those of the Elk of the Old World. From its custom of browsing on trees and eating the bark, the Indians gave this animal the name of *musee* or "wood-eater," whence comes our word Moose.

The young Moose can be easily tamed, and learn in a few days to know their keeper and follow him with confidence. But as they grow older, they grow worse in temper, and become savage and dangerous.

Audubon, who had the gratification of bringing a Moose down with his rifle and examining it in detail, states that to him he appeared awkward in his gait, clumsy in his limbs, and inelegant in form. The head, he adds, is long and clumsy, the snout is long and almost prehensile, the eye deep-seated, and small in comparison with the jackass-like head.

During the winter several of these animals associate together and form groups of two, three, or four, and make what is technically called "a yard" by beating down the snow; in such places they feed on all the branches they can reach, stripping the trees of their bark, and breaking boughs as thick as a man's thigh. When obliged to run, the male goes first, breaking the way, the others tread exactly in his tracks, and when their path runs through other "yards," they all join together, still going in Indian file.

The seasons for hunting the Moose are March and September. The sun then melts the snow on the surface, and the nights being frosty form a crust which greatly impedes the animal's progress. It is necessary for the hunter to have two or three small curs that can run upon the snow without breaking the crust; their use is to annoy the Moose by barking at its heels. The males when thus pressed stop, and the hunter can come up and dispatch them while their attention is thus engaged. Another method of hunting is described by Audubon: "In September two persons in a canoe paddle by moonlight along the shore of the lake, imitating the call of the male. He answers the call, and rushes down to meet his rival. The man in the bow of the canoe fires; if the animal is only wounded he makes for the shore, and can be tracked by his blood to the place where he has lain down, and where he is generally found unable to proceed further." Sometimes hunters find out the beaten tracks of the Moose leading to some spring, and bend down a sapling with a strong noose over his path. If his branching antlers pass through the dangling snare, he makes a struggle which disengages the rope holding down the sapling. It springs up, and the Moose is strangled. At other times they are "pitted," but their legs are so long that this method of securing them seldom succeeds. The skill of a moose-hunter is mostly tried in the winter; he must track the creature by its footmarks in the snow, keeping always to the leeward of the chase. The difficulty of approach is increased by a habit which the Moose has of making daily a sharp turn in its route; the hunter therefore forms his judgment, from the appearance of the country, of the direction it is likely to have taken. When he has discovered by the footmarks and other signs that he is near the chase, he approaches cautiously; if he gets close without being seen, he breaks a small twig, which noise alarms the animal. It starts up, stops, and offers a fair mark to the hunter. The method of "calling" is efficacious in the rutting season. A sportsman writes of it as follows: "Calling is the most fascinating, disappointing, exciting of all sports. You may be lucky at once, and kill your Moose the first night you go out, perhaps at the very first call you make. You may be weeks and weeks, perhaps the whole calling season, without getting a shot. Moose calling is simple enough in theory; in practice it is immensely difficult of application. It consists in imitating the cry of the animal with a hollow cone made of birch-bark, and endeavoring by this means to call up a Moose near enough to get a shot at him by moonlight, or in the early morning. He will come straight up to you, within a few yards—walk right over you almost—answering, 'speaking,' as the Indians term it, as he comes along, if nothing happens to scare him; but that is a great if. So many unavoidable accidents occur."

In December Moose-deer cast their horns; by April their successors begin to sprout; by the end of June full form is developed, but not till many weeks later are they denuded of velvet. At one year the bull-calf throws out a brace of knots an inch in length; in the second season these are about six inches long; the third year they increase to nine or ten inches with a fork; in the fourth season palmation is exhibited at many points. "Twenty-three," writes Mr. Parker Gilmore," is the greatest number of points I have seen on one head, and the weight of the horns exceeded seventy pounds.

The "Moose Wood" or "Pennsylvania Maple," *Acer Pennsylvanicus*, derives its name from the fact that the Moose browse on it by preference. The Indians, we may add, believe that after a feast on Moose-flesh they can travel three times as far as after any other food.

CHAPTER XIX

THE REINDEER AND THE CARIBOU.

THE REINDEER—ITS LIFE IN NORTHERN EUROPE—ITS LIFE IN SIBERIA—ITS LIFE WHEN DOMESTICATED—ITS VALUE—THE CARIBOU—MODES OF HUNTING IT.

IN the case of the Reindeer, as with the Elk, naturalists are not agreed as to whether the American variety attains to the dignity of a separate species or not. Wallace, and other high authorities assign but *one* species to the genus which we are now about to describe.

GENUS TARANDUS.

In this genus the animals of both sexes possess antlers, the arrangement of which is very characteristic. From the principal stem, which is round and short, spring two branches of flattened shape, the longest of which tends upward with various twists terminating in an indefinite number of points; the other, more moderate in size, stretches horizontally over the muzzle. The hoofs are very broad and long; the head is wide and ox-like, but there is no muffle, and the nostrils open in the midst of the hair. The legs, although finely made, are less slender than those of the stag. The feet are covered all over with stiff hair, even on the under parts, an arrangement which facilitates the animal's tread on ice, or icy snow. Its coat is rough, of a grayish-brown color, and is long under the throat, becoming in the winter woolly in texture, and often white in color. The eye of the Reindeer is protected from the blinding glare of the snow by a third nictitating eyelid, which, at the animal's will, covers the whole eye.

The Reindeer may be described as the most valuable of the *Cervidæ*. Whole tribes depend on it for their existence and well-being, and it is more necessary to them than the horse or ox to ourselves.

THE REINDEER.

The REINDEER, *Tarandus rangifer* (Plate XL), is a native of the icy deserts of the Arctic regions. It is found in Spitzbergen, Greenland, Lapland, Finland, and the whole of Northern Russia, as well as in Siberia and Tartary. In Russia it sometimes migrates southward as far as the range of the Caucasus.

The ancients were well acquainted with the Reindeer. Cæsar describes it, Pliny confounds it with the elk. Ælian relates that the wild Scythians used deer instead of horses. But the first good account of the animal is found in the work of Scheffer, published 1673, and in the next century the great Linnæus, who had personally seen and observed it, gave a complete description of the animal.

"There is no animal," writes Brehm, "in which the burden of servitude, the curse of slavery, is so evident as in the Reindeer. No two creatures of the same family can be so different as the tame and the wild Reindeer. The former is a miserable slave of a poor miserable master, the latter is a lord of the mountain, as agile as the chamois. When we see a troop of wild and a herd of tame Reindeer, it is almost impossible to believe that they are members of the same genus."

The Reindeer is naturally a child of the mountains, and loves the wide, treeless, mossy plateaus which the natives of Scandinavia call "Fjelds." The barren expanses where a few Alpine plants grow between the rocks, or the heaths covered with Reindeer-moss, are its favorite abode. Woods and woodlands it usually avoids. In the north of Siberia, according to Pallas, it is sometimes found in the forests making annual journeys from the woods to the hills and back again, according to the season. Their chief object in leaving the forests in the summer months appears to be their hope of escaping the continual attacks of mosquitoes and other insect-pests that are found in such profusion about forest land. The principal plague of the Reindeer is one of the gadflies, peculiar to the species, which deposits its eggs in the animal's hide, and subjects it to great pain and continual harassment. Even in the domesticated state the Reindeer is obliged to continue its migrations, so that the owners of the tame herds are perforce obliged to become partakers in the annual pilgrimages, and to accompany their charge to the appropriate localities.

Toward the end of May, the wild Reindeer leave the forests and proceed toward the northern plains. They are thin, and covered with wounds caused by the flies, but when they return in autumn, they are healthy and fat, and supply an excellent article of food. In some years this migration consists of thousands which, although divided into herds of two or three hundred, follow close after each other, so that the whole form an enormous column. Their track is always the same. In Siberia a body of them has been seen which required two hours to pass the point of observation. Equally long are the migrations of the Reindeer in the Western Hemisphere. They leave the continent of America in spring, and, using the frozen sea as a bridge, appear in Greenland, sojourning there till the end of October, when they retrace their steps. In Norway they do not migrate to the same extent, merely changing one mountain range for another, ascending to the glaciers and snow-fields when summer is burning. The Wild Reindeer always live in companies, which are larger than the herds of other deer, and more resembling the herds of antelopes in South Africa. The writer has seen herds of twelve to twenty in summer in the snows of the Fille Fjeld near Nystuen, but in winter they gather to the number of three or four hundred.

The Reindeer is admirably adapted by the conformation of its hoofs for those northern regions which in summer are a morass, and in winter a field of snow. Their gait is a quick walk, or a trot. At every step a peculiar crackling is heard like the noise produced by an electric spark. This noise is not caused by collision of the hoofs, or by one part of the hoof striking against the other. Even tame reindeer can produce the sound without lifting a foot from the ground, by merely swaying the body. It seems to be produced in the interior of the limb, just like the noise caused by pulling one's fingers. The Reindeer is an excellent swimmer, and the wild ones never hesitate to plunge into any stream that crosses their path.

All the senses of the Reindeer are good; its power of smell is remarkable, it can hear as keenly as a stag, and its sight is so sharp that a hunter, even coming against the wind, has to conceal himself most carefully. They are, according to the testimony of all sportsmen, shy and cunning in the highest degree. During the summer their food consists of Alpine plants, in winter they scrape away the snow with their feet, and eat the lichens on the rocks.

The chase of the Reindeer is of the highest importance to the North ern tribes. Many of those in Siberia depend entirely on the Reindeer for food, clothing, conveyance, and shelter. The chase of the Reindeer decides whether there will be famine or prosperity, and the season when these animals migrate is the harvest-time. The hunters attack them when crossing a river, and the slaughter made on these occasions can be best described as immense.

The Reindeer is domesticated by the Lapps and Finns, as well as by the Samoyede tribes, the Ostzaks, Tunguses, and others in Siberia. According to Norwegian statistics, the Lapps in that kingdom possess seventy-nine thousand reindeer. It is the support and pride, the joy and riches, the plague and torment of the Laplander. He is the slave of his Reindeer; where they go, he must follow. He has to be out for months, tormented in summer by the mosquitoes, half-killed in winter by the cold, and with no other companion than his dog. The latter is an indispensable auxiliary; watchful, sagacious, reliable, it obeys every sign of its master, and will for days together keep the herd together by its own independent action. The uses to which the tame Reindeer is put are manifold. The Lapps use it for driving, the Tunguses mount and ride on their backs. On even ground it can travel seven or eight miles an hour, but its ordinary pace is four or five miles.

The mode of harnessing and driving the Reindeer is most simple. A collar of skin is fastened round its neck, and from this a trace hangs down, which, passing under the belly, is fastened into a hole bored in the front of the sledge. The rein consists of a single cord fastened to the root of the animal's antlers, and the driver drops it on the right or left side of the back, according to the side to which he wishes to direct the animal. The vehicle being very light, traveling may be rapidly performed in this equipage, but not without running some risk of breaking your neck; for, to avoid being upset, one must be very skilful in this sort of locomotion. The Laplander becomes by practice a perfect master of this art.

We have not yet mentioned the most important articles this Ruminant of the Arctic regions yields man. The female produces milk superior to that of the cow, and from it butter and cheese of excellent quality are made. Its flesh, which is nutritious and sweet, forms a precious alimentary resource, and almost the only one in the polar regions. Its coat furnishes thick and warm clothing, and its skin is converted into strong and supple leather. The long hairs on the neck of this animal are also used for sewing, while out of its tendons string is manufactured. From the old antlers of the Reindeer various utensils are made, such as spoons, knife-handles, etc., and when the horns are young, gelatine is extracted from them by submitting them to a severe course of boiling. Their excrement, when dried, is formed into bricks, which serve for fuel. Many tribes even turn to advantage the cropped lichens contained in the stomach of a slaughtered animal. The Esquimaux and Greenlanders add to these lichens chopped meat, blood, and fat; when this is smoke-dried, they are extremely fond of it. The Tunguses, or nomadic inhabitants of Siberia, add wild berries to the above northern delicacy, then make it into cakes, which rank high among the articles of their *cuisine*.

THE CARIBOU.

The CARIBOU, or American Reindeer, is considered by some naturalists as identical with the species above described. We shall take the liberty of regarding it as a separate species for the purpose of description.

The CARIBOU, *Tarandus caribou* (Plate XL.), is larger than the Reindeer, has smaller antlers and a darker color, and lives mostly in the forests. It has never been brought under the sway of man, nor used for any domestic purposes. The Caribou lives in herds varying from ten to three hundred in number, and is an object of chase for both white and red hunters. The Esquimaux take it in an ingeniously constructed pitfall. A hole, about five feet deep, and capable of holding several deer, is dug out. It is then covered with a slab of ice or frozen snow, which is balanced on two pivots, so that when the Caribou treads upon it, it gives way, and precipitates him into the pit. Other tribes are said to make a large inclosure into which the deer is driven. In the inclosing fence numerous narrow gaps are left, and in each gap a strong running noose is suspended. The Caribou, in its attempts to escape, makes for one of these treacherous outlets, and is caught by the fatal noose. Some Indians go in couples to the chase. The sportsman who goes first carries in one hand a Caribou's antler; the other sportsman, who follows his leader closely, bears a bundle of twigs, which he rubs against the antler borne by his partner. When the herd perceives the approach of this remarkable object, it stands still in astonishment. The Indians creep up to the gazing Caribou till they are within range, then both fire their guns at once, then run toward the herd, loading as they run, and discharge a second couple of shots. But these shots must be well directed in order to kill. In most cases, however, the pursuit of the Caribou is a long affair, for it is not only a very strong but a very enduring animal, and often leads its pursuers a chase of four or five days. A small herd of these animals was chased continually for a week, when the original hunters were tired out, and gave up the pursuit to a new party. Whenever practicable, the Caribou makes for the frozen lakes, and then it is sure to escape, although in a very clumsy and ludicrous fashion. Rushing recklessly forward, the Caribou will be suddenly startled by some object in front, and falls on the ice in a sitting posture, in which attitude it slides for a considerable distance before it can recover its feet. As soon as it does so, it rushes off in another direction, and with the same results. Still, the speed attained is so great that the hunters always give up pursuit when the animal gets upon ice.

During the greater part of the year, the flesh of the Caribou is dry and tasteless, and when eaten seems to have no effect in satiating hunger. There is, however, a layer of fat, sometimes two or three inches in thickness, that lies under the skin of the back and croup in the male, and is technically termed the *dépouillé*. This fatty deposit is so highly esteemed that it outweighs in value the remainder of the carcass, including skin and horns. The marrow is also remarkably excellent, and is generally eaten raw. When pounded together with the *dépouillé* and the dried flesh it makes the best pemmican, a substance which is invaluable to the hunter. Even the horns are eaten raw, while they are young and soft. The skin is very valuable, especially when taken from the young animal; and when properly dressed, it is an admirable defence against the cold and moisture of the inclement North.

CHAPTER XX.

THE TRUE DEER.

THE TRUE DEER—THE WAPITI—THE RED DEER OF EUROPE—THE VIRGINIA DEER OR CARCAJOU—THE PERSIAN DEER—THE INDIAN STAGS—THE BARASINGA—THE AXIS DEER—THE SAMBUR—THE MANED STAG—THE HOG DEER—THE SOUTH AMERICAN STAGS—THE PAMPAS DEER—THE RED DEER OR GUASUPITA.

THE True Deer range over the whole of the great continents, except that they do not go beyond 57° north in America, and a little further in Europe and Asia. In South America they extend over Patagonia, and even to Tierra del Fuego. They are found in the north of Africa, and over the whole of the Oriental region and beyond it, as far as the Moluccas and Timor.

GENUS CERVUS.

In this genus it is only the male animals which carry antlers. These antlers are round, and bear more or less numerous tines. The lachrymal grooves attain a considerable size. Many of the *forty* species comprised in this genus differ from each other so slightly that we give only those in which the difference is strikingly marked.

THE WAPITI.

The WAPITI, *Cervus Canadensis* (Plate XLI), is the largest of all the True Deer. The adult male measures nearly five feet in height at the shoulders, and about eight feet from the nose to the tail. It is very commonly known by the name of Elk.

The herds of Wapiti vary in number from ten or twenty to three or four hundred; but each one is always under the command of an old leader. When it halts, the herd halts; when it moves on, the herd follows; they all wheel right or left, advance and retreat with almost military precision when it commands. The proud position of ruler is gained by dint of many a fight; and the combats are unusually fierce, often indeed ending in the death of one of the rivals. Sometimes both perish miserably; their branching horns become inextricably locked, and the two adversaries, united in a common fate, slowly succumb to hunger and thirst. When the antagonists meet, they do not push with their horns, but, backing from each other for about twenty feet, with blazing eyes, hair turned the wrong way, and heads lowered, rush together like knights in the tournay, with tremendous speed. At the moment of contact there is a snort of defiance, then a crash of horns, and then each backs off for a new start. This combative nature is retained even in captivity. Audubon relates the following anecdote: "A gentleman in the interior of Pennsylvania, who kept a pair of Wapiti in a large woodland pasture, was in the habit of taking pieces of bread or a few handfuls of corn with him when he walked in the inclosure, to feed these animals, calling them up for the amusement of his friends. Having occasion to pass through his park one day, and not having furnished himself with bread and corn for his pets, he was followed by the buck, who expected his usual gratification. The gentleman, irritated by its pertinacity, turned round and hit it a sharp blow, upon which, to his astonishment and alarm, the buck, lowering his head, rushed at him and made a furious pass with his horns. Luckily, the man stumbled as he attempted to fly, and fell between two prostrate trunks of trees where the Wapiti was unable to injure him, although it butted at him repeatedly and kept him prisoner for more than an hour." On the other hand, General Dodge says that, "in a close encounter with either man or dog, he is not to be compared for a moment, as a dangerous animal, with the common red deer."

These deer are great travelers, and when not molested walk in single file, but when disturbed, they run together like a flock of sheep; and then, when they again advance, form a wedge-shaped mass, the leader at the point. If this leader is shot, all the followers stop, huddle together, and seem to consult. Then another deer steps to the front, and acts as leader. This peculiarity of having to stop and select new leaders when one is killed, enables a good marksman to bring down several deer. "The gravest objection to this style of shooting," writes General Dodge, is that, nine times out of ten, the leader is a doe. The buck rarely takes the lead, and if he brings up the rear, it is because his fat prevents him from running faster. His favorite position is in the middle of the herd, surrounded by admiring females." It is therefore very difficult to get a good shot at him.

The Wapiti is a good swimmer, and even when very young, will fearlessly breast the current of a wide and rapid river. Like many of the larger animals, it is fond of submerging itself under water in the warm weather for the sake of cooling its body and of keeping off the troublesome insects. It is also a good runner, and although burdened with its large and widely-branched horns, can charge through the forest haunts with perfect ease. In performing this feat, it throws its head well back, so that the horns rest on the shoulders, and shoots through the tangled boughs like magic. Sometimes a Wapiti will make a slight miscalculation in its leap; Mr. Palliser saw one strike a small tree so forcibly with its forehead that the recoil of the trunk threw the Wapiti on its back upon the ice of a stream which it had just crossed.

The food of the Wapiti consists of grass, wild pea-vine, various branches, and lichens. In winter it scrapes the snow with its fore-feet, so as to lay bare the scanty vegetation below. When alarmed or excited, it utters a peculiar whistling sound which may be heard at the distance of a mile on a clear quiet day. The buck's call for the doe is a deep bellow, said to be the natural E of the organ.

The flesh is in great favor among hunters, and the skin is also very valuable, being employed in the manufacture of mocassins, belts, thongs, and other articles in which flexibility and strength are required. The teeth are employed by the Indians in decorating their dresses, and a robe thus adorned, which belonged to Audubon, was valued as worth thirty horses. The horns also are used for various purposes, and it is said that in no two individuals are the horns precisely alike.

THE RED DEER.

The RED DEER or STAG, *Cervus Elaphus* (Plate XLII), is one of the nearest kindred of the Wapiti. It is one of the most noble and stately of animals, and exceeds in size all others of this genus except the Wapiti. It is found in nearly every country of Europe except the far North, and in a great part of Asia, extending southward to the Caucasus and the mountains of Mantchooria. In all thickly inhabited regions it has been either exterminated or very much reduced in numbers. It is still abundant, however, in Poland and Hungary, Styria, Carinthia, and the Tyrol. It prefers mountainous to level districts, and loves to frequent long stretches of forest. With regard to it in Great Britain, Wood writes: "In the olden days of chivalry and Robin Hood, the Red Deer were plentiful in every forest; and especially in that sylvan chase which was made by the exercise of royal tyranny at the expense of such sorrow and suffering. Even in the New Forest itself the Red Deer is seldom seen, and those few survivors that still serve as relics of a bygone age, are scarcely to be reckoned as living in a wild state, and approach nearly to the semi-domesticated condition of the Fallow Deer. Many of these splendid animals are preserved in parks or paddocks, but they no more roam the wide forests in unquestioned freedom. In Scotland, however, the Red Deer are still to be found, as can be testified by many a keen hunter of the present day, who has had his strength, craft, and coolness thoroughly tested before he could lay low in the dust the magnificent animal, whose head with its forest of horns now graces his residence."

The Red Deer forms troops of various sizes, divided according to sex or age. The females and calves usually keep together; the older stags form smaller bands, but the master-stags live alone till the breeding season comes on. At all times the herd, when traveling, follows a doe; the buck appears last of all. If we see in a herd several stout bucks, we can with certainty look for a still stouter one some five hundred paces behind. In winter the Red Deer comes down from the mountains, and when its horns are soft, it avoids the forests. The color varies slightly according to the time of year. In summer its coat is a warm reddish-brown, but in winter the ruddy hue becomes gray. The young, which are born about April, have their fur mottled with white upon the back and sides, the white marking gradually fading as they increase in size. The young deer, for a short time after its birth, is very helpless, and crouches close to the ground till it looks like a block of stone when it has been warned by its mother that danger is nigh.

All the movements of the stag are full of grace and dignity, and its speed, when it is in full gallop, is incredibly swift. Immense leaps are executed with sportive lightness, all obstacles surmounted, and lakes or streams crossed by swimming. Its senses of hearing, smell, and sight are highly developed; it can scent a man perhaps about six hundred yards off, and hears the slightest rustle made by its pursuer. Like many animals it seems to have a love for some kinds of music; the notes of a flute will attract it, or at least bring it to a standstill. The stag does not seem possessed of much intelligence; it is shy, but not cautious; it acts without reflection when its passions are aroused. Although it has several times been partially tamed, and even trained to run in harness, the stag is a very unsafe servant, and at certain seasons becomes dangerous. In attacking, it uses its fore-feet with terrible effect, the hard, sharp-edged hoofs being formidable weapons.

Formerly, the stag was placed in Europe under the protection of the severest penalties, its slaughter being visited with capital punishment on the offender if he could be known and arrested. Indeed, a man who murdered his fellow might hope to escape retribution except by the avenging hand of some relation of the slain man, but if he were unfortunate or daring enough to dip his hands in the blood of a stag, he could hope for no mercy if he were detected in the offence.

THE VIRGINIA DEER.

The VIRGINIA DEER or CARCAJOU, *Cervus Virginianus* (Plate XLI), is a very beautiful species, remarkable for its peculiar horns, which are of moderate size, bent boldly backward, and then suddenly curved forward. Its color is a light reddish-brown in spring, slaty-blue in autumn, and dull brown in winter. The abdomen, throat, and chin are white. It is considerably smaller than the species which we have described above, seldom exceeding five feet and a half in length, or three feet and a quarter in height. It is found everywhere in North America from Canada to Mexico, and from the Atlantic Ocean to the Rocky Mountains. It has a strong attachment to certain localities, and if driven from its resting-place on one day, it will surely be found on the next day within a few yards of the same spot. Sometimes it chooses its lair in close proximity to some plantation, and after feasting on the inclosed vegetables, leaps over the fence as soon as its hunger is satiated, and returns to the spot which it had previously occupied. The animal, however, does not often lie in precisely the same bed on successive nights, but always couches within the compass of a few yards. It is a very good swimmer, and loves to immerse itself in rivers to get rid of ticks and mosquitoes. When swimming, only the head appears above the surface; and the creature moves so fast as to be hard to overtake by a boat. Audubon gives the following anecdote: "We recollect an occasion when, on sitting down to rest on the margin of the Santee River, we observed a pair of antlers on the surface of the water, near an old tree, not ten steps from us. The half-closed eye of the buck was upon us; we were without a gun, and he was therefore safe from any injury we could inflict upon him. Anxious to observe the cunning he would display, we turned our eyes another way and commenced a careless whistle, as if for our own amusement, walking gradually toward him in a circuitous route, until we arrived within a few feet of him. He had now sunk so deep in the water that an inch only of his nose and slight portions of his prongs were seen above the surface. At length, we suddenly directed our eyes toward him, and raised our hands, when he rushed to the shore, and dashed through the rattling cane-brake in rapid style." It has been seen crossing broad rivers, and, when hard pressed by dogs, has even swum boldly out to sea.

In those parts of the country where it is unable to visit the plantations, the Carcajou feeds on the young grasses of the plains, being fastidiously select in choosing the tenderest herbage. In winter it finds sustenance on various buds and berries, and in autumn it finds abundant banquets under the oaks, chestnuts, and beeches; but, excepting in the months of August, September, and October, the Carcajou is in very poor condition. It is then, however, very fat, and the venison is of remarkably fine quality. It is in October and November that the buck becomes so combative, and in a very few weeks he has lost all his sleek condition, shed his horns, and retired to the welcome shelter of the forest.

The sight of the Carcajou does not seem to be very keen, but its senses of scent and hearing are wonderfully acute.

It is a thirsty animal, requiring water daily, and generally visiting some stream or spring at nightfall. It is remarkably fond of salt, and resorts in great numbers to the saline springs, or "salt-licks," as they are popularly termed. The Deer do not drink the briny water, but prefer licking the stones at the edge where the salt has crystallized from the evaporation of the water.

The Virginia Deer has been often tamed. A pair kept as pets by Audubon were most mischievous creatures. They would jump into his study-window, and when the sashes were shut would leap through glass and woodwork like harlequin in a pantomime. They ate the covers of his books, nibbled his papers, and scattered them in sad confusion, gnawed the carriage-harness, cropped all the garden-plants, and finally took to biting off the heads of his ducklings and chickens.

The skin of the Carcajou is peculiarly valuable to the hunter, for when properly dressed and smoked, it becomes as pliable as a kid glove, and does not shrivel or harden when subjected to the action of water. Of it are formed the greater part of the Indian's apparel, and it is also employed for articles of civilized raiment.

As the Carcajou feeds, it always shakes its tail before it lowers or raises its head. So by watching the movement of the tail, the hunter knows when he may move toward his intended prey, and when he must lie perfectly quiet. So truly indicative of the animal is this habit, that when an Indian wishes to signal to another that he sees a Carcajou, he moves his fore-finger up and down. This sign is invariably understood by all the tribes of North American Indians.

THE PERSIAN DEER.

The PERSIAN STAG, *Cervus Wallichii*, differs from the Red Deer of Europe only in its heavier build, and by possessing a stronger development of mane on the neck. It is a magnificent animal, but calls for no further remark.

THE DEER OF INDIA.

The *five* specimens of the genus which we are now about to describe, are natives of the Eastern regions of the Old World, and each of them is regarded by some naturalists as the representative of a sub-genus.

THE BARASINGA.

The BARASINGA, *Cervus barasinga*, is the type of the so-called *Recurvus* group. The head is short and pointed, the ear large and very broad, the eye large and beautiful, the legs long but powerful. The antlers, although they cannot be called properly palmate, yet approximate to the form of the elk's horns, and are characterized not only by their breadth, but by their repeated ramifications. The coat is rich and thick, each hair being long and fine, but the general appearance is that of a rough fur. The color in summer is a golden-brown, with a dark streak along the back between two rows of small golden spots. Long bristles are scattered over the muzzle and around the eye.

The Barasinga is found in Farther India, but it is not known whether it prefers the mountains or the lowlands. It is lively and courageous, and, unlike most deer, it "bells" at all seasons of the year; its voice being a short bleat, very like that of a young goat.

THE AXIS DEER.

The AXIS DEER, *Cervus axis* (Plate XLI), is a species well known in India and Ceylon. The horns are placed on long footstalks, and simply forked at their tips. The color of the animal is a rich golden-brown with a dark stripe along the back between two series of white spots. The sides are covered with white spots arranged in oblique curved lines. It is nocturnal in its habits, and is not so active or restless as many other deer. It loves to lie in the low jungle-lands near a stream, where it sleeps all through the hot day. As far as marking and color go, the Axis Deer is the most beautiful of its race.

THE SAMBUR.

The SAMBUR, *Cervus Aristotelis*, is often regarded as the type of a sub-genus to which the name *Rusa*—from a Malay word meaning "stag" —has been given. It is a large, powerful animal, quite as large as the Red Deer, and equally active. Its horns, like those of the Axis Deer, are set on a long foot-stalk, with a prong projecting forward just above the crown, and the tip forked. The color is a sooty-brown, with a patch of tan over the eyes. It is a savage and ill-tempered beast, and is found throughout India, and in Sumatra and Malacca.

THE MANED STAG.

The MANED STAG, *Cervus hippelaphus*, is scarcely inferior to the Red Deer in size, and possesses the usual characteristics of the Rusine group in the formation of its horns. The color of its coat varies with the seasons. In summer it consists of rough, sparse hair of a brownish-fawn color which is hard to describe. Both sexes are of the same color, even the young ones do not possess the dappled coats which distinguish the young of most of the *Cervidæ*. The buck is conspicuous by a strong mane which develops itself on the lower part of the neck and chin.

As far as is known, this deer is confined to the islands of Java, Sumatra and Borneo, and in India, where it is found in abundance, but in small herds. It lies in the long jungle-grass during the heat of the day, and at nightfall begins to browse. It is very fond of water, drinking frequently. It runs with great swiftness and endurance, but interrupts its extended gallop by frequent short leaps or bounds. When walking, its pace is stately, like the so-called Spanish pace sometimes taught in the riding-school. It lifts its foot carefully, stretches it out before it and sets it gracefully down, accompanying each step with a corresponding movement of the head.

The chase of this deer is a favorite amusement of the Javanese princes, who use only sword and spear to kill it, or running nooses to capture it alive.

THE HOG-DEER.

The HOG-DEER, *Cervus hydaphus*, is the representative of a subgenus *Hydaphus*, and is one of the clumsiest-built species. Its body is thick, the legs, neck, and head short; the antlers are short, thin, and three-pronged, and stand wide apart. The usual color is a beautiful coffee-brown, darkening in the male to deep black-brown. In all the Hog-deer some traces of dappled marks can be discovered, but in the young these spots are larger and clearer. In Bengal it is often hunted on horseback, and killed by the stroke of a sword. The venison is said to be excellent.

SOUTH AMERICAN DEER.

South America is the home of a species of deer which is by some naturalists regarded as the type of a genus *Blastoceros*, which is characterized by upright horns with three or five prongs.

The PAMPAS DEER, *Cervus campestris*, is the best known variety. It attains a length of four feet and a half, and a height of about three feet. Its coat is thick, rough, and shining, of a reddish-brown color on the back, but gray on the abdomen; the tip of the tail is pure white, and a white ring surrounds the eye, and white spots mark the upper lip.

The greater part of South America is the home of this very common deer. It loves open and dry plains, and even when chased, avoids swamps and woodlands. Its senses are very acute, and its movements active, so that if it has a start, the best horse cannot overtake it. By day it hides itself in the long grass, and lies so still that one may ride close past it without rousing it. It is usually taken by the lasso. The hunters form a half-circle, and wait till another party with dogs drive the game towards them. The lasso is then thrown over the horns or around the feet. The native name of this deer is Gua-zu-y.

The GUASUPITA, *Cervus rufus*, may be regarded as a representative of a group to which the title *Subulo* has been given, and which is characterized by moderate size, slender build, and two short, often rudimentary horns. These horns are thick at the root, but end in a sharp point. All the animals of this group have a rather long, well-haired tail. The Guasupita stands about two feet high, and attains a length of three feet and a half. The neck is short and slender, the ears large but not especially long, the eyes small and lively, the tear-grooves almost invisible, the limbs tall, slender, and very gracefully built. The prevailing color is a yellowish-brown, which becomes grayish on the forehead, and quite gray on the lower side of the body.

This deer inhabits Guiana, Brazil, and Peru. They live on the plains as well as on the mountains, preferring everywhere forest-land, and avoiding the open country. They do great injury to any plantations in their neighborhood, destroying the melon-shoots, the young corn, and the beans. They display great caution and timidity in commencing their forays, looking carefully around, and at the slightest alarm plunging back into the woods. Their pace is swift, but they have little endurance, and can be ridden down by a well-mounted horseman.

CHAPTER XXI.

THE FALLOW DEER, ROE DEER, AND MUSK DEER.

THE GENUS DAMA—FALLOW DEER—GENUS CAPREOLUS—ROE DEER—GENUS CERVULUS—MUNTJAK OR KIDANG—GENUS MOSCHUS—MUSK DEER—ITS ABODE—HABITS—THE MUSK.

WE conclude our account of the *Cervidæ* with some Old World genera which have no representatives either in North or South America. The animals contained in them are less in size than those we have been describing, but if they are less stately, they are quite as beautiful and graceful as the magnificent animals we have mentioned in our previous chapters.

GENUS DAMA.

The *only* species of this genus is distinguished from the Stag by its spreading palmated horns and its spotted coat. In the latter respect it resembles the Axis; in the former, the Reindeer; and some naturalists place it next to the *Tarandus* in their classification.

THE FALLOW DEER.

The FALLOW DEER, *Dama vulgaris* (Plate XLII), seems to have spread over Europe from the shores of the Mediterranean Sea. It extends to the south of those inland waters as far as the northern limits of the desert of Sahara; it is found in Tunis, the Greek islands, Sardinia, and Spain. It does not love the severe climates of the North, and does not venture beyond Southern Norway and Sweden. It is, at the present day, most numerous in England; it is not even there in a wild state, but

FALLOW MOUSE DEER
ROEBUCK STAG
PLATE XLII UNGULATA

kept in the parks of the large landowners, where it adds much to the beauty of the scene. Soft slopes of undulating land, where the grass is short and thick, alternating with shady dells, where the fern grows high, are its favorite haunts. There is hardly a more interesting sight than a herd of these graceful and active creatures, either lying calmly under the shadow of a broad clump of trees, or tripping along the sward under the guidance of their leaders, the old and sober proceeding at their peculiarly elastic trot, and the young fawn exerting all kinds of fantastic gambols by way of expressing the exuberance of youthful spirits. The color of the animal is a reddish-brown, spotted with white, with two or three white lines on the body, and dark rings round the muzzle and the eyes. The legs are shorter and less powerful than those of the stag, and the body is proportionately stouter, the neck is shorter, and the tail considerably longer. There is often great variety of color in the Fallow Deer; specimens of a pure white color are not uncommon, but black ones are very rare. In its habits it resembles the stag, but as it is always seen in a state of comparative domestication, it is less shy and timid. It runs very fast, and is an excellent leaper. It forms larger or smaller herds, in which there is always one master-deer, who often couches alone in solitary state, apart from the rest of the herd, and accompanied by a few chosen does whom he honors with his preferences. In his absence, the herd is commanded by the younger bucks, but they take care to keep out of his way when he condescends to join the community.

The food of the Fallow Deer consists chiefly of grass, but it is very fond of bread, and will sometimes display a very curious appreciation of unexpected dainties. They have often been seen to eat ham-sandwiches in spite of the mustard, and enjoy them so thoroughly that they pushed and scrambled with each other for the fragments as they fell on the ground. At Magdalen College, Oxford, where many deer are kept, it used to be a common amusement to tie a crust to a piece of string, and let it down to the deer out of a window. The animals would nibble the bread, and as it was gradually drawn aloft by the string, would raise themselves on their hind-legs, in order to reach it. But when the master-deer loomed in the distance, all retired, leaving him to eat the bread in solitary state. It was curious to see how a single deer would contrive to take into her mouth the entire side of a "half-quartern" loaf, and though it projected on each side of her jaws, would manage, by dint of patient nibbling, to swallow the whole crust without ever letting it drop out of her mouth.

The Fallow Deer display great susceptibility to music; even the wildest will come near and listen to the notes of a horn. Sportsmen in Germany sometimes avail themselves of this taste for melody to allure the poor creature within range of the gun.

It is from the Fallow Deer that the best venison is procured, that of the stag being comparatively hard and dry. The skin is well known as furnishing a valuable leather, and the horns are manufactured into knife-handles and other articles of common use. The shavings of the horns are employed for the purpose of making ammonia, which has therefore been long popularly known under the name of hartshorn. The height of the adult Fallow Deer is about three feet at the shoulders. It is a docile animal, and can be readily tamed. Indeed, it often needs no taming, but becomes quite familiar with strangers in a very short time, especially if they should happen to have any fruit, bread, or biscuit, and be willing to impart some of their provisions to their dappled friends.

GENUS CAPREOLUS.

The Roe Deer inhabits all Temperate and Southern Europe to Syria, with a distinct variety in Northern China. We will confine our remarks to the best known of the *two* species.

THE ROE DEER.

The ROE DEER, *Capreolus vulgaris* (Plate XLII), is smaller than the Fallow Deer, being only two feet and a quarter in height at the shoulder. The antlers rise up straight, without any prong projecting forward over the eye, and fork rather than branch, throwing out one prong in front, and one or two behind, according to age. The antler, from the base to the first fork, is thickly covered with wrinkles. The head is short, the neck slender and longer than the head, the fore-quarters powerful, the legs long and slender; the eyes are large and lively, with long lashes on the upper lid, the lachrymal groove very small; the ears are of moderate length, and stand wide apart. The thick coat of the Roebuck alters according to the season. In summer the back and sides are of a reddish-brown, in winter of a brownish-gray color, but the lower side of the body is always lighter colored. The fawns, as is usual with nearly all deer, display white or yellow dapplings. Milk-white Roes are by no means unfrequent, and in Germany raven-black specimens have been seen by some sportsmen.

The Roe Deer is found in all Europe, except the high North, and in the greatest part of Asia. It prefers to frequent the larger forests, whether they are on the plain, or on the mountains, provided that the underwood is thick, and that there is plenty of shade. It seeks the higher grounds in summer, and descends in winter to the low lands. In Siberia this change of dwelling assumes the character of a regular migration. It possesses a much greater love of liberty than the Fallow Deer does, and seeks change of abode, of food, and of society. In its motions it is active and graceful. It can execute astonishing leaps, crossing without apparent exertion wide ditches and high hedges. It swims well. All its senses are keen, and it is very cautious and shy. It is more irritable and worse-tempered than the stag, and in old age becomes dangerous and not to be trusted.

The Roe Deer never forms large herds like the Red Deer. During the greater part of the year it lives in family fashion, usually with only one doe; but in winter occasionally several of these families unite and live together. The food of the Roe Deer is nearly the same as that of the Stag, but it selects the more tender plants. Leaves and young shoots, green corn, and the like, constitute its favorite diet. Pure water is a necessity, and it is fond of licking salt.

Speaking of this animal, Mr. St. John makes the following remarks. After stating that when captured young it can readily be tamed, he proceeds to say:

"A tame buck becomes a dangerous pet, for after attaining to his full strength, he is very apt to make use of it in attacking people whose appearance he does not like. They particularly single out women and children as their victims, and inflict severe and dangerous wounds with their sharp-pointed horns. One day, at a kind of public garden, I saw a beautiful but small Roebuck in an enclosure fastened by a chain which seemed strong enough and heavy enough to hold down an elephant. I asked the reason for this cruel treatment of the poor animal. The keeper of the place informed me that small as the Roebuck was, the chain was quite necessary, as he had attacked and killed a boy of twelve years old a few days before, stabbing the poor fellow in fifty places with his sharp-pointed horns. Of course I had no more to urge in his behalf."

GENUS CERVULUS.

The *four* species of this genus are found in all the forest districts of the Oriental region, from India and Ceylon to China, as far north as Ningpo and Formosa, and as far south as the Philippine Islands, Borneo, and Java. They are all small animals, and have very short imperfect horns. The distinguishing characteristics of the genus consist in the possession of two large tusk-like canines in the upper jaw, and in an extraordinary development of the cranial bones which form elongated pedestals for the support of the two pronged horns; the forehead is marked with three deep vertical folds of skin.

THE MUNTJAK.

The MUNTJAK or KIDANG, *Cervulus muntjak*, the best known species, is about the size of the Roebuck. It is rather slenderly built, with a compact body, moderately long neck, short head, long and fine limbs, and a short hairy tail. The coat is short, smooth, and thick; the color a yellowish-brown, darkening into chestnut on the middle of the back. The inner sides of the limbs, as well as the chin and throat, are white. The antlers are cream color.

Sumatra, Java, Borneo and Banca, and the Malay Peninsula, are the homes of the Muntjak. Horsfield writes: "The Muntjak selects for its resort certain districts to which it forms a peculiar attachment, and which it never voluntarily deserts. Many of these are known as the favorite resort of the animal for several generations. They consist of moderately elevated grounds diversified by ridges and valleys tending toward the acclivities of the more considerable mountains or approaching the confines of extensive forests." "The Kidang," the same author observes, "is impatient of confinement, and is not fitted for the same degree of domestication as the stag. It is, however, occasionally found in the enclosures of natives and Europeans, but requires a considerable range to live comfortably. It is cleanly in its habits, and delicate in the choice of food. The flesh affords an excellent venison. The natives eat the males, but have an aversion to use the females as food." The Muntjak is monogamous, and whenever a troop of them is seen, it proves to consist of the members of a single family.

The Kidang is regularly hunted by the Javanese. It leaves a good scent, and can be easily followed by hounds. When it sees itself pursued, it does not, like the stag, run straight onward, but, after a short burst at its highest speed, it slackens its pace and describes a large curve calculated to bring it back to its original starting-point. The natives assert that it is a weak, lazy creature. If it is followed up perseveringly, it will finally push its head into a thick bush, and stand there motionless, as if it were in complete security. The dogs used in the chase are commonly the half-wild Paria dogs. The Muntjak, when brought to bay, knows how to make good use of his small antlers, and makes many a hound bite the dust. In the island of Banca the favorite method of taking the Kidang is by driving it into a space between two long hedges which gradually come nearer to each other, and are furnished at the narrow exit with nooses suspended from the trees. The Kidang is driven in by dogs, and as he attempts to escape, is caught by the horns.

GENUS MOSCHUS

The members of this hornless genus are sometimes classed as a distinct family, but they differ in no important points of organization from the rest of the *Cervidæ*. They are found in Central Asia from the Amoor River to the Himalayas and the Siamese mountains, above eight thousand feet elevation.

THE MUSK DEER.

The MUSK DEER, *Moschus moschiferus* (Plate XL), is the *only* species known. It is about the size of the Roebuck, but stands lower in the front than the hind quarters. Its legs are slender, the neck short, the head rounded at the muzzle, the eyes of moderate size with long lashes, and the ears are oval. The hoofs are small and narrow, but can be extended by means of a fold of skin between the two parts so that they form, in connection with the false-hoofs which reach to the ground, a sure support for the animal on the snow-fields and glaciers. The coat is thick and close, becoming longer at each side of the breast, and on the throat; the hairs forming it are stiff, long and twisted, and display the most perfect cellular structure of all kinds of hair. The color of these deer is various; hardly two specimens can be found alike in this respect. Some are dark, some a dirty-white, others reddish-brown, others yellowish-brown, while others have longitudinal lines of light-colored spots on the back. The canine teeth project, in the male, outside of the mouth, and are curved backward; these tusks are sometimes as much as three inches in length. Both sexes are devoid of horns.

The popular as well as scientific name of this deer is derived from the possession of a powerfully odorous secretion. The musk-pouch lies near the navel, and communicates with the air by two small openings. Small glands placed in the interior of this pouch secrete the musk; the average quantity found in an animal when at maturity is thirty grammes, or four hundred and fifty grains Troy. The secretion when dried becomes a granular mass, which at first is reddish-brown, but finally darkens to a coal-black shade. The odor decreases as the musk becomes darker, and is quite destroyed by mixture with sulphur or camphor. If burnt, it gives out an offensive smell.

Neither the Greeks nor the Romans, although they were curious about all kinds of odoriferous substances, knew anything about this animal. On the other hand, the Chinese have used this musk for thousands of years. Our first knowledge of it came from the Arabs. Aboo Senna writes that the best musk came from Thibet, and was produced by an antelope-like creature which had two projecting tusks. Marco Polo, the celebrated Venetian traveler, describes the animal, and calls the musk "the finest balsam known by man."

The Musk Deer is found most abundantly on the Tibetian slopes of the Himalaya, near Lake Baikal, and in the mountains of Mongolia. Here they are killed by hundreds. In the Western Himalaya they are found in the lower part of the range, never in herds, but usually in pairs. They love grassy slopes that are near thickets, in which they can hide by day, for they only venture at twilight to visit the treeless feeding-grounds. Although the deer is eagerly pursued on account of its musk, yet it is by no means timid, and seldom runs away, unless molested. Its gait is a series of short leaps followed by a brief pause. When it is at liberty, it has never been heard to utter a sound; it does not even call for its mate, but when captured, it utters a kind of scream. Its tracks are very remarkable, as both the false hoofs leave a clear impression.

All its movements are quick; it runs like an antelope, leaps like a goat, and climbs like a chamois; it swims broad streams, and crosses, almost without leaving a trace, fields of snow in which any dog will sink.

In Siberia the chase of the Musk Deer is an important branch of industry; the usual method is by fixing running nooses in places through which the deer is driven. The flesh is uneatable by Europeans. It is the musk-pouch which rewards the hunter. Official reports give the numbers of Musk Deer killed in Siberia every year, as fifty thousand, of which nine thousand are males. The best musk, however, is not from Siberia, but from China. It is very much adulterated by the Chinese dealers before it reaches the markets of Europe.

GENUS HYDROPOLIS.

This is a new genus, of *one* species, discovered by Mr. Swinhoe. It inhabits China from the Yang-tse-Kiang northward. Its nearest affinities are with *Moschus*. Other new forms inhabiting North China are *Lophotragus*, which is hornless, and *Elaphodus*, which has horns about an inch long.

CHAPTER XXII.

THE GIRAFFE.

THE CAMELOPARDALIDÆ OR GIRAFFES—ITS SIZE AND APPEARANCE—ITS HABITAT—ITS ADAPTATION TO ITS LOCATION—ITS MOVEMENTS—ITS FOOD—ITS SENSES—GIRAFFES IN LONDON AND PARIS—MODES OF HUNTING—MEANING OF THE WORD "GIRAFFE."

THE family CAMELOPARDALIDÆ or Giraffes, now consist of but a *single* species of a *single* genus, which ranges over all the open country of Africa. It is almost entirely absent from West Africa, which is more especially a forest district. During early epochs of the world's history, these animals had a wider range. Extinct species have been discovered in Greece, in the Siwalik hills of Western India and in the island of Perim in the Red Sea, while an extinct genus *Helladotherium*, more bulky but not so tall as the Giraffe, ranged from the South of France to Greece and Northwestern India.

GENUS CAMELOPARDALIS.

The name of this genus was given to the remarkable animals that bear it because the ancient writers saw in it a "mixture of the Camel and the Leopard." It is one of those strange forms which men may be excused for having believed to be fabulous. Even at the present day, neither education nor experience diminishes the astonishment with which we view this extraordinary creature. Tallest of all the dwellers upon earth, it raises its stately head far above any animal that walks upon the surface of our globe. It seems, indeed, to be rather the fancied form of something devised by the brain of an eccentric artist, than an animal who lives, moves, and has its being among the creations with which we are familiar. Its singular proportions and the peculiarity of its gait, as well as its variegated coat, have always excited curiosity.

THE GIRAFFE.

The GIRAFFE, *Camelopardalis girafa* (Plate XLIII), is the *only* representative of the family. The enormously long neck, the tall legs, the finely formed head with large, beautiful clear eyes, and two peculiar cranial excrescences covered with skin, are its leading characteristics. Its total length is about seven feet and a half, but its height to the shoulder is fully ten feet, and to the head eighteen to twenty feet. The tail, including the tuft of hair at its termination, is about forty inches long. From the muzzle to the root of the tail the distance is nearly fourteen feet. The weight is about one thousand one hundred pounds. These proportions alone are striking, but its construction is as remarkable as its size. The Giraffe seems made out of portions of different animals. The head and body resemble those of the horse; the neck and shoulders, those of the camel; the ears, those of the ox; the tail, that of the ass; the legs are imitations of those of the antelope; the color and markings are borrowed from the panther. The result naturally is, a certain want of symmetry. The short body is out of proportion with the neck and legs; the sloping back is no beauty; even its height does not give it grace. The head is beautiful, the eyes wonderful, the markings agreeable; but the whole is peculiar.

The eyes of the Giraffe are large, vivacious, brilliant, and yet soft; they are *spiritual* eyes. The ears are well-formed, and very movable. Between the two cranial developments which are commonly called horns is a round protuberance, almost like a third horn. The neck is as long as the fore-legs. It is thin, and adorned with a mane, but possesses only seven vertebræ, the same number as is found in ordinary animals. The body is broad at the chest, the withers much higher than the rump; the almost vertical shoulder-blades are very distinct. The legs are of almost equal length; it is to the elongation of the shoulder-blades that the height of the fore-quarters is due. The hoofs are small and fine. On the knee-joints are callosities like those of the camel. The skin is very thick and very smooth, except the mane, the tail-tuft, and two tufts of dark hair, which adorn the "horns." The ground color of the coat is a sand-yellow, becoming darker on the back, and passing into white on the belly; on this ground are placed large, irregular, angular spots of a brownish color, so closely arranged that the yellow tints appear like network. These marks are smaller on the neck and limbs.

The mane is streaked fawn-color and brown, the front of the ears is white, the back brownish; the tufts of hair are black. Males, when they grow old, become darker; the females lighter, as if bleached.

The Giraffe is a silent animal, and has never been heard to utter a sound even when in the agonies of death. In its native land it is so strongly perfumed with the foliage on which it feeds, that it exhales a powerful odor, which is compared by Gordon Cumming to the scent of a hive of heather honey.

In the present day, the Giraffe inhabits Africa, between the 17th degree north latitude, and 24th degree south latitude, commencing with the southern border of the Sahara, and ending at the Orange River. It is found in Abyssinia, Darfour, Kordofan, and on the White Nile, but does not appear in Senegambia, or on the Congo. Its home is on the wide plains, and is defined by the growth of various kinds of mimosa. It is never seen in mountainous or forest-clad districts.

In its native haunts, the marvellous adaptation of the animal to its abode is clearly perceptible. "When one sees a herd of Giraffes," writes Gordon Cumming, "in a grove of the picturesque spreading mimosas which ornament the plains, and of which it can reach the topmost twigs, one must be lost to all sense of natural beauty if the sight is not an attractive one." All other observers agree in this remark. "No animal in the whole world," says Baker, "is so picturesque as the Giraffe in its native dwelling-place." The harmony in color between the animal and the trees on which it browses is striking. "Often have I," says Cumming, "been in doubt respecting the presence of a whole herd of Giraffes till I examined them with a telescope; even my native attendants confessed that their sharp trained eyes were, at times, deceived; they often took the tree stems for Giraffes, and Giraffes for old trees." When they are seen on the treeless plain, at the edge of the horizon, when the evening sunlight is shining, they seem supernatural creatures. Usually they are found in small troops of six or eight, but occasionally, where it feels itself safe, in larger numbers. Gordon Cumming met bands of thirty and forty, but gives the average number as sixteen. Baker saw herds of sixty to hundred. Brehm never saw more than three together.

The movements of the Giraffe are peculiar. It is seen to most advantage when walking quietly; it is then dignified and graceful. It advances with a slow, measured pace, moving the two legs of the same side. When galloping it is by no means graceful. "The Giraffes when

GIRAFFE

running away," writes Liebtenstein, "presented such an extraordinary sight, that with astonishment and laughter I forgot the whole chase. With the disproportion between the fore and hind quarters, and between the height and the length, quick movement presented great difficulties. It cannot trot; it can only gallop. And this gallop is so awkward and clumsy that, judging by the slowness with which the limbs are moved, one would fancy that a man could overtake it on foot. But this slowness is compensated for by the length of the stride, for each stride measures fourteen to eighteen feet. The peculiar formation of its forequarters compels it to throw back its neck in order to throw the centre of gravity away from the fore-legs. Every movement of the neck is accompanied by a spring of the hind-legs. Thus the neck keeps swinging to and fro, like the mast of a ship rolling." At the same time it lashes its sides with its long tail, and often turns its head to look with its lovely eyes whether its pursuers are drawing nigh.

Most extraordinary is the position assumed by the Giraffe when it has to lift anything from the ground, or to drink. It spreads its fore-legs out wide apart from each other, till it can reach the ground with its long neck. When it lies down, it first sinks on its fore-knees, and then draws up its hind-legs, finally resting on its breast like the camel. It sleeps lying on one side, with its head on its hind legs. Its sleep is light, and only of short duration. It can go for several days without sleep, and seems to repose standing. The Giraffe is not adapted to eat grass, but to strip leaves from the trees. For this purpose its uncommonly flexible tongue is of as great service as the trunk is to the elephant. It can take up with it the smallest object, and pluck the most tender leaf. "In our Zoological Gardens," Owens writes, "many a lady has been robbed by the Giraffes of the artificial flowers in her bonnet. The animal is guided in the selection of food less by the smell than by the eye, and hence it is often deceived when it has seized artificial flowers with its pliant tongue. In its wild state it consumes chiefly twigs, buds, and leaves of the mimosa. In South Africa the "wait-a-bit" thorn and camel-thorn constitute most of its diet; in North Africa the Karrat-mimosa or the climbing plants which cover the trees are eagerly devoured by it. As the trees just mentioned are not taller than the Giraffe, it easily procures its food, for its lips and tongue are as insensible to the needle-like thorns as the camel's. It seldom eats grass, but does not despise it when it is green. When its food is juicy, it can do without water for a long time, and hence when the leaves are fresh, it is met with in places where there is no water for miles. But in the dry season, when the foliage is parched, it goes long distances to pools or ponds in the river beds. The Giraffe stands up while it chews the cud; it chiefly ruminates by night, and does not spend much time in the process.

The senses of the Giraffe, especially those of sight and hearing are highly developed. Its intellectual qualities are not inferior. It is prudent and cautious, very amiable and good-natured. It is a peaceful, gentle creature, that lives in harmony not only with its own kind but with other animals. In case of need it defends itself with courage, striking powerful blows with its long, sinewy legs. It is in this fashion that the males fight with each other, and that the female defends her young from the crouching *Felidæ*; she has been known to knock down even a lion. The keepers of menageries have often to be on their guard, although they generally are on good terms with the Giraffe. It is full of curiosity, and seems to be gratified by the presence of visitors, whom it investigates with an air of great interest. It was a little more than fifty years since Giraffes were first seen alive in Europe, after a very long interval. The Pasha of Egypt heard that the Arabs of Senaar had brought up a pair of Giraffes on camel's milk; he ordered them to be brought to Cairo, and resolved to send them as presents to the kings of England and France. They arrived at their destinations in 1827, and in Paris, fashions *à la giraffe* became in vogue. Since then, numerous specimens have been imported, most of them being brought from the country between the Blue Nile and the Red Sea. At present, the Arabs do a regular and profitable trade in catching and selling these animals. Twenty-four were sent to Germany alone in 1874. The Giraffes require great attention when kept in captivity. They are attacked with a disease in the bones, which is named "Giraffe-sickness." It arises from want of exercise and proper food. Plenty of room and a warm surface to stand on are indispensable requisites for its health. Hay, carrots, onions and different vegetables form its principal diet in captivity, but to this provender the keeper ought to add some tannic acid, a substance in which the mimosa is very rich.

The Giraffe is an object for the native as well as the European hunter. The former hunt it on horseback, and when they overtake it, cut the tendon Achillis. They esteem its flesh very highly. Europeans usually stalk it—a difficult task, as the animal's enormous length of neck enables it to see for a great distance. The natives employ the pitfall also. For this purpose a very curiously constructed pit is dug, being about ten feet in depth, proportionably wide, and having a wall or bank of earth extending from one side to the other, and about six or seven feet in height. When the Giraffe is caught in one of these pits, its fore-limbs fall on one side of the wall, and its hind-legs on the other, the edge of the wall passing under its abdomen. The poor creature is thus balanced, as it were, upon its belly across the wall, and in spite of all its plunging, is unable to obtain a foothold sufficiently firm to enable it to leap out. The pitfalls for the capture of the hippopotamus and the rhinoceros are furnished with a sharp stake at the bottom; but it is found by experience that, in the capture of the Giraffe, the transverse wall is even more deadly than the sharpened stake. The Giraffe is easily traced by its "spoor," or footmarks, which are eleven inches in length, pointed at the toe and rounded at the heel. The pace at which the animal has gone is ascertained by the depth of the impression, and by the scattering of disturbed soil along the path.

The slain Giraffe is used for many purposes. Its skin is made into leather, its tail-tuft forms a fly-flapper, its hoofs are worked up like horn. Yet it must be said, for the credit of the natives of Africa, that they pride themselves more on possessing living than dead Giraffes. In the villages of the interior, the traveler will often see a pair of Giraffes raising their gentle heads high over some garden-wall, or else walking about the streets like cows. When Brehm was traveling on the Blue Nile, a Giraffe came to the boat as if to welcome him. It approached with perfect confidence, and eat bread out of his hand as if it had been an old acquaintance. It came every day afterward to be caressed. The name Girafe, the naturalist adds, is a corruption of the Arabic Seráfe, which means the "lovable," and fitly designates this noble creature.

CHAPTER XXIII.

THE HOLLOW-HORNED RUMINANTS.

THE BOVIDÆ—THE THIRTEEN SUB-FAMILIES—THE BOVINÆ—THE OXEN, ETC.—THE DOMESTIC OX—THE WILD CATTLE—THE CATTLE OF THE PAMPAS—CATTLE OF AMERICA—DOMESTIC CATTLE—THE HIGHLAND CATTLE—THE DURHAM—THE ALDERNEY.

THE large and important family Bovidæ includes all the animals commonly known as oxen, buffaloes, antelopes, sheep and goats. Some naturalists have classed them in three, some in four or five distinct families. Zoologically, they are all briefly and satisfactorily defined by the words with which this chapter is headed, "hollow-horn ruminants," and although they present wide differences in external form they pass so insensibly into one another, that no satisfactory definition of the smaller family-groups can be found. As regards the distribution of the family it may be said that, as a whole, they are almost confined to the great Old World continent, only a few forms being found in North America. Different types prevail in different regions; thus, antelopes prevail in Africa, sheep and goats in the northern parts of Europe and Asia, and oxen are perhaps best developed in the far East.

Following the arrangement adopted by Wallace from Sir Victor Brooke, we subdivide the family into *thirteen* sub-families. The first of these sub-families BOVINÆ is one of the best marked groups in the family. It comprises the Oxen and Buffaloes, and their allies, and has a distribution very nearly the same as that of the whole family. The animals comprehended in it are large and strong; their chief characteristic being more or less round and smooth horns, a broad muzzle, with nostrils wide apart, and a long and tufted tail. The udder of the females has four teats. The skeleton displays a thick and powerful form. The skull is broad at the brow, and slightly narrowed in at the muzzle. The orbits of the eyes are round and project laterally; the processes on which the horns are placed grow from the back part of the skull. The smooth and round horns have, in many cases, wrinkles about the base. The coat is usually short and smooth, but in some species forms a mane-like growth on certain parts.

Slow and awkward as the *Bovine* appear, they are capable of quick movements. They are all good swimmers, and cross broad streams without hesitation. Their strength is extraordinary, and their endurance wonderful. Their sense of smell is good, so also is that of hearing; but their powers of sight are not highly developed. The wild species display more intelligence than the domesticated ones. In general, they exhibit a gentle and confiding disposition toward all animals which do not annoy or threaten them, but show the highest courage and fierceness toward beasts of prey, and usually use their terrible horns so effectively as to come off victorious. The sub-family BOVINÆ contains *six* genera, the first of which contains our domestic ox.

THE OX.

There has been considerable dispute concerning the origin of these patient and useful creatures. Ratimeyer asserts that three different wild stocks have given rise to the forty or fifty varieties of domestic oxen. He supposes that the Broad-faced Ox is the progenitor of the Norwegian Mountain-Ox; the Long-faced Ox, of the cattle which existed in Switzerland during the Stone Age, and which the German naturalist believes were introduced into Britain by the Romans; while the *Bos primigenius*, or Original Ox, has produced the cattle found on the continent of Europe.

GENUS BOS.

The DOMESTIC OX, *Bos taurus*, has been so modified in form, habits, and dimensions by long intercourse with mankind, that it has developed into many permanent varieties. We regard the genus as containing only *one* species. We will begin our description with the European varieties, from which our own domestic cattle are descended, and reserve an account of the Domestic Ox of India for a subsequent chapter (Chap. XXV), in which we speak of the Indian Wild Cattle of the genus *Bibos*.

The variety which approaches most nearly to the original species is still preserved in a half-wild state in some parts of England and Scotland. The most celebrated herds are those of Lord Tankerville at Chillingham, and of the Duke of Hamilton at Hamilton Palace.

THE WILD CATTLE.

The WILD CATTLE, as they are called, are of moderate size, strongly but not clumsily built; their hair is short and thick, becoming longer and curly on the head and neck, and with a slight indication of a mane as far as the withers. The color is milk-white, with the exception of the muzzle, the ears, the horns, and hoofs. The color of the ears varies according to the breed. The cattle of Chillingham have the interior of the ear red; those of Hamilton have it coal black; the former have brown muzzles, the latter black. Both have black hoofs, and black tips to their horns. These distinctions of color in the two breeds are kept up by a strict process of weeding out all calves that do not conform to the standard. In Hamilton, all calves with brown marks are put to death; in Chillingham, a black muzzle and ear is the creature's death-warrant. The herds at Gisburne in Yorkshire, and Chartly in Staffordshire, resemble the oxen of Chillingham; those that used to be at Drumlanrig and Cumbernauld were of the Hamilton variety. In all cases a superstition prevails that danger threatens the owner's house and family if a black calf is born in the herd.

According to Lord Tankerville, the cattle of Chillingham have all the peculiarities of wild animals. They hide their young, feed by night, and sleep by day. In summer they are seldom seen, preferring at that season the shade of the wood; in winter they come to the places where they are accustomed to be fed, and then a man on horseback can ride almost into the middle of the herd. When they are alarmed or provoked at the intrusion of a strange human being within the limit of their territories, they toss their heads wildly in the air, paw the ground, and steadfastly regard the object of their dislike. If he should make a sudden movement, they scamper away precipitately, gallop round him in a circle, and come to another halt at a shorter distance. This process is continually repeated, the diameter of the circle is shortened at every fresh alarm, till the angry animals come so near the spectator that he is glad to escape as best he can. They are usually shot when they are six years old. The flesh is not different from that of the common ox. The herds are maintained out of family pride, and the annual expense is heavy. They present a majestic sight as they gallop between the gray, gnarled trunks of the primeval oaks of the old Caledonian forest.

THE CATTLE OF THE PAMPAS.

We have already spoken of the herds of horses descended from the domesticated horse which roam over the wide plains of South America. A similar phenomenon is presented in the enormous herds of cattle which are found there. At the discovery of this continent no cattle existed in South America. Columbus imported some on his second voyage into San Domingo, and in 1540 some Spanish bulls and cows were landed in the southern parts of the continent. Circumstances favored their rapid increase, the herds became too large to be always watched, and soon wandered about in perfect liberty. Within a hundred years of their introduction, they were roaming over the Pampas in hundreds and thousands, and were hunted by the natives as the Northern Indians hunt the Bison of the Plains. At present the plains on both sides of the River Platte and its tributaries are swarming with cattle. They all have owners. Vast establishments named "Estancias" are scattered over the Pampas; and thirty thousand cattle, five thousand horses, and twelve thousand sheep are moderate numbers for the animals belonging to one owner. The cattle of each proprietor are branded with his mark, and are looked after by Gauchos, who display incredible courage, patience, and skill in their occupation, collecting the herds when necessary, or catching those that have to be killed or sold. The cattle are drilled, as far as possible, to assemble on the appearance of the herdsman at a certain spot situated at a convenient distance from the corral, and it is no unusual sight to see thirty-five thousand thus assembled. The proportion of men employed is very small, when compared with the numbers of the oxen. The usual allowance is four men to every five thousand head; thus an extent of two hundred square miles may have only fifty inhabitants.

Those that remain in a half-wild state, are for the most part taken with the lasso, and sold to the drovers in troops of five hundred each. When a five-year-old ox is lassoed by the horns, and he turns out a Tartar, after a few ineffectual shakes of the head to throw off the lasso, he directly darts at the horse, who immediately starts off at full speed, the foaming ox close at his heels, and fast to the saddle with twenty-five yards of lasso. The horse must take all that comes in his way: patches of long grass that reach up the stirrups, the burrows of the viscachas, and every other obstacle. Should the ox give up the chase suddenly, the rider must immediately check the speed of his horse, otherwise the jerk would break the lasso, or what is worse, it would draw the saddle back to the flanks of the horse, or break the girths; in which case the man would be brought to the ground, and be at the mercy of the furious animal, still with the lasso on his horns, but no longer fast to the horse.

The troops of oxen when formed, are driven at the rate of nine to twelve miles a day to the *Saladero* or Salting establishment.

The hide is the most valuable part of the animal, and the preparation of it is carefully attended to. The workmen lay each hide on the flat of their left hands, scrape off all the beef and fat which may be adhering to the inner coating with a knife in the right hand, trim the edges, and then stretch out the hides by means of stakes driven into the ground, if the skins are to be dried. If they are to be salted, a pile is made of them with layers of salt. Dried hides require much more time and skill than when they are only salted. In the latter case they are packed in casks for exportation, in the former, when shipped, they are tied up in bundles. Hides form the chief export from the River Platte.

AFRICAN CATTLE.

In Africa, the cattle are not only employed for the yoke, but are also educated for the saddle, and are taught to obey the bit as well as many horses. The bit is of very primitive form, being nothing more than a stick which is passed through the nostrils, and to which the reins are tied. One end of the stick is generally forked to prevent it from falling out of its place, and in guiding the animal, the rider is obliged to draw both reins to the right or left side, lest he should pull out the wooden bit. The saddle oxen are not very swift steeds, their pace being about four or five miles an hour; and as their skin is so loosely placed on their bodies that the saddle sways at every step, their rider has no very agreeable seat. In breaking the Ox for the saddle, the teachers avail themselves of the aid of two trained oxen, between whom the novice is tied, and who soon teach it the proper lesson of obedience.

The horns of this variety of the Ox are of marvellous length, having been known to exceed thirteen feet in total length, and nearly nine feet from tip to tip. The circumference of these enormous horns was more than eighteen inches, measured at their bases. One such horn is capable of containing upwards of twenty imperial pints. These weapons are not only long, but are sharply pointed, and are of so formidable a nature that a lion has been kept at bay during a whole night, not daring to leap upon an animal so well defended. As these horns might prove dangerous to the rider in case of the animal suddenly jerking its head, or flinging him forward by a stumble, the natives are in the habit of trimming them in various fashionable modes, by which the danger is avoided.

Their chief employment at the Cape of Good Hope is in drawing the wagons of the Boers or farmers. Ten or twelve yoke are frequently employed to drag a single vehicle over the tracks of the wild country.

THE SANGA.

The SANGA, which at present is found in all Interior Africa, is remarkable for possessing a hump on the withers. It is a strong, powerfully built animal with long legs and tail. The horns are placed close together at the root, and measure a yard in length. The hair is smooth, fine, and usually of a chestnut color.

DOMESTIC CATTLE.

It is only in America that the Ox has escaped from the thraldom of man. Elsewhere, it is his most valued possession. In general, high honor is paid to it. The Egyptians worshipped their god Apis in the form of an ox, and their Isis, like the Greek Io, is represented with cow's horns. According to the traditions of India an Ox keeps the gates of heaven, and the clouds are said to be the cows which Indra drives to their pasture. The nomads of the Soudan possess countless herds which they keep solely for their milk. The whole of South Russia swarms with cattle. In Hungary the herds are so wild that men dare not approach them; and even in Italy, in the Maremma, the cattle are half-wild. A remarkable contrast is presented by the condition of these animals in Switzerland and Holland. The Swiss lives with his cows in a perpetual exchange of reciprocal acts of kindness; they are never ill-treated or beaten. The best of them are adorned with bells, the leading cow—a black one—bearing the largest. If she is deprived of this honor she manifests her disgrace by lowing incessantly, and losing condition, or by attacking her fortunate rival. In Spain the bull is the hero of the people. The cattle live like wild ones, never entering a stall, or coming under a roof. Not large, but beautiful and powerful, they have pretty long, very sharp, outward-curving horns. They are revengeful, and never forget a blow. As they roam over the hills of Southern Spain, no wolf, no bear dares to assail them, and the prudent traveler keeps out of their way. The most fierce and active of the bulls are reserved for the bull-fights to which the Spanish nation is so passionately addicted. The bloody and brutalizing spectacle has been often described, and we need not repeat it here.

There is, at the mouth of the Rhone, stretching from the town of Arles to the Mediterranean, a vast extent of marshy land, intersected by woods. This tract has been formed by successive deposits of the river, and is called the Camargue. Large herds of cattle live in an almost wild state in these humid plains and solitary woods. The bulls of the Camargue are all black, of a moderate size, with long tapering horns. Their wild nature, agility, and exceptional strength render them very dangerous to the traveler who intrudes on their domain.

The most ancient documents of historic ages describe the Ox, the Horse, the Dog, and the Sheep, as associated with man. The humble and patient Ox forms the most useful assistant of the small farmer, and also constitutes the main performer of the most important agricultural operations. It helps to till the ground; it drags immense and heavily laden wagons; it takes a part in all the labors of the farm; and after fifteen or sixteen years of a well-spent life, it yields up for the benefit of man its flesh, bones, fat, skin, horns, hoofs, and blood—all of them products which supply with material a host of useful manufactures.

The Ox is neither so dull nor so stupid as is popularly supposed; but, on the contrary, is endowed with a degree of intelligence which, in certain countries, man has developed and turned to his profit; for some of the tribes of South Africa entrust to oxen the care of their flocks—duties which the sagacious Ruminants fulfil with a zeal and intelligence worthy of all praise. Prudence and a quick perception of danger are also qualities possessed by the Ox. If, either by his own fault or that of his guide, he finds himself in a dangerous place, he develops resources for extricating himself quite surprising.

When we are considering the advantages which society derives from them, domestic cattle may be looked at in four different aspects: as beasts of burden, that is, producers of mechanical force applicable to the cultivation of the soil; as supplying milk; as furnishing meat; and lastly, as makers of manure or fertilizing matter. Allowing all this, the question arises, is it possible to manage the breeding and rearing of the Ox so as to ensure the maximum result of all these four requirements? All the agriculturists who have had any experience in breeding cattle give a negative reply to this question. Qualities so different in their nature as muscular vigor, abundance of milk, fitness for fattening, and richness of fertilizing residuum, cannot, they say, be the attribute of one animal, or one breed; in fact, they exclude one another, and one quality can only be encouraged at the expense of the others. A good breed for work can hardly at the same time be a good breed for the butcher. If therefore, any one quality is to be specially developed, the others must, to some extent, be sacrificed. By this plan perfection may, at all events, be arrived at in one point, while by a different course of procedure nothing but mediocrity can be attained. This is the principle which ought to guide the agriculturist in the choice and breeding of his cattle, whether for the dairy, the market, or the farm.

For the butcher it is required to produce, as quickly and as economically as possible, an animal excelling in the quality and quantity of its meat. Such are the short-horned breeds. Next to meat, milk is the most valuable product, as it is not only universally consumed in its natural state, but supplies us with cheese and butter. A French farmer, named Guenon, professes to have discovered a method of determining, by examination of the cow, both the quantity and the quality of its milk. He remarked that in cows the hairs on the hinder face of the udders are turned upward, and added to this, these hairs extend more or less over the region of the perinæum, so as to form a figure which he describes under the name of an escutcheon. By a multiplicity of observations, he became assured that a cow's power of giving milk varied in proportion to the size of this escutcheon, and he divided cows into orders and classes accordingly. A commission appointed by the French government made an investigation, and confirmed his hypothesis that, the longer and wider the escutcheon is, the greater are the milking qualities. As far as regards the richness of milk, Guenon considers that it finds its maximum in those cows which have the skin of their udders of a yellowish hue, freckled with black or reddish spots, furnished with fine and scanty hair, and covered with a greasy substance, which becomes detached when it is scratched on the surface.

The breeds of cattle in Great Britain are almost as various as the soils of the different districts. They have, however, been conveniently classed into Middle Horns, Short Horns, Long Horns, and Polled, or Hornless Cattle. The Middle Horns are represented by the Devon, Hereford, and the Ayrshire; the Long Horns, by the Lancashire, as improved by the famous breeder Bakewell of Dishley; the Short Horns, by the Durham; and the Polled, by the Suffolk and Angus breeds. From one or other of these breeds our domesticated cattle are mainly descended.

THE SHORT HORNS.

The DURHAM (Plate XLIV), is the progenitor of the modern Short Horn. The breed had always existed in the North of England, especially on the banks of the Tees, but did not attain general celebrity till Mr. Collings astonished the world by showing its capacity for producing animals for the butcher. Under his care the modern or Improved Short Horn came into general favor in England. The first great importation of Short Horns to America took place in 1834, when a combination of farmers brought some to the Scioto Valley in Ohio. In 1853, Mr. Thorne, of Dutchess County, New York, brought here several of the famous "Duchess" and "Oxford" strains. His example was followed by Mr. Cornell of Ithaca, General Wadsworth of the Genesee Valley, and Mr. Alexander of Kentucky. By their enterprise, America, in 1856, possessed specimens of the stock better than any to be found in England. At various sales in this and other States, enormous prices have been paid for American-bred stock for reshipment to England. As a beef-producing animal the Short Horn is unrivalled. It lays on flesh in places where other cattle fail to give it. At the same time the cows, if bred with a view to giving milk, are very valuable, but naturally they become lean during the period in which they are being milked.

THE HIGHLAND CATTLE.

The SCOTCH BULL (Plate XLIV). These animals are small, usually black in color, with a small head, thin ears, and fine muzzle; the face is broad, the eyes prominent, the countenance placid. The horns taper to a point, are of a waxy color, and widely set at the root. The neck is

ZEBU OR BRAHMIN BULL
ALDERNEY COW
SCOTCH CATTLE
DURHAM COW
PLATE XLIV UNGULATA

fine, the breast wide, the shoulders broad, the back straight and flat. The legs are short, straight, and muscular. The whole body is covered with a thick, long coat; thick tufts of hair hang about the face and horns, and that hair is not curly.

THE ALDERNEY.

The ALDERNEY (Plate XLIV) is a breed of small and very elegant cattle exported from the Islands of Jersey and Alderney, in the English Channel. They are essentially a breed for the pail, not for the butcher. The milk they yield is remarkable for its richness and deep yellow color, as well as for the quantity of cream and butter it supplies. The quantity given is not great, eight to twelve quarts a day being a maximum. The Alderney is exceedingly handsome, the head particularly being indicative of blood, and reminding one of the head of the American Elk. It is very gentle and kindly in disposition, loving to be petted. The color is usually light-red, or fawn, occasionally smoky-gray, rarely black. A roan color indicates a cross with the Short Horn. In their native island, the principal food given them is parsnips, and to this diet is attributed the richness of the milk.

CHAPTER XXIV.

THE BISONS.

THE BONASSUS OR EUROPEAN BISON—CALLED ALSO THE AUROCHS—THE REAL AUROCHS EXTINCT—THE FOREST OF BIALOWICZ—DESCRIPTION OF THE BONASSUS—THE BISON OF THE CAUCASUS—THE AMERICAN BISON OR BUFFALO—ENORMOUS NUMBERS—TERRIBLE DESTRUCTION—ESTIMATE OF NUMBERS KILLED—THE MOUNTAIN BUFFALO—DEATH OF A BULL.

IF we are to believe the writings of ancient and mediæval naturalists, there were scattered over a great part of Europe two distinct kinds of wild cattle. Pliny speaks of the *Bonassus*, which he says was distinguished by its rich mane; and of the *Urus*, which was characterized by its huge horns. Both these animals are repeatedly mentioned under the names of "Bison" and "Aurochs" by a series of writers down to the year 1669. Brehm considers we are justified in trusting these accounts of the difference between the Aurochs and the Bison. The former is now extinct. A painting of the first quarter of the sixteenth century represents it as a rough-coated, maneless animal with a large head, thick neck, thin dewlap, and long horns turned out forward and upward. The coat of the creature was black.

GENUS BISON.

This genus comprises *two* species, one found in small numbers in Europe, the other abounding on the prairies of North America. Its generic characteristics are small round horns directed upward and forward, a broad convex forehead, soft long hair, and a large number of ribs. The European variety has fourteen, the American fifteen pairs of ribs.

THE BONASSUS.

The BONASSUS, *Bison bonassus*, is often called the Aurochs, more rarely the Zubr. It is found only in the Russian province of Grodno, where it is preserved by stringent laws in the forest of Bialowicz, a genuine northern forest about ten miles long by seven broad. No one, except foresters and game-keepers, dwell within its limits. The trees are of enormous age, great height and size; the whole wood, indeed, looks to-day as it looked thousands of years ago. Here this, the largest of European mammals, lives undisturbed. In 1857, the number of Bisons in Białowicz was estimated at eighteen hundred and ninety, but in 1863 an official count was made which reduced these figures to eight hundred and sixty-four.

The Bison or Bonassus has, perhaps, diminished in size during the lapse of centuries, but it is still a powerful beast. It stands six feet high, is eleven feet long, and weighs from thirteen hundred to seventeen hundred pounds. The head is large and well-formed, the forehead high and broad, the muzzle broad, the ear short and rounded, the eye rather small than large, the neck very powerful. The body is, in the fore-quarters, strongly developed, but slopes to the rump; the tail is thick and short; the legs powerful but not short, the hoofs are rather long, and the false hoofs small. The horns are placed on the sides of the head, first curve outward, then backward, so that the points are over the roots. A thick, rich covering of long, curly hairs and felt-like wool is spread over the whole body, but lengthens on the back of the head into a mane which falls over the brow, and hangs over the temples, while it forms on the chin a long beard, which extends over the whole throat down to the chest. The usual color is light brown, but the beard is dark brown, and the tuft on the tail is black. The cows are noticeably smaller than the bulls, with weaker horns, and less mane.

In summer and autumn the Bisons live in the moister parts of the forest of Białowicz, but in winter choose drier and higher quarters. The old males live solitary, the younger cattle forming herds of fifteen or twenty in summer, and thirty or fifty in the winter. Each herd ranges in its own limits, and returns to the same spot. The Bonassus feeds chiefly in the mornings and evenings; grass, leaves, buds, and bark constitute its nourishment; they peel the trees as far as they can reach, and bend down the younger ones in order to get at the crown. The movements of the animal are lively enough. They walk quickly, their run is a heavy but effective gallop, during which the head is lowered, and the tail raised. They wade and swim easily. Their disposition varies with their age. The younger ones are good-natured, the older ones irritable and malicious. In general they leave men alone, but the slightest disturbance can excite their anger and render them terrible. The angry Bonassus puts out his dark-red tongue, rolls his red eyes, and dashes with fury at the object of his wrath. An old bull ruled for a long time over the road running through the forest of Bialowicz, and did much mischief. He stopped carriages, or sleighs, especially those laden with hay. If the peasants threatened him, he charged and threw the sleigh over. Horses were terrified at the sight of him, and seemed to lose their senses. The Bonassus and the domestic ox similarly display a mutual repugnance, and even the calves which are brought up by tame cows exhibit no change in this respect.

It has been disputed whether the Bonassus occurs in the Caucasus. The balance of evidence seems to indicate that they are found in Abkasia from the Kuban to the source of the Psib. At a dinner given to General Rosen by a Caucasian chief, sixty silver-mounted bonassus-horns served as drinking-cups. Hunters have shot them on the Great Selentshuga, and say that they extend upward to the snow line.

THE BISON.

The AMERICAN BISON, *Bison Americanus* (Plate XLV), is usually called the Buffalo. "As Buffalo, he is known everywhere," writes General Dodge, "as Buffalo he lives, as Buffalo he dies, and when, as will soon happen, his race has vanished from earth, as Buffalo he will live in story and tradition." It is a giant among our mammals; its bulk, shaggy mane, vicious eye and sullen behavior give it a ferocious appearance, but it really is a mild, inoffensive beast, unwieldy, sluggish, and stupid. A few years ago, the numbers of these animals was past all counting. General Dodge, in 1871, drove from Fort Zara to Fort Larned, a distance of thirty-four miles. At least twenty-five miles of this distance was through an immense herd of buffalo. "The whole country appeared one mass of buffalo; but the apparently solid mass was really an agglomeration of innumerable small herds of from fifty to two hundred animals. When I had reached a point where the hills were only a mile from the road, the buffalo on the hills started at full speed directly toward me, stampeding and bringing with them the numberless herds through which they passed, pouring down upon me in one immense compact mass of plunging animals, mad with fright, and irresistble as an avalanche. When they arrived within fifty yards, a few well-directed shots split the herd, and sent them left and right in two streams." In like manner, whole herds would charge the trains on the Atchison and Santa Fé railroad, and often threw the cars from the track. This senseless obstinacy characterizes the animal even when it encounters natural obstacles. A herd of four thousand tried to cross the South Platte in 1867, when the water was low. The leaders were soon stuck in the mud. Those behind, pressed forward by the column, trampled over their struggling companions, and soon the whole bed of the river, nearly half a mile wide, was filled with dead or dying buffalo. Usually the Buffalo is very careful of the roads by which he passes from one creek to another, and shows an antipathy to steep grades. But when alarmed, he will with impunity climb banks, or plunge down precipices where no horse can follow. Contrary to the statements usually given, General Dodge declares that there is very little fighting among the bulls, and that the small herds have each generally more bulls than cows, seemingly all on the very best terms with each other. These small herds are in no sense families; they are merely instinctive, voluntary and accidental, as regards their individual components. Herds merge together, and herds break up gradually during the process of grazing. The only perceptible change that takes place is that the bulls work themselves out to a new circumference, so as to have the cows and calves within the circle. The bulls have the duty of protecting the calves. An army surgeon asserts that he saw six or eight bulls protecting from wolves a poor little calf which could scarcely walk.

Vast quantities of Bisons are killed annually, whole herds being sometimes destroyed by the cunning of their human foes. The hunters, having discovered a herd of Bisons at no very great distance from one of the precipices which abound in the prairie-lands, quietly surround the doomed animals, and drive them ever nearer and nearer to the precipice. When they have come within half a mile or so of the edge, they suddenly dash toward the Bisons, shouting, firing, waving hats in the air, and using every means to terrify the intended victims. The Bisons are timid creatures, and easily take alarm, so that on being startled by the unexpected sights and sounds, they dash off, panic-struck, in the only direction left open to them, and which leads directly to the precipice. When the leaders arrive at the edge, they attempt to recoil, but they are so closely pressed upon by those behind them that they are carried forward and forced into the gulf below. Many hundred of Bisons are thus destroyed in the space of a few minutes.

A much fairer and more sportsmanlike method of hunting these animals is practised by red and white men, and consists in chasing the herds of Bisons and shooting them while at full speed. This sport requires good horsemanship, a trained steed, and a knowledge of the habits of the Bisons, as well as a true eye and steady hand. The hunter marks a single individual in the herd, and by skilful riding contrives to separate it from its companions. He then rides boldly alongside the flying animal, and shoots it from the saddle. In this method of shooting, the hunter requires no ramrod, as he contents himself with pouring some loose powder into the barrel, dropping a bullet from his mouth upon the powder, and firing across the saddle without even lifting the weapon to his shoulder. The Indians are very expert in this sport, and, furnished only with their bows and arrows, will do a good day's work.

"The difficulty in this style of hunting," to quote again from General Dodge, "is the cloud of dust which prevents very careful aim—the explosion of the pistol creates such confusion among the flying herd, that it is impossible to shoot at any individual buffalo more than once. The danger arises not from the buffalo, but from the fact that neither man nor horse can see the ground which may be rough or perforated with prairie-dog or gopher-holes."

Such slaughter, however, made little or no impression on the numbers of buffalo. Unfortunately, in 1872, it was discovered that their hides were merchantable; buffalo-hunting became an organized trade. The slaughter would seem incredible, but the figures are taken from official sources. In 1872, buffalo were around Fort Dodge in such numbers as to interfere with other game, and sportsmen paid no heed to them. "In 1873, where there were myriads of buffalo the year before, there were now myriads of carcases. The vast plain which a short twelvemonth before was teeming with animal life was a dead, solitary, putrid desert." In 1874, General Dodge remarks that there were more hunters than buffalo. The hunting-parties engaged in this occupation consist of four men—one shooter, two skinners, and a cook. One shooter, from one spot, in less than three-quarters of an hour, killed one hundred and twelve buffalo within a radius of two hundred yards. During the three years for which General Dodge gives statistics—1872, 1873, 1874—there were 1,378,350 hides sent to market, representing a slaughter of 3,158,730

AMERICAN BUFFALO

PLATE XLV. UNGULATA.

buffaloes. Add to this 1,215,000 killed by Indians, and we have the number of nearly four and a half millions. "Nor is this all," he adds. "No account has been taken of the immense numbers killed by hunters who took the skins away in wagons, nor of the numbers sent to St. Louis, Memphis, and elsewhere by other railroads than the Santa Fé and Topeka line. No wonder that men fear that the Buffalo will soon cease to exist. Congress has talked of interfering, but only talked. But in fact, the extinction of the Buffalo is inevitable. Civilization cannot spare room for its ranges. When industry and skill have turned the rich prairies into smiling fields and rich pastures, the Bison will be confined to small reservations, or seen only as curiosities in the Zoological Gardens." The government of the United States has shown a laudable desire to preserve the curiosities of our country. It has rescued from destruction the giant trees of California, and saved from utter annihilation the Sea Lions of Santa Barbara. We need not fear that a nation which has set apart the whole Yellowstone Valley as a National Park, will ever refuse to give to the remnants of our buffaloes—when the day comes that their wild freedom is incompatible with the progress of civilization—a district compared with which the Russian forest which shelters its European congener will appear narrow and confined.

THE MOUNTAIN BUFFALO.

In the so-called Parks of the Rocky Mountains, there is found an animal which old frontiersmen call by the name of Bison. It is to the Buffalo of the plains what a mountain pony is to a well-built horse. It is admirably adapted for its dwelling-place. Its body is lighter, and its legs shorter, thicker, and stronger than those of the Buffalo. It is rare and very shy, seeking out the most retired glens and passes, and scrambling with wonderful agility over the craggy sides of almost inaccessible mountains.

General Dodge writes: "The deep gorges which intersect the mountains that join the Parks are the favorite haunt of the Mountain Buffalo. Early in the morning he enjoys a bountiful breakfast of the rich nutritious grasses, quenches his thirst with the finest water, and, retiring just within the line of jungle whence, himself unseen, he can scan the open, he reposes in comfort till appetite calls him to dinner late in the evening. He does not, like the buffalo, stare stupidly at the intruder. At the first symptoms of danger they disappear like magic in the thicket, and never stop until far removed from even the apprehension of pursuit."

Old mountaineers give marvellous accounts of the number of these animals "many years ago," and ascribe their present rarity to the great snow-storm of 1844, which wrought such havoc with the buffaloes of the plains. The shyness of the Bison renders it difficult to shoot. Two, however, were killed in one afternoon in the Tarrgall range of mountains between Pike's Peak and the South Park. The following story from the pages of General Dodge will give a good idea of the haunts and habits of this animal:

"One of my friends determined on the possession of a bison's head, and, hiring a guide, plunged into the mountain wilds that separate the Middle from the South Park. After several days, fresh tracks were discovered. The sportsmen started on foot on the trail; for all that day they toiled and scrambled with the utmost caution, now up, now down, through deep and narrow gorges and pine thickets, over bare and rocky crags. Next morning they pushed on, and when both were well-nigh disheartened, their route was intercepted by a precipice. Looking over, they descried on a projecting ledge several hundred feet below, a herd of twenty bison lying down. The ledge was about two hundred feet at widest, about one thousand feet long. Its inner boundary was the wall of rock on the top of which they stood; its outer seemed to be a sheer precipice of at least two hundred feet. This ledge was connected with the slope of the mountain by a narrow neck. My friend selected a magnificent head, that of a fine bull, young but full grown, and both fired. At the report, the bisons all ran to the far end of the ledge and plunged over. The precipice was so steep that the hunters could not follow them. At the foot lay a bison. A long, fatiguing detour brought them to the spot, and my friend recognized his bull—his first and last Mountain Buffalo. The remainder of the herd were never seen after the grand plunge, down which it is doubtful if even a dog could have followed without risk of life or limb."

CHAPTER XXV.

EASTERN CATTLE.

THE DOMESTIC CATTLE OF INDIA—THE ZEBU—THE WILD CATTLE OF INDIA—GENUS BIBOS—THE GAYAL—THE GAUR—THE BANTENG—GENUS POEPHAGUS—THE YAK—THE TROUGH YAK—HUNTING THE YAK—GENUS ANOA—THE CHAMOIS-BUFFALO OF CELEBES—ITS FIERCENESS.

As this chapter is devoted to Indian Cattle, we have included in it the Domestic Cattle of Hindostan, although they belong to the genus *Bos*, and ought properly to be treated of in Chapter XXIII.

THE ZEBU.

The Domestic Cattle of India is commonly known by the name of Zebu, and is, like the Sanga, conspicuous for the curious fatty hump which projects from the withers. These animals are further remarkable for the heavy dewlap which falls in thick folds from the throat, and which gives to the fore part of the animal a very characteristic aspect. The limbs are slender, and the back, after rising toward the haunches, falls suddenly at the tail. The ears are long and drooping, the horns very short, the color of the coat is various, but usually a reddish-brown, or dun.

The Zebu is a quiet and intelligent animal, and is capable of being trained in various modes for the service of mankind. It is a good draught animal, and is harnessed either to carriages or ploughs, which it can draw with great steadiness, though with but little speed. Sometimes it is used for riding, and is possessed of considerable endurance, being capable of carrying a rider for fifteen hours in a day, at an average rate of five or six miles per hour. The Nagore breed is specially celebrated for its capabilities as a steed, and is remarkable for its peculiarly excellent action. These animals are very active, and have been known to leap over a fence which was higher than our five-barred gates, merely for the purpose of drinking at a certain well, and, having slaked their thirst, to leap back again into their own pasture. As a beast of burden, the Zebu is in equally great request, for it can carry a heavy load for a very great distance.

The Zebu race has a very wide range of locality, being found in India, China, Madagascar, and the eastern coast of Africa. It is believed, however, that its native land is India, and that it must have been imported from thence into the other countries. There are various breeds of Zebu, some being about the size of our ordinary cattle, and others varying in dimensions from a large ox to a small Newfoundland dog.

The well-known BRAHMIN BULL (Plate XLIV.), is the most familiar of these varieties. It is so called, because it is considered to be sacred to Brahma.

The more religious among the Hindoos, scrupulously observant of the letter of a law which was intended to be universal in its application, but to which they give only a partial interpretation, indulge this animal in the most absurd manner. They place the sacred mark of Siva on its body, and permit it to wander about at its own sweet will, pampered by every luxury, and never opposed in any wish or caprice which it may form. A Brahmin Bull will walk along the street with a quaintly dignified air, inspect anything and anybody that may excite his curiosity, force every one to make way for himself, and if he should happen to take a fancy to the contents of a fruiterer's or greengrocer's shop, will deliberately make his choice, and satisfy his wishes, none daring to cross him. The indulgence which is extended to this animal is carried to so great a height, that if a Brahmin Bull chooses to lie down in a narrow lane, no one can pass until he gets up of his own accord.

Bishop Heber, in his well-known journal, mentions the Brahmin Bulls, and the unceremonious manner in which they conduct themselves, and remarks that they are sometimes rather mischievous as well as annoying, being apt to use their horns if their caprices be not immediately gratified.

THE WILD CATTLE OF INDIA.

The Wild Cattle ranging the woods and jungles of a great part of the Oriental region from Southern India to Assam, Burmah, the Malay Peninsula, Borneo, and Java form the third genus of the sub-family BOVINÆ, and are divided into *three* species.

GENUS BIBOS.

The GAYAL, *Bibos frontalis*, was first scientifically described by Lambert in 1802, from a specimen which was sent alive to England. These animals are often tamed by the Hindoos and used as domestic cattle, or for improving the breed of the latter. They are large, powerful animals, standing five feet high, and measuring nearly eleven feet, including two feet and a half of tail. No animal better deserves its specific title, *frontalis* or "broad-browed." The unparalleled breadth of the forehead at once strikes the eye. Its body is well proportioned, compact, and powerful; the whole appearance gives an impression of strength and beauty. The head, with its thick muzzle, forms almost a truncated pyramid, the base of which would pass through the roots of the horns and the angles of the under jaw. But the base is not a regular square; the side between the horns is much longer than the others. The breadth of the forehead is about two-fifths of the length of the head. The horns are very thick, and curve slightly outward and backward. The eyes are small, and placed deep in their sockets; the ears are upright, and large, and pointed. A small triangular double dewlap springs behind the chin, ending at each jaw. Three or four deep folds of skin separate the head from a thick, hump-like protuberance which covers the whole neck, withers, and half of the spine. The rest of the body is fleshy, the legs strong but well-formed, the hoofs are short, and straight in front, the tail is thin, and ends in a long tuft. A short, thick and shining coat covers the whole body; it is slightly longer on the throat, and becomes a rich tuft on the tail, and on the fetlocks of the forelegs. The predominant color is deep black, with some white on the throat. The inside of the ear is flesh-color, the horns gray, with black tips. The cow is smaller, and more slightly built than the male, and has much shorter horns.

Some Hindoo tribes regard the Gayal as a sacred animal. It is not slain, but driven into the holy groves. Other tribes, on the contrary, eat its flesh without any compunction, and employ newly-caught ones in bull-fights. The people of Silhoot, Chittagong, and Tepoor possess tame herds of the Gayal, and the English have attempted to introduce it into Bengal. The Gayal lives chiefly in the woody hills which separate Bengal from Arracan. It is a mountain animal, climbing well, and being very sure-footed. It never attacks mankind, but defends itself against beasts of prey with courage, repulsing even the panther and tiger. The Kuki tribe tame many of them. They place a ball of salt and earth in some place where the wild Gayals pass, and drive some tame ones to the same spot. Both love salt, and remain together for months, new balls of salt being continually supplied. The wild Gayals become friends with their tame brethren, and gradually lose their dread of man. They finally submit to be stroked and handled, and end in becoming domesticated and useful animals.

THE GAUR.

The GAUR, *Bibos gaurus*, is the largest of all the existing members of the Ox tribe, and it may be easily recognized by the extraordinary elevation of the spinal ridge and the peculiarly white "stockings." The general color of the Gaur is a deep-brown, verging here and there upon black, the females being usually paler than their mates. The dimensions of the Gaur are very considerable, a full-grown bull having been known to measure six feet ten inches in height at the shoulders. The great height of the shoulder is partly owing to the structure of the vertebrae, some of which give out projections of sixteen inches in length.

The Gaur associates in little herds of ten, twenty, or thirty in number, each herd generally consisting of a few males and a comparatively large number of the opposite sex. These herds frequent the deepest recesses of the forest. During the heat of noonday, the Gaurs are buried in the thickest coverts, but in the early morning, and after the setting of the sun, they issue from their place of concealment, and go forth to pasture on the little patches of open verdure that are generally found even in the deepest forests. The watchfulness of this animal is extremely remarkable, as, independently of placing the usual sentries, the Gaurs are said to arrange themselves in a circle while at rest, their heads all diverging outward, so as to preserve equal vigilance on every side. They may, however, be readily approached if the spectator be mounted on an elephant, as they seem to regard these huge animals without any suspicion or fear. In all probability, the imperturbable indifference with which they look upon the elephant is caused by the fact that the elephant is never used in Gaur-hunting, and, unless accompanied by human beings, never attempts to attack these animals.

This creature is found in all the large connected woodlands of India from Cape Comorin to the Himalaya, selecting always as its dwelling the thickest parts of the forest, and deep shady glens. In Bahar and the Western Ghauts, they are especially numerous. The country here is a series of abrupt hills with deeply cleft valleys, covered with an impenetrable growth of underwood, thorns, and tree ferns. Here the Gaur has lived since time immemorial, and has compelled the savage carnivora to leave him the undisputed monarch of the region. Like the Gayal, it is eaten by some tribes, and avoided by others, who regard it as akin to the sacred cow. Its flesh is said to be much more delicate to the palate than that of the domestic ox.

THE BANTENG.

The BANTENG, *Bibos Sondaicus*, is a most beautiful species, equalling many of the antelopes in grace, and being distinguished also by its color. It extends over Java, Borneo, and the Eastern part of Sumatra. It loves moist, or well-watered portions of the forests. It lives in small companies, consisting of a leader and six or seven cows. The softest and most juicy grasses which cover the slopes and young leaves and twigs constitute its food, but it prefers, above everything, as an article of diet, the tender shoots of the bamboo.

The timidity of the Banteng renders the chase of it both difficult and dangerous. Although it usually flies, yet when driven to bay, it uses its horns with skill and effect. The cows which are nursing calves are especially to be dreaded, and the Javanese sportsman who ventures to the chase armed only with a forest-knife, runs no slight peril. Adult Bantengs cannot be tamed; calves, however, become completely domesticated, and live in harmony with other cattle. The mild and gentle disposition of the Banteng is in harmony with the beauty of its form. The second or third generation of captives willingly submit to the rule of man, learn to know and love their keeper, and act like our common domestic animal.

GENUS POEPHAGUS.

The YAK, *Poephagus grunniens*, (Plate XLVI), is the *only* species of the genus. The body is stout and powerful, the head large and very broad; the eye, small and dull; the ear, small and round; the horns, round in front and angular behind; these horns project horizontally and then turn upward and forward, the point curving outward; the neck is short, the withers are elevated into a bison-like protuberance, the tail is long, and covered with bushy hairs that reach the ground; the legs are short and powerful, the hoofs wide, and the false-hoofs well developed. The coat consists of fine, long hair which, on the brow, falls in curling tufts over the face, and on the neck forms a thick waving mane; the legs from the knee downward are covered with smooth, short, close hair. The adult Yak is of a beautiful deep black color, with a silver-gray stripe along the spine. It is from the long tail of the Yak that the Mongols and Turks form the so-called "Horse-tails" which are carried before high officers of state, their number indicating the rank. The tail is used in India as a fly-flapper, under the name of chowrie. The chowries used by princes, or in religious ceremonies, are mounted in gold or silver, and cost large sums, the white ones being especially valuable. The Chinese dye the tails bright red, and wear them as tufts in their caps.

In Thibet the Yak has been domesticated, and is used for riding and as a beast of burden. It always retains, however, a good deal of its original wildness, and is subject to paroxysms of rage. It lives on good terms with the common domestic cattle, and a cross-breed of great robustness is held in high esteem. This variety, often called the "Plough Yak," is altogether a more plebeian-looking animal, humble of deportment, carrying its head low, and almost devoid of the magnificent tufts of long silken hairs that fringe the sides of its more aristocratic relation. Their legs are very short in proportion to their bodies, and they are generally tailless, that member having been cut off and sold by their avaricious owner. There is also another variety which is termed the Ghainorik. The color of this animal is black, the back and tail being often white. When overloaded, the Yak is accustomed to vent its displeasure by its loud, monotonic, melancholy grunting, which has been known to affect the nerves of unpractised riders to such an extent that they dismounted, after suffering half an hour's infliction of this most lugubrious chant, and performed the remainder of their journey on foot.

The Yak is found on the highlands of Thibet between twelve to fifteen thousand feet above the sea. It forms small troops in all parts of Northern Thibet, and large herds where pasture is plentiful. In summer they appear on the grassy meadow-lands near the streams, often to the number of a thousand. They disperse to feed, but collect again to repose, or when storms are threatening. At any sign of danger they range themselves into a compact body with the calves in the centre, while one or two old bulls go out to reconnoitre the enemy. At the sound of a gun, they gallop away, heads down, tails raised aloft. Clouds of dust envelope them as they fly along, and the earth echoes with the stamping of their hoofs. When resting, the Yak prefers some cool, deep glen, where it can avoid the sunshine, for it dreads heat more than cold, and even when in the shade, lies in the snow. It climbs, like a goat or a chamois, over the most rugged and steep rocks with complete sureness of foot. "The most noteworthy peculiarity of the Yak," writes Przewalski, "is its laziness. Morning and evening it feeds; the rest of the day it devotes to repose, which it indulges in, either standing or lying. The chewing of the cud is the only sign of life; otherwise it remains as motionless as a statue of stone."

This traveler gives an account of hunting the Yak. In Siberia the sportsman usually fires from a rest consisting of a long forked stick. Holding erect the forked rest, and covered with a sheep-skin jacket with the wool outward, he creeps near the game. The latter, probably regarding the moving figure as a kind of antelope, remains stationary. The hunter when near enough fires. The wounded Yak either turns and flies, or charges the assailant. Usually after charging a few yards, the animal pauses, thus offering a steady mark for the shooter. Another ball is put into him. Another charge of a few yards succeeds, and so on till the Yak falls. Its tenacity of life is remarkable. Przewalski followed one till nightfall; he found it dead next morning, with three balls in its head, and fifteen in its breast. To kill a Yak at the first shot is a rare occurrence. The Mongols usually fire in small platoons of six or eight, and from behind a blind, and then follow the trail of blood left by the wounded animal till it—after a day or two—dies from loss of blood.

The Yak has been transported to Europe, and stands the change of climate well. As it supplies excellent wool, savory meat, and rich milk, and as it is indefatigable in labor, and satisfied with commoner food, it would be a valuable addition to our stock of domesticated Ruminants.

GENUS AÑOA.

The AÑOA, *Anoa depressicornis*, is the *solitary* species known. For a long time it was placed by naturalists among the antelopes, then it was removed to the buffaloes, with which indeed it is closely related. Modern classifiers raise it to the rank of a genus. It is the dwarf of the race, measuring only fifty inches in height at the shoulders, with a length of five feet and a half. The head is broad at the brow, but pointed at the muzzle, the eyes are large, and dark brown; the horns are set wide apart, and form nearly a straight line with the face. There is no lachrymal sinus. The hair is of moderate length and thickness, and is usually of a dark-brown color.

Nothing is known of the wild life of the Anoa. It seems to be confined to the island of Celebes, where it is an inhabitant of the mountains. It has been often captured and brought to Europe. It gives the impression of a small cow, is lazy and indisposed to move, standing for hours on the same spot, either eating, or chewing the cud. Like the buffaloes, it is remarkable for its silence; it very rarely utters a short low, more like a groan. It resembles the buffaloes also in its predilection for water and moist places, and its love for water-plants as food. Though so small, it is very fierce. Some of them which were kept in confinement, killed in one night fourteen stags placed in the same inclosure. Brehm has proposed for the animal the name of Chamois-Buffalo; the word *Anoa* means in the Malay language "forest-cow."

MUSK OX CAPE BUFFALO YAK
PLATE XLVI UNGULATA

CHAPTER XXVI.

THE BUFFALOES.

THE GENUS BUBALUS—THE CAPE BUFFALO—DRAYSON'S ACCOUNT—BUFFALO SHOOTING—THE INDIAN BUFFALO—BUFFALO AND TIGER FIGHTS—WILLIAMSON'S ACCOUNT—THE KERABAU—THE DOMESTICATED BUFFALO—ITS HABITS—ITS USES.

THE Buffaloes are heavily built animals with a clumsy figure, short, powerful, and thick legs, a short neck, a broad convex forehead, a large bare muzzle, dull malicious-looking eyes, large, broad, hairy ears, and a rather long tufted tail.

GENUS BUBALUS.

Of the *five* species into which the Buffaloes are divided, three are African, ranging over all the continental parts of the Ethiopian region; one is Indian; one is domesticated in South Europe and North Africa. The animal is subject to considerable modifications in external aspect, according to the climate, or the particular locality in which it resides; and has, in consequence, been mentioned under very different means. In all cases, the wild animals are larger and more powerful than the domesticated ones, and in many instances the slightly different shape and greater or lesser length of the horns, or the skin denuded of hairs have been considered as sufficient evidence of separate species.

THE CAPE BUFFALO.

The CAPE BUFFALO, *Bubalus Caffer* (Plate XLVI) is formidable for its strength, and terrible in aspect. The heavy bases of the horns, that nearly unite over the forehead, and under which the little fierce eyes twinkle with sullen rays, give to the creature's countenance an appearance of morose, lowering ill-temper, which is in perfect accordance with its real character.

Owing to the enormous heavy mass which is situated on the forehead, the Cape Buffalo does not see very well in a straight line, so that a man may sometimes cross the track of a buffalo within a hundred yards, and not be seen by the animal, provided that he walks quietly, and does not attract attention by the sound of his footsteps. This animal is ever a dangerous neighbor, but when it leads a solitary life among the thickets and marshy places, it is a worse antagonist to a casual passenger than even the lion himself. In such a case, it has an unpleasant habit of remaining quietly in its lair until the unsuspecting traveler passes closely to its place of concealment, and then leaping suddenly upon him like some terrible monster of the waters, dripping with mud, and filled with rage. When it has succeeded in its attack, it first tosses the unhappy victim in the air, then kneels upon his body, in order to crush the life out of him, then butts at the dead corpse until it has given vent to its insane fury, and ends by licking the mangled limbs until it strips off the flesh with its rough tongue.

Many such tragical incidents have occurred, chiefly, it must be acknowledged, owing to the imprudence of the sufferer; and there are few coverts in Southern Africa which are not celebrated for some such terrible incident. Sometimes the animal is so recklessly furious in its unreasoning anger, that it absolutely blinds itself by its needless rush through the formidable thorn-bushes which are so common in Southern Africa. Even when in company with others of their own species, they are liable to sudden bursts of emotion, and will rush blindly forward, heedless of everything but the impulse that drives them forward. In one instance, the leader of the herd being wounded, dropped on his knees, and was instantly crushed by the trampling hoofs of his comrades, as they rushed over the prostrate body of their chief.

The flesh of the Cape Buffalo is not in great request even among the Kaffirs, who are in no wise particular in their diet. The hide, however, is exceedingly valuable, being used for the manufacture of sundry leathern implements where great strength is required without much flexibility.

The Cape Buffalo is little larger than an ordinary ox. The strangely shaped horns are black in color, and so large that the distance between their points is not unfrequently four to five feet. They are very wide at the base, and form a kind of bony helmet which is impenetrable by an ordinary musket-ball. Captain Drayson gives the following description of the animal:

"The hide of this animal is a bluish-black in color, and is so very tough that bullets will scarcely penetrate it if they are fired from a distance, or are not hardened by an addition of tin in the proportion of one to eight. It is of a fierce, vindictive disposition, and from its cunning habits is esteemed one of the most dangerous animals in Southern Africa. The Cape Buffalo is naturally a gregarious animal, but at certain seasons of the year the males fight for the mastery; a clique of young bulls frequently turn out an old gentleman, who then seeks the most gloomy and retired localities in which to brood over his disappointments. These solitary skulkers are the most dangerous of their species; and although it is the nature of all animals to fly from man, unless they are badly wounded, or are intruded upon at unseasonable hours, these old hermits will scarcely wait for such excuses, but will willingly meet the hunter half-way and try conclusions with him.

"Although frequently found in large herds on the plains, the buffalo is principally a resident in the bush; here he follows the paths of the elephant or rhinoceros, or makes a road for himself. During the evening, night, and early morning, he roams about the open country and gorges; but when the sun has risen high, or if he has cause for alarm, the glens and coverts are sought, and amidst their shady branches he enjoys repose and obtains concealment. The 'spoor' of the buffalo is like that of the common ox, the toes of the old bulls being very wide apart, while those of the young ones are close together; the cow buffalo's footprints are longer and thinner than the bull's, and smaller. As these animals wander in the open ground during the night, and retreat to their glens during the day, their spoor may be taken up from the outside of the bush, and followed until the scent leads to the view. When the hunter comes near to his game, of which he should be able to judge by the freshness of the footprints, he should wait and listen for some noise by which to discover their position. Buffaloes frequently twist and turn about in the bush, and do so more especially just before they rest for the day.

"I knew a Kafir who carried about him the marks of a buffalo's power and cunning. He was hunting buffaloes one day in the bush, and came upon a solitary bull, which he wounded; the bull bounded off, but the

Kaffir, thinking him badly hurt, followed after at a run, without taking sufficient precautions in his advance. Now, dangerous as is a buffalo when untouched, he is still more to be dreaded when hard hit, and should therefore be followed with the utmost caution. The Kaffir had hurried on through the bush for a hundred yards or so, and was looking for the spoor, when he heard a crash close to him, and before he could move himself, he was sent flying in the air by the charge of the buffalo. He fell into some branches and was thus safe, for the buffalo was not satisfied with this performance, but wished to finish the work which he had so ably begun. After examining the safe position of his victim, he retreated. The Kaffir, who had two or three ribs broken, reached his home with difficulty, and gave up buffalo-shooting from that day. It appeared that this cunning animal had retraced its steps after retreating, and had then backed into a bush, and waited for the Kaffir to pass.

"A great sportsman at Natal, named Kirkman, told me that he was shooting buffaloes when he was across the Sugela river on one occasion, and having wounded a bull, he was giving him his quietus, when the creature sent forth a sort of moan. Now the buffalo always dies game, and rarely makes any other noise when hard hit. This moan was probably a signal; and as such it was translated by the herd to which this animal belonged, as they suddenly stopped in their retreat, and came to the rescue. Kirkman dropped his gun and took to some trees, where he was in safety. Fortunate it was for him that timber happened to be near, as the savage herd really meant mischief, and came round his tree in numbers. When they found that he was safe from their rage, they retreated."

The other African species differ but slightly from the Cape species, the specific distinction being in the shape of the horns.

THE INDIAN BUFFALO.

According to Brehm, the Indian Buffalo ought to be regarded as the ancestor of the domesticated species which is found in Italy and Egypt. He considers that we are not yet in a position to define the species very accurately, and regards the "Arni" and the "Bain" as, at most, varieties of the Wild Buffalo which is dispersed through the greater part of India, Ceylon, and South-eastern Asia.

The INDIAN BUFFALO, *Bubalus buffelus*, measures in length nine to ten feet, including the tail, and stands nearly five feet high. The head is short and broad, the neck thick, with folds in front, but without a dewlap, the body full and round, and raised at the chine, the tail pretty short, and the legs have long broad hoofs, capable of being widely outspread. The eye has a wild defiant look, the horizontal ear is long and broad. The long, strong, smooth horns, thick and broad at the roots, have a triangular section; they rise close together, and at first curve outward and downward, then backward and upward, and finally turn inward and forward. The sparse, stiff, bristle-like hair is rather longer on the forehead and shoulders, while the hind-quarters, breast, belly, and most part of the leg are bare. Thus the color of the dark-gray or black hide is more prominent than that of bluish, brown, or reddish hair. White and piebald specimens are rare.

This animal frequents wet and marshy localities, being sometimes called the Water Buffalo on account of its aquatic predilections. It is a most fierce and dangerous animal, savage to a marvellous degree, and not hesitating to charge any animal that may arouse its ready ire. An angry buffalo has been known to attack a tolerably-sized elephant, and by a vigorous charge in the ribs to prostrate its huge foe. Even the tiger is found to quail before the buffalo, and displays the greatest uneasiness in its presence.

The buffalo, indeed, seems to be animated by a rancorous hatred toward the tiger, and if it should come inadvertently on one of the brindled objects of its hate, will at once rush forward to the attack. Taking advantage of this peculiarity, the native princes are in the habit of amusing themselves with combats between tigers and trained buffaloes. The arena is always prepared by the erection of a lofty and strongly-built palisade, composed of bamboos set perpendicularly, and bound together upon the outside. The object of this contrivance is, that, the surface of the bamboo being hard and slippery, the tiger's claws may find no hold in case of an attempted escape.

The tiger is first turned into the arena, and generally slinks round its circumference, seeking for a mode of escape, and ever and anon looking up to the spectators, who are placed in galleries that overlook the scene of combat. When the tiger has crept to a safe distance from the door, the buffalo is admitted. On perceiving the scent of the tiger, it immediately becomes excited, its hairs bristle up, its eyes begin to flash, and it seeks on every side for the foe. As soon as it catches a glance of its enemy it lowers its head toward the ground, so that the tips of its horns are only a few inches above the earth and its nose lies between its fore-legs, and then it plunges forward at the shrinking tiger. Were the latter to dare the brunt of the buffalo's charge, the first attack would probably be the last; but, as the tiger is continually shifting its position, the force of the onset is greatly weakened. Usually the buffalo is victorious, for the tiger, even if successful in the first onset, does not follow up his advantage, and allows his adversary to recover his breath. The buffalo, on the other hand, delivers charge after charge, never giving the tiger time to rest. At last the wearied feline is off his guard, and with a grand rush the buffalo impales him on his horns, and hurls him into the air to fall crushed and lifeless to the earth.

Captain Wilhamson, in his work on "Oriental Field Sports," describes the Buffalo, and its mud-loving propensities:

"This animal not only delights in the water, but will not thrive unless it have a swamp to wallow in. Then rolling themselves, they speedily work deep hollows, wherein they lie immersed. No place seems to delight the Buffalo more than the deep verdure on the confines of *jeels* and marshes, especially if surrounded by tall grass, so as to afford concealment and shade, while the body is covered by the water. In such situations they seem to enjoy a perfect ecstasy, having in general nothing above the surface but their eyes and nostrils, their horns being kept low down, and consequently hidden from view.

"Frequently nothing is perceptible but a few black lumps in the water, appearing like small clods, for the buffaloes being often fast asleep, all is quiet; and a passenger would hardly expect to see, as often happens, twenty or thirty great beasts suddenly rise. I have a thousand times been unexpectedly surprised in this manner by tame buffaloes, and once or twice by wild ones. The latter are very dangerous, and the former are by no means to be considered as innocent. The banks of the Ganges abound with buffaloes in their wild state, as does all the country where long grass and capacious *jeels* are to be found. Buffaloes swim very well, or, I may say, float. It is very common to see droves crossing the Ganges and other great rivers at all seasons, but especially when the waters are low. At a distance one would take them to be large pieces of rock or dark-colored wood, nothing appearing but their faces. It is no unusual thing for a boat to get into the thick of them, especially among reedy waters, or at the edges of jungles, before it is perceived. In this no danger exists; the buffaloes are perfectly passive, and easily avoid being run down, so the vessel runs no danger."

The KERABAU is a variety of buffalo found in Ceylon, Borneo, Sumatra, Java, and the Philippine Islands. Its horns attain a monstrous size, the body is nearly devoid of hair, but a tuft springs out between the horns. The Kerabaus that are found wild are not to be considered as wild cattle; properly speaking, they are domesticated cattle that have run wild. They are, however, dangerous to meet, especially for European travelers. Even the tame ones, which are docile enough to a native, evidently dislike a European.

THE DOMESTICATED BUFFALO.

The BUFFALO, *Bubalus vulgaris*, is found in Egypt and in Italy. It was introduced into the former country by the Mohammedan conquerors, and first appeared in the latter in the reign of the Lombard prince, Agilulf. It prefers warm and marshy districts; it thrives in the Delta of the Nile, in the poisonous Campagna of Rome, in the marsh lands of Apulia and Calabria, and in the Maremma of Tuscany. In the Italian fens it is the only representative of the family, as all others quickly perish; in Lower Egypt it is the domestic animal from which milk and butter is procured. Every Egyptian village has a pond for it to wallow in. It is less often seen in the fields than in water, where it lies or stands with only its head and part of the back visible. The period of inundation, when the Nile overflows, is to the Buffalo a time of enjoyment. It swims from place to place, cropping the grass on the dykes, and eating the long reed-grasses; large herds of them are seen playing with each other in the water; and they return home only when the cows want to be milked. It is a pretty sight to see a herd of these creatures crossing the swollen stream, and bearing on their backs women and children. The swimming powers of the Buffalo are wonderful. They act as if water were their natural element; they sport together, dive, roll, and drift at pleasure. They pass at least eight hours a day in the water, and become unruly and disquieted if they cannot reach it. If long deprived of it, they gallop furiously as soon as a stream or swamp comes in sight, in order to plunge into it. Many accidents result, for the animals drag into the flood the wagons to which they are yoked. The

Buffalo is more awkward on land. Its gait is clumsy, and its gallop very heavy and labored.

The first aspect of these buffaloes, as the traveler sees them on the desolate Roman marshes, is calculated to cause alarm. Their look is defiant and savage, their eyes are full of deceit and viciousness. But in Egypt they are good-tempered, and can be managed by a child. Complete indifference to everything except water and fodder is their leading characteristic. They draw the plow or the wagon, they can be led or driven, they will bear a rider or a load. They are easily satisfied, selecting the driest, hardest, most tasteless plants, and this provender suits them, for they give rich, good milk, from which excellent butter is made.

The Buffalo, having the swine-like propensity for rolling in the mud, is often very dirty. Very orthodox Mussulmans suspect it of being too near akin to the unclean pig, and, as it charges furiously at the red standard of Mohammed, the Turks consider it an accursed creature. The Egyptians, looking to its usefulness, piously believe that the Almighty will pardon its acts of impiety. The Buffalo seldom bellows, but the cow lows to call up her calf; the sound is an unpleasant one, midway between a grunt and a bellow.

The flesh of the Buffalo is tough, and is disagreeable to Europeans on account of its musky odor. The hide supplies thick strong leather, and the horns are fashioned into cups.

CHAPTER XXVII.

THE ANTELOPES.

THE ANTELOPES—THE ELAND—THE KOODOO—THE BOSCH-BOK—THE NYLGHAU—THE PASSAN—THE PEESA—THE SABRE ANTELOPE—THE ADDAX—THE SABLE ANTELOPE—THE LEAP-BOK.

THE *eleven* sub-families of Hollow-horned Ruminants which we are about to describe in this and the subsequent chapter, are often grouped together as *Antilopinæ*, and called Antelopes. But the number of Antelopes is so great, and the differences in their form so immense, that a further subdivision is necessary. Some of the Antelopes are as clumsy as cows, others as agile as the roe, some approach the horse in appearance, others the musk-deer, already mentioned. The horns present nearly every form: they are curved, forward, backward, downward, upward, they rise straight, or sweep gracefully aloft like a lyre; they are round, or angular, or, as in one species, forked. The anatomy of them all is very like that of the deer. With one exception, the Prong-horn, they are all denizens of the Eastern Hemisphere, and are especially abundant in Africa.

The first sub-family, TRAGELAPHINÆ, represents to a certain extent the connecting link with the Ox-tribe. The body of the animals contained in it are heavy, thick, and strong, the neck short, the head large, the tail like an ox-tail; the neck has a dewlap, the horns are common to both sexes, and are formed like a screw. The sub-family contains *three* genera.

GENUS OREAS.

The ELAND, *Oreas canna* (Plate XLVII), of which *two* species are known, is the largest of all the African Antelopes, and attains the dimensions of a large ox: it stands six feet high, and is stout in proportion. Its color varies with age, but usually is a pale grayish-brown. The horns are straight, and spirally twisted. The old males have a strong musky odor. Their food consists of the fragrant herbs which cover the plains. They require very little water, living without it for months, even when the herbage is so dry that it crumbles in the hand.

The heavy build of the Eland renders it an easy prey to the hunter, who can run it down by a steady pursuit. A heavy Eland will weigh over a thousand pounds; the fat around the heart alone weighs fifty pounds. The flesh is cut off as soon as the animal is killed, dried, or salted, then packed in skins for future consumption. The fat is made into candles, and the hide cut up into thongs. The flesh is like beef, but has an unpleasant taste when fresh, which it loses when smoked. The "Bil tongues" are eaten raw, and are a great delicacy. They consist of the tongue cut out its whole length, slightly smoked, and sliced very fine.

The Eland has a wide range. It is abundant near the Cape of Good Hope, and is not uncommon on the White Nile. Its favorite haunts are the grassy plains, where clumps of mimosas rise up like islands in a sea. It is occasionally, however, found in mountainous districts. The herds usually consist of eight or ten animals, and at a distance may be mistaken for cows grazing. Some walk slowly about, others bask in the sun, others lie chewing the cud in the shadow of the mimosa-shrubs. In going to pasture they follow in close order the guidance of an old male, and trot at a good pace. Elands have been brought to Europe, where they have propagated, and many of them may be seen in the parks of wealthy land-owners. They are, however, such large feeders, that the expense of keeping them destroys all hope of profit.

A species called the STRIPED ELAND has been shot in Southern Africa. Except in the color, it resembles perfectly the common Eland.

GENUS TRAGELAPHUS.

The animals comprised in this genus are classed in *eight* genera. They are of graceful form, and attain the size of the Roe Deer. The head is slender, the eyes large, the ear large and broad. Only the males have horns, which are slightly spiral, and set in a line with the face. The hair is thick, and forms a ridge along the spine. The hide is dappled, and curiously marked in the different species.

GNOO or HORNED HORSE   KOODOO
ELAND

The Koodoo, *Tragelaphus kudu* (Plate XLVII), is not much inferior to the elk in size, and is of a very imposing appearance. It is widely dispersed over Africa, where it frequents the thorny scrub-forests of the interior of the continent.

This truly magnificent creature is about four feet in height at the shoulder, and its body is rather heavily made, so that it is really a large animal. The curiously twisted horns are nearly three feet in length, and are furnished with a strong ridge or keel, which extends throughout their entire length. It is not so swift or enduring as the Bless-bok, and can be run down without difficulty, provided that the hunter be mounted on a good horse, and the ground be tolerably fair and open. Its leaping powers are very great, for one of these animals has been known to leap to a height of nearly ten feet without the advantage of a run.

The Bushmen have a curious way of hunting the Koodoo, which is generally successful in the end, although the chase of a single animal will sometimes occupy an entire day. A large number of men start on the "spoor," or track, one taking the lead and the others following leisurely. As the leading man becomes fatigued he drops into the rear, yielding his place to another, who takes up the running until he too is tired. A number of women bearing ostrich egg-shells filled with water accompany the hunters, so that they are not forced to give up the chase through thirst. As the chase continues, the Koodoo begins to be worn out with continual running, and lies down to rest, thereby affording a great advantage to its pursuers, who soon come within sight, and force it to rise and continue the hopeless race. At last it sinks wearied to the earth, and falls an unresisting prey to its foes.

The flesh of the Koodoo is remarkably good, and the marrow of the principal bones is thought to be one of Africa's best luxuries. So fond are the natives of this dainty, that they will break the bones and suck out the marrow without even cooking it in any way whatever. The skin of this animal is extremely valuable, and for some purposes is almost priceless. There is no skin that will make nearly so good a "fore-slock," or whip-lash, as that of the Koodoo; for its thin, tough substance is absolutely required for such a purpose. Shoes, thongs, certain parts of harness, and other similar objects are manufactured from the Koodoo's skin, which, when properly prepared, is worth a sovereign or thirty shillings, say five or six dollars, even in its own land.

The Koodoo is very retiring in disposition, and is seldom seen except by those who come to look for it. It lives in little herds or families of five or six in number, but it is not uncommon to find a solitary hermit here and there, probably an animal which has been expelled from some family, and is awaiting the time for setting up a family of his own. As it is in the habit of frequenting brushwood, the heavy spiral horns would appear to be great hindrances to their owner's progress. Such is not, however, the case, for when the Koodoo runs, it lays its horns upon its back, and is thus enabled to thread the tangled bush without difficulty. Some writers say that the old males will sometimes establish a bachelor's club, and live harmoniously together, without admitting any of the opposite sex into their society.

It is a most wary animal, and is greatly indebted to its sensitive ears for giving it notice of the approach of a foe. The large, mobile ears are continually in movement, and serve as admirable conductors and condensers of sound. From the conduct of a young Koodoo that was captured by Mr. Anderson, and reared by him, the disposition of the animal appears to be gentle, playful, and affectionate. The little quadruped, which was taken at so tender an age that it was fed with milk from a bottle, became strongly attached to its owner, and was a most active and amusing little creature. Domestication to any extent, is, however, not very practicable, as the animal is, in common with the gnoo and the zebra, liable to the terrible horse-sickness, which destroys so many of those useful animals.

The color of the Koodoo is a reddish-gray, marked with several white streaks running boldly over the back and down the sides. The females are destitute of horns.

The BOSCH-BOK, *Tragelaphus sylvaticus*, is upward of three feet in height, and five in length. It is gracefully, yet stoutly built. The horns are a foot long, nearly straight, and wrinkled at the base. The general color is dark chestnut, with a white streak along the back, and some white spots about the body. The female is hornless, smaller, and lighter colored. The animal is extremely watchful, and requires the perfection of bush-craft to be surprised.

These beasts are generally found in couples, male and female, although sometimes an old ram leads a hermit life. The Kaffirs frequently caution the hunter about these solitary animals, but they never display any signs of a ferocious disposition except when brought to bay, and under such circumstances even a rat will fight. It is said that the tiger-bosch-katte (the Serval) has been found dead in the bush, pierced by the horns of the Bosch-bok.

The districts from the Cape of Good Hope to Delagoa Bay and some distance inland are the resorts of this antelope. Although frequently passing from three to four days per week in the bush, the sportsman rarely sees more than a dozen black Bosch-boks, even though their spoor is imprinted on the ground in all directions, thus proving that they are numerous. Seldom by fair stalking can this crafty and wary antelope be slain. The Kaffirs frequently form large hunting parties, and by "spooring" their tracks and surrounding the bush in which they are concealed, drive them out and despatch them with assagais. This is, however, but a butcherly proceeding, and one which no true sportsman would follow. The Bosch-bok is so wary, so rare, and so beautiful an antelope, that any man may feel delighted if he can fairly procure one or two specimens during his sporting career.

The STRIPED ANTELOPE, *Tragelaphus scriptus*, is characterized by a coat of three colors. The head is tawn-gray, the neck and back dark-gray, while the flanks are reddish. The breast is dark-brown, the ridge of hair along the back of the neck is nearly black, but that along the spine consists of hairs with white tips. There is a white spot under the eye, and another near the ear. The front of the legs from the knee to the hoof is white, and the sides are marked with a moderately broad longitudinal stripe, and several narrow vertical ones; these are crossed by transverse stripes, and contain between them oval spots. Oval spots of white are very numerous on the thigh.

These graceful creatures are not rare in Zoological Gardens; they live on common provender, and give little trouble.

GENUS PORTAX.

The NYLGHAU, *Portax pictus*, is the *solitary* species known. It is an inhabitant of the thickly wooded districts of India. It is about four feet high at the shoulders, and nearly seven feet in length. In its general appearance it seems like a hybrid between the deer and the ox. The general color is a slaty-blue. The face is marked with brown, the long neck is furnished with a bold dark mane, and a long tuft of coarse hair hangs from the throat. The female is smaller than her mate, and hornless. Her coat is generally a reddish-gray, instead of partaking of the slate-blue tint which colors the form of the male. The hind-legs of this animal are rather shorter than the fore-legs. Its name, Nylghau, is of Persian origin, and signifies " Blue Ox."

It does not seem to be of a social disposition, and is generally found in pairs inhabiting the borders of the jungle. There are, however, many examples of solitary males. It is a shy and wary animal, and the hunter who desires to shoot one of these antelopes is obliged to exert his bush-craft to the utmost in order to attain his purpose. To secure a Nylghau requires a good marksman as well as a good stalker, for the animal is very tenacious of life, and if not struck in the proper spot will carry off a heavy bullet without seeming to be much the worse at the time. The native chiefs are fond of hunting the Nylghau, and employ in the chase a whole army of beaters and trackers, so that the poor animal has no chance of fair play. These hunts are not without their excitement, for the Nylghau's temper is of the shortest, and when it feels itself aggrieved, it suddenly turns upon its opponent, drops on its knees, and leaps forward with such astounding rapidity, that the attack can hardly be avoided, even when the intended victim is aware of the animal's intentions and is prepared for the attack.

Even in domesticated life the Nylghau retains its hasty and capricious temper, and though there may have been several successive generations born into captivity, the young Nylghaus display the same irritable temper as their parents. Its disposition is very uncertain, and not to be depended upon. It takes offence at trifles, and instantly attacks the object of its dislike. The flesh is coarse and insipid, and the animal is of no great value commercially.

The next sub-family, ORYGINÆ, comprises two genera, which are distinguished chiefly by the formation of the horns. Both genera are depicted on the oldest monuments of Egypt and Nubia. In the Great Pyramid one of these animals is represented with one horn, and hence it is supposed that the legend of the Unicorn has arisen.

GENUS ORYX.

The *four* species of ORYX belong to the largest and heaviest class of antelopes, but in spite of their stout figure, they give the spectator an impression of majesty. Both sexes have horns which are very long and thin; quite straight in some species, boldly curved in others. They have no lachrymal sinus.

The PASSAN or CAPE CHAMOIS, *Oryx capensis*, stands about five feet high. The horns, which in the female are thinner and much longer than in the male, exceed a yard in length; they rise straight from the head in a line with the face; the lower portion is marked with thirty or forty rings, the upper part is smooth and pointed. The smooth coat consists of short, stiff hair. The neck, back, and sides are yellowish-white, the head, ears, and the legs from the knee are dazzling white. A streak on the brow, a broad patch on the nose, a line running from the eye to the chin are black, while black lines separate the white and yellowish colors on the rest of the body. The mane-like ridge on the neck and the tuft on the tail are black-brown. The Passan is only found in Southern Africa, and is called by the Boers of the Cape, the Gems-bok. In Gordon Cumming's work on Southern Africa may be found the following notes concerning this animal:

"The Gems-bok was intended by nature to adorn the parched karroos and arid deserts of South Africa, for which description of country it is admirably adapted. It thrives and attains high condition in barren regions where it might be imagined that a locust could not find subsistence; and burning as is the climate, it is perfectly independent of water, which, from my own observation and the repeated reports both of Boers and aborigines, I am convinced it never by any chance tastes. Its flesh is deservedly esteemed, and ranks next to that of the eland. At certain seasons of the year they carry a great quantity of fat, at which time they can more easily be ridden into.

Owing to the even nature of the ground which the Gems-bok frequents, its shy and suspicious disposition, and the extreme distances from water to which it must be followed, it is never stalked or driven to an ambush like the antelopes, but is hunted on horseback, and ridden down by a long, severe, tail-on-end chase. Of several animals in South Africa which are hunted in this manner, the Gems-bok is by far the swiftest and the most enduring."

The long horns are terrible weapons, and the Gems-bok wields these natural bayonets in a manner which makes it a match for most of the Carnivora. It has been known to beat off the lion itself. Even when the lion has overcome the Gems-bok, the battle may sometimes be equally claimed by both sides, for in one instance the dead bodies of a lion and a Gems-bok were found lying on the plain, the horns of the antelope being driven so firmly into the lion's body, that they could not be extracted by the efforts of a single man. The lion had evidently sprung upon the Gems-bok, which had received its foe upon the points of its horns, and had sacrificed its own life in destroying that of its terrible and redoubtable adversary.

The BEISA, *Oryx beisa*, is the Oryx of the ancients. It is equal to the Pas-an in size, but its color is lighter, and the white markings are somewhat differently arranged. It inhabits at present the coast of Abyssinia, and southward to the country of the Somali.

The SABRE ANTELOPE, *Oryx leucoryx*, has long horns which are considerably bent, and sweep backward in a noble curve; they are ringed at the base thirty or forty times, and very sharp-pointed. This Antelope, which the Arabs call the "Desert Cow," extends over the northern part of Africa. It is not uncommon in Senaar and Kordofan, as well as in the Soudan.

In habits it resembles the Gems-bok, and can use its cimeter-like horns with equal effect.

GENUS ADDAX.

Like the genus just described, the Addax is represented on the monuments of Egypt. The horns which adorn the heads of kings or priests in ancient Egypt are the horns of this antelope.

The ADDAX, *Addax nasomaculatus*, the *only* species of the genus, is found in many parts of Northern Africa, and is formed by nature for a residence among the vast plains of arid sand which are spread over that portion of the globe.

These animals are not found living together in herds, but in pairs, and their range of locality seems to be rather wide. As they are intended for traversing large sandy regions, the feet are furnished with broad, spreading hoofs, which enable them to obtain a firm foothold upon the dry and yielding sand. The horns of this animal are long, and twisted after a manner that reminds the spectator of the Koodoo, an antelope which has been described and figured. Measured from the tip to the head in a straight line, the horns are about two feet three inches in length; but if the measurement is made to follow the line of the spiral, the length is obviously much greater. The distance between the tips is about the same as that from the tip to the base. From their roots to within a few inches of their extremities, the horns are covered with strongly marked rings, arranged in an oblique manner, and some of them partially double. The spiral of the horns is as nearly as possible two turns and a half in its whole length.

Upon the forehead there is a bunch or tuft of long hair, and the throat is also covered with a rather heavy mane of long hair, but there is no mane on the back of the neck. The muzzle and nose are rather peculiar, and bear some resemblance to the same parts of a sheep or goat. The general color of the Addax is a milk-white, with the exception of the black patch of hair on the forehead, a brown-black mane, and a reddish shade on the head and shoulders. Both sexes have horns.

The next sub-family, HIPPOTRAGINÆ, comprises only *one* genus, but the animals embraced in it are the stateliest and noblest of the tribe. The name *Hippotragus*, or "Horse-antelope," is derived from the horse-like form which all the *three* species possess.

GENUS HIPPOTRAGUS.

The SABLE ANTELOPE, *Hippotragus niger* (Plate XLVIII), is a magnificent creature, very shy, and therefore seldom seen by the colonists of South Africa. Gordon Cumming gives the following description of it:

"Cantering along through the forest, I came suddenly in full view of one of the loveliest animals which graces this fair creation. This was an old buck of the Sable Antelope, the rarest and most beautiful animal in Africa. It is large and powerful, partaking considerably of the nature of the ibex. Its back and sides are of glossy black, beautifully contrasting with the belly, which is white as driven snow. The horns are upward of three feet in length, and bend strongly back with a bold sweep, reaching nearly to the haunches."

It lives in herds of no very great size, consisting mostly of ten or twelve does led by a single buck. As a general fact, the buck takes matters very easily, and trusts to the does for keeping a good watch and warning him of the approach of an enemy. Owing to the jealous caution of these female sentinels, the hunter finds himself sadly embarrassed when he wishes to enrich his museum with the horns of their leader, and if any of them should happen to take alarm, the whole herd will bound over the roughest ground with such matchless speed that all pursuit is hopeless, and soon abandoned.

In the native dialect, the Sable Antelope is known under the name of Potaquaine. It is very tenacious of life, and will often make good its escape even though pierced entirely through the body with several bullets. It therefore fully tests all the powers of the hunter.

The BLAU-BOK, *Hippotragus leucophæus*, is as large as its sable congener. It was formerly quite common at the Cape of Good Hope, but has within the last sixty years been exterminated in the colony.

It is a gregarious animal, living in little herds not exceeding ten or twelve in number, and preferring hills and slopes to level ground. Like the preceding animal, it exhales a powerful odor, which penetrates throughout its entire body, and which renders its flesh so unpalatable that it is never eaten as long as other food can be obtained. It is a swift and active creature, being remarkable for its speed even among the swift-footed antelopes. There is a variety of this animal, called the Docoi, which is found by the Gambia, and which is not quite of the same color. The natives assert that the female never produces more than a single young one during her lifetime, for the mother's horns grow so rapidly after the birth of the offspring, that they penetrate into her back and kill her. The Blau-bok is about four feet in height, and the horns are nearly thirty inches in length.

CHAPTER XXVIII.

THE GAZELLES.

THE GAZELLE—ITS BEAUTY AND GRACE—THE ARIEL GAZELLE—THE JAIROU—THE SPRING-BOK—ITS IMMENSE NUMBERS—THE DSEREN—THE SASIN—THE PALLAH—THE SAIGA—THE SUB-FAMILY ANTILOCAPRINÆ—THE PRONG-HORN.

THE sub-family GAZELLINÆ—comprising *six* genera—consists of a group of small or moderate-sized animals, remarkable for their graceful forms, their lyre-shaped horns, their long-pointed ears, and small false hoofs. The tear-bag below the eyes is distinct.

GENUS GAZELLA.

The Gazelle, as it ranges the desert, presents a figure so attractive, that the poets of the East have from time immemorial been eloquent in singing its grace, agility, and beauty. The old Egyptians dedicated it to their goddess Iris, the Queen of the gods; it is the "roe or the young hart" to which Solomon compares his spouse, it is the "roe or the hind of the field" by which he conjures the daughter of Jerusalem. The highest expression of beauty which an Eastern bard can utter is the comparison—"she is like a gazelle as it browses beneath the roses." The Arabs find no words to depict it adequately; their oldest poems praise it, the wandering minstrels of to-day still sing its loveliness.

THE GAZELLE.

The GAZELLE, *Gazella dorcas* (Plate XLVIII), is not quite as large as the Roe-deer, but is much more finely and slenderly built, and much more prettily marked. A full-grown male stands about a yard high, and measures nearly a yard and a quarter in length. The legs are extremely delicate, and the hoofs small and pretty. Its home is the northeastern district of Africa. It extends from the Barbary States to the Desert of Arabia, and from the Mediterranean to the plains of Central Africa. Wherever vegetation is found in the desert, the Gazelle is seen in large numbers, but it is rare either in the rich river-bottoms, or in the mountain ranges. A rolling, sandy district, where the mimosas grow thickly, form its favorite haunts. In Kordofan herds of fifty have been observed, but usually the troop consists of less than ten members. But, although they are moving about all day, except during the burning noon-tide hours, the Gazelles are not easily discovered. Their coat resembles closely the color of the soil, and renders it difficult for any but a native of the desert to detect them. At the first sign of danger the herd bounds away as if in sport. Every movement is graceful. They may be seen in play leaping up from one to two yards from the ground over each other's backs, or skipping over stones in their path. All their senses are acute, their scent is remarkable, their sight keen, their intelligence great. Harmless and timid as they are, they are by no means so devoid of courage as is often supposed. Combats take place even in their herds. With all other animals they are willing to be friends; but the leopard and the lion do not reciprocate with good feeling. The Gazelle makes no resistance to such enemies; against weaker ones the herd defends itself by forming a circle with horns pointed out. They are well aware of the advantages of association, and seem to be inspired with feelings of natural attachment to each other.

The eye of the Gazelle is large, soft, and lustrous. The color of the coat is a light-tawn, deepening into a dark-brown band on the flanks which forms a line of demarcation between the tawn color of the back and the pure white of the abdomen. The face is marked with a dark-brown and a white streak, running from each horn to the muzzle.

There are *seventeen* species recognized in the genus Gazella, and almost as many varieties not entitled to the dignity of species. Among them, the most beautiful is the Ariel Gazelle.

The ARIEL is much darker than the Dorcas Gazelle. It is found in Syria and Arabia, and as it is not only a most graceful and elegant animal in appearance, but is also docile and gentle in temper, it is held in great estimation as a domestic pet, and may be frequently seen running about the houses at its own will. So exquisitely graceful are the movements of the Ariel Gazelle, and with such light activity does it traverse the ground, that it seems almost to set at defiance the laws of

CHAMOIS
GAZELLE   SPRINGBOK
PRONGHORN SABLE ANTELOPE
PLATE XLVIII UNGULATA

gravitation. When it is alarmed, and runs with its fullest speed, it lays its head back so that the nose projects forward, while the horns lie almost as far back as the shoulders, and then skims over the ground with such marvellous celerity that it seems rather to fly than to run, and cannot be overtaken even by the powerful, long-legged, and long-bodied greyhounds which are employed in the chase by the native hunters.

When the Gazelle is hunted for the sake of the sport, the falcon is called to the aid of the greyhound, for without such assistance no one could catch an Ariel in fair chase. As soon as the falcon is loosed from its jesses, it marks out its intended prey, and overpassing even the swift limbs by its swifter wings, speedily overtakes it, and swoops upon its head. Rising from the attack, it soars into the air for another swoop, and by repeated assaults bewilders the poor animal so completely that it falls an easy prey to the greyhound, which is trained to wait upon the falcon, and watch its flight.

When, however, the Gazelle is hunted merely for the sake of its flesh and skin, a very different mode is pursued. Like all wild animals, the Gazelle is in the habit of marking out some especial stream or fountain, whither it resorts daily for the purpose of quenching its thirst. Near one of these watering-spots the hunters build a very large inclosure, sometimes nearly a mile and a half square, the walls of which are made of loose stones, and are too high even for the active Gazelle to surmount. In several parts of the edifice the wall is only a few feet in height, and each of these gaps opens upon a deep trench or pit. A herd of Gazelles is quietly driven toward the inclosure, one side of which is left open, and being hemmed in by the line of hunters, the animals are forced to enter its fatal precincts. As the pursuers continue to press forward with shouts and all kinds of alarming noises, the Gazelles endeavor to escape by leaping over the walls, but can only do so at the gaps, and fall in consequence into the trenches that yawn to receive them. One after another falls into the pit, and in this manner they perish by hundreds at a time. The flesh of the Ariel Gazelle is highly valued, and is made an article of commerce as well as of immediate consumption by the captors. The hide is manufactured into a variety of useful articles. The Ariel is a small animal, measuring only about twenty-one inches in height.

The JAIROU, or common Gazelle of Asia, which is so celebrated by the Persian and other Oriental poets, is ascertained to be a different species from the Dorcas, and may be distinguished from that animal by the general dimness of the marking, and the dark brown streak on the haunches. Several other species belong to the genus Gazella, among which we may mention the MOHR of Western Africa, the ANDRA of Northern Africa, and the KORIN, or KEVEL, of Senegal.

THE SPRING-BOK.

The SPRING-BOK, *Gazella euchore* (Plate XLVIII), is the representative of the genus in South Africa. It derives its name from its extraordinary agility. It can rise to a height of seven or eight feet without any difficulty, and can reach, on occasions, a height of twelve or thirteen feet. It will never cross a road, if it can avoid doing so; when forced, it clears it at a bound. The color of the Spring-bok is a warm cinnamon-brown above, and pure white on the abdomen; a broad band of reddish-brown parting the two colors.

In the vast plains of Southern Africa, the Spring-bok roams in literally countless herds. "For two hours before the day dawned," Gordon Cumming writes, "I had been lying awake listening to the grunting of the bucks within two hundred yards of me. On rising and looking about me, I beheld the ground to the northward actually covered with a dense living mass of Spring-boks marching slowly and steadily along, extending from an opening in a long range of hills on the west, through which they poured like the flood of some great river, to a ridge about a mile to the east, over which they disappeared. The breadth of the ground they covered might have been somewhere about half a mile. I stood upon the fore-chest of my wagon for nearly two hours, lost in wonder at the novel and beautiful scene which was passing before me, and had some difficulty in convincing myself that it was reality which I beheld, and not the wild and exaggerated picture of a hunter's dream. During this time, their vast legions continued streaming through the neck in the hills, in one unbroken compact phalanx."

The wonderful density of these moving herds may be imagined from the fact, that a flock of sheep have been inextricably entangled among a herd of migrating Spring-boks, and carried along with them without the possibility of resistance or even of escape. Even the lion himself has been thus taken prisoner in the midst of a mass of these animals, and has been forced to move in their midst as if he belonged to their own order. Want of water is said to be the principal cause of these migrations, for they have been always observed to depart as soon as the district in which they live has been deprived of water, and to return as soon as the genial rains have returned moisture to the earth, and caused the green herbage to make its appearance. Dr. Livingstone, however, doubts whether the Spring-bok is a sufficiently thirsty animal to be driven into these migrations only by want of water, and thinks that there must be other causes also at work.

They are extremely fond of the short tender grass as it springs from the earth, and the Bakalahari Kaffirs, taking advantage of this predilection, are in the habit of burning large patches of dry stubbly herbage for the sake of attracting the Spring-boks, who are sure to find out the locality, and to come and feed upon the short sweet grass that always makes its appearance on the site of burnt vegetation. Spring-boks are very seldom seen in the deep, rank grass that is so plentiful in their native country, for they would not be able to raise their head above the tall blades, and to perceive the lion, leopard, or other enemy that might be crawling toward them under its shelter.

While engaged in these pilgrimages, the Spring-bok suffers sadly from many foes, man included, who thin their numbers along the whole line of march. Various beasts of prey, such as lions, leopards, hyænas, and jackals, hang around the skirts of the herd, and are always ready either to dash boldly among the moving mass and to drag out some unfortunate animal which may happen to take their fancy, or to prowl in a crafty manner about the rear of the troop, in hopes of snapping up the weakly or wounded animals as they fall out of the ranks. The black and white inhabitants of Southern Africa also take advantage of the pilgrimages, and with guns and spears, which may be used almost indiscriminately among such multitudes of animals, without any particular necessity for a careful aim, destroy myriads of the Spring-boks, and load themselves with an ample supply of hides and meat.

There is a curious provision of nature for preserving the herds in proper condition. It is evident that as the animals move in a compact mass, the leaders will eat all the pasture, and those in the rear will find nothing but the bare ground, cut to pieces by the hoofs of their predecessors. The rearward animals would therefore soon perish by starvation, did not matters arrange themselves in a rather remarkable manner. The leading Spring-boks, having the choice of the best pasture, soon become so satiated and overloaded with food, that they are unable to keep pace with their eager and hungrily active followers, and so are forced to drop into the rear. The hindermost animals in the meantime are anxiously pushing forward in search of food, so that there is a continual interchange going on as the herd moves onward, those in front dropping back to the rear, while those in the rear are constantly pressing forward to take their place in front.

In size the Spring-bok is rather superior to the Dorcas gazelle, but may be immediately distinguished from that animal by means of a curious white patch of long hairs on the croup. Although the animal is so marvelously agile, the body is rather clumsily formed, and seems to be disproportionately large when contrasted with the slight and delicate limbs on which it is supported. While standing at rest, the Spring-bok may be recognized by the peculiar line of the back, which is more elevated at the croup than at the shoulders. The horns of this animal are much larger in the adult male than in the young or the female, and when full grown, are marked with eighteen or twenty narrow complete rings. The lyrate form of the horns is not so perceptible in the young Spring-bok as in the older animal, for until the creature has attained its full growth, the tips of the horns point forward.

GENUS PROCAPRA.

The *two* species of this genus inhabit Mongol-Tartary and the steppes between China and Thibet.

The DSEREN, *Procapra gutturosa*, stand about two feet and a half high, and measures about four feet and a half in length. The body is slender, the head short and thick; the throat of the males has a large protuberance, the tail is short, and its upper part covered with curly hair; the legs are slender and graceful; the horns, which are borne only by the males, are placed close together on the skull, but gradually widen out, curving backward and inward, and marked at the base with about twenty prominent rings. The ears are large and pointed. The color of the coat varies according to the season. In summer the throat and the thighs below the tail are pure white, and the rest of the body is cream-color, with some brownish marks on the brow. The winter coat is uniformly lighter. The hair is very thick, and longer on the hind-quarters than in front.

These animals are usually hunted in the winter season, when they visit the frozen streams, obtaining water by breaking the ice with their hoofs. If they are surprised on the ice, they can be easily killed, as they cannot keep their feet. A lucky sportsman will kill two hundred head in a favorable winter, for the Dseren keep together in such large herds, that a single ball will often kill two or three of them.

GENUS ANTILOPE.

The SASIN, or INDIAN ANTILOPE, *Antilope bezoartica*,—the *only* species—is generally found in herds of fifty or sixty together, each herd consisting of one buck and a large harem of does.

It is a wonderfully swift animal, and quite despises such impotent foes as dogs and men, fearing only the falcon, which is trained for the purpose of overtaking and attacking them, as has already been related of the gazelle. At each bound the Sasin will cover twenty-five or thirty feet of ground, and will rise even ten or eleven feet from the earth, so that it can well afford to despise the dogs. As its flesh is hard, dry, and tasteless, the animal is only hunted by the native chiefs for the sake of the sport, and is always chased with the assistance of the hawk or the chetah, the former of which creatures overtakes and delays it by continual attacks, and the other overcomes it by stealthily creeping within a short distance, and knocking over his prey in a few rapid bounds. It is a most wary animal, not only setting sentinels to keep a vigilant watch, as is the case with so many animals, but actually detaching pickets in every direction to a distance of several hundred yards from the main body of the herd.

The young Sasins are very helpless at the time of their entrance into the world, and are not able to stand upon their feet for several days, during which time the mother remains in the covert where her little one was born. As soon as it has attained sufficient strength, she leads it to the herd, where it remains during its life, if it should happen to be a doe, but if it should belong to the male sex, it is driven away from its companions by the leading buck, whose jealousy will permit no rivals in his dominions. Forced thus to live by themselves, these exiles become vigilant and audacious, and endeavor to attract mates for themselves from the families of other bucks.

The horns of this animal are large in proportion to the size of their owner; their form is spiral, and they diverge considerably at their tips. From the base to the last few inches of the points, the horns are covered with strongly marked rings. In color, the Indian Antelope is grayish-brown or black on the upper parts of the body, and white on the abdomen, the lips, breast, and a circle round the eyes. The outer sides of the limbs, together with the front of the feet and the end of the tail, are nearly black. Some of the oldest and most powerful males are so deeply colored that their coats are tinted with the two contrasting hues of black and white, the fawn tint being altogether wanting. The height of this animal is about two feet six inches at the shoulder.

GENUS ÆPYCERUS.

The PALLAH, *Æpycerus melampus*, is the *only* species of the genus. It is found in enormous herds in South Africa. It is a remarkably fine animal, three feet in height at the shoulder, and possesses elegantly shaped horns, and a beautifully tinted coat. The predominant color is bay, fading into white on the abdomen, and the peculiar patch of lighter colored hair which surrounds the root of the tail. A black semi-lunar mark on the croup serves as a visible distinction from the other antelopes; and the hoofs are black. It is less timid than the Spring-bok. When alarmed, it walks away in the quietest and most silent manner imaginable, lifting its feet high from the ground. When on a journey, they walk in Indian file, and when they have once settled the direction in which they intend to go, they cannot be turned aside even by the presence of a human being.

The name *Æpycerus* or "High-horned," indicates the characteristic of the genus. The horns are nearly two feet long, slender, and lyre-formed, and marked by rough rings at the base. The females are hornless.

GENUS SAIGA.

The SAIGA, *Saiga tartarica*, a *solitary* species, in form and habits reminds one of the sheep. The body is thick, the legs slender and short, the hair remarkably long and thick. The nose is arched and broad, and terminates in a snout. This snout is very peculiar, it projects over the jaw, is mobile, and has in the center two naked nostrils, so that it resembles a real proboscis. The horns are lyrate, thin, and transparent; the ears are short, and almost hidden in its rough coat. In summer the color is a grayish-yellow; in winter the color becomes lighter and the hairs are nearly three inches in length.

The Saiga inhabits the steppes of Eastern Europe and Siberia, from the frontiers of Poland to the Altai. It lives in herds, forming in autumn bands of several thousands, which perform regular migrations. They are very watchful, and never all repose at once, and Pallas observed that the sentinels were regularly relieved. They are very fleet, so that not even a greyhound can take them; they leap well, but without the grace of the antelope. For food, they prefer the herbage that grows on the dry steppe near salt-springs. They walk backward when feeding, as the projecting snout is, otherwise, in their way. The flesh, owing to the nature of the herbage on which the Saiga feeds, has a sharp, balsamic odor; but in spite of this, the natives are eager hunters. In addition to the usual methods of the chase, the Black Eagle is employed.

THE PRONG HORN.

The sixth of our sub-families of the BOVIDÆ is constituted by the ANTILOCAPRINÆ, consisting of only *one* species of a single genus. It inhabits both sides of the Rocky Mountains, extending north to the Saskatchewan and the Columbia rivers, west to the coast-range of California, and east to the Missouri. It seems to represent a transition between the families which have solid and deciduous horns, and those possessing hollow and permanent ones, for the horns, although hollow, like those of the antelope, are shed annually, like those of the deer.

GENUS ANTILOCAPRA.

The PRONG HORN, *Antilocapra Americana* (Plate XLVIII), derives its name from the character of its horns. They are from eight to fourteen inches in length, and at about two-thirds of their height become palmated, and give out a short prong. The tips are bent inward. They are shed every year, in a peculiar manner. They do not fall off entirely like the horns of the elk; the pith remains, and the hard horny shell comes off the pith like the shell from a crab. This shell becomes loose in May, and the Prong-horn retires into seclusion. When it has dropped off, it leaves a spongy, white fleshy substance, sparsely covered with short black bristles pointing upward. This pith grows rapidly, and becomes larger than the old horn. The outside, in the short time of two or three days, hardens again into horn.

The hair is very thick, brittle, and nearly two inches long. The color on the back is reddish, turning to yellow on the sides and white on the belly. Around the tail the hair is dazzlingly white, long, and capable of being erected like a fan when the animal is alarmed. The Prong-horn is chiefly found in the treeless plains. Its run is even and regular, and it is the fleetest of all the animals on the plains. General Dodge says that they often display a strange combination of curiosity and terror. "They become beside themselves at the appearance of any unusual object. A wagon train will attract every herd within the range of vision. They rush at it with every indication of extreme terror, and passing within a few yards, will make a complete circuit, and go off in the direction from which they came." When the leader goes, the rest follow like sheep; if the leader leaps up in the air, all the herd leap in the like manner. As a rule, except during the month of April, these antelopes are in herds. The does are very motherly, and will give suck to other young beside their own; thus the orphan-fawns whose dams have been killed seldom die.

Their food is the short succulent grass of the prairies, and they love saltish water or pure salt, and remain at the salt licks till hunger drives them away. They are good swimmers, and easily pass broad streams. They are hunted alike by Indians and white men.

CHAPTER XXIX.

THE LESSER ANTELOPES.

THE OUREBI—THE KLIPSPRINGER—THE WATER BUCK—THE BLUE-BUCK—THE MUSK ANTELOPE—THE DUYKER BOK—THE RHOODE BOK—THE CHICKARA—THE HARTEBEEST—THE SASSABY—THE GNU—THE CHAMOIS—THE GORAL—THE MOUNTAIN GOAT OF THE ROCKY MOUNTAINS.

THE sub-family CERVICAPRINÆ contains *five* genera. They are all confined to the Southern and Tropical regions of the African continent.

GENUS CERVICAPRA.

The OUREBI, *Cervicapra urebi*, the first of the *four* genera, is thus described by Captain Drayson:

"While many animals of the Antelope kind fly from the presence of man, and do not approach within a distance of many hundred miles of his residence, there are some few which do not appear to have this great dread of him, but which adhere to particular localities as long as their position is tenable, or until they fall victims to their temerity. It also appears as if some spots were so inviting, that as soon as they become vacant by the death of one occupant, another individual of the same species will come from some unknown locality, and reoccupy the ground. Thus it is with the Ourebi, which will stop in the immediate vicinity of villages, and on hills and in valleys, where it is daily making hair-breadth escapes from its persevering enemy—man.

"When, day after day, a sportsman has scoured the country, and apparently slain every Ourebi within a radius of ten miles, he has but to wait for a few days, and upon again taking the field he will find fresh specimens of this graceful little antelope bounding over the hills around him. It is generally found in pairs, inhabiting the plains, and when pursued, trusts to its speed, seeking no shelter either in the bush or the forest. Its general habitation is among the long grass which remains after a plain has been burned, or on the sheltered side of a hill, among rocks and stones."

Its mode of progression, when alarmed or disturbed, is very beautiful. It gallops away with great rapidity for a few yards, and then bounds several feet in the air, gallops on, and bounds again. These leaps are made for the purpose of examining the surrounding country, which it is enabled to do from its elevated position in the air. Sometimes, and especially when any suspicious object is only indistinctly observed in the first bound, the Ourebi will make several successive leaps, and it then looks almost like a creature possessed of wings, and having the power of sustaining itself in the air. If, for instance, a dog pursues one of these antelopes, and follows it through long grass, the Ourebi will make repeated leaps, and by observing the direction in which its pursuer is advancing, will suddenly change its own course, and thus escape from view. In descending from these leaps the Ourebi comes to the ground on its hind feet.

The Ourebi stands about two feet high, and is four feet in length. The horns of the males are about five inches long, straight, pointed, and ringed at the base. The female is hornless. The color of the animal is pale-tawny above, white below.

The KLIPSPRINGER, *Cerviapra saltatrix*, is called by Gordon Cumming a darling little antelope. It is peculiarly formed for rocky ground, its hoofs being small, hard, and sharply pointed. It stands like the chamois, with its feet drawn close together. When alarmed, it bounds up the most precipitous rocks with such rapidity that it is soon beyond all danger. Its color is dark brown, sprinkled with yellow. Each hair is yellow at the tip, then brown, then gray to the base. It measures about twenty-one inches in height when fully grown. The female is like the female Ourebi, destitute of horns.

GENUS KOBUS.

The WATER BUCK, *Kobus ellipsoprymnus*, is the most striking of the *six* species of this genus.

It is a peculiarly timid animal, and when alarmed rushes at once toward the nearest river, into which it plunges without hesitation, and which it will cross successfully even when the stream is deep, strong, and rapid. The animals are probably induced to take to the water by their instinctive dread of the lion and leopard, which will never voluntarily enter the water, except under peculiar circumstances. The Water Bucks are generally found in small herds, which never wander far from the banks of some large river. The horns of this species are remarkable for their formation, being somewhat lyrate, bent back, and thrown forward at their extremities. The tail is rather long, and is covered with long hairs toward its termination. The flesh of this animal is very powerfully scented, and is of so bad a flavor that none but a hungry Kaffir will eat it, and even he will not do so until forced by dire hunger. This peculiar scent is probably variable in potency according to the season of the year, as is the case with all perfumed animals. Captain Harris says that those which he has killed have been totally uneatable, not even the native palate being proof against the rank flavor. The scent extends to the skin, which exhales so powerful an odor that when Captain Harris was engaged in cutting off the head of one, he was compelled to desist. The calves, however, according to Schweinfurth, are very good eating.

GENUS NEOTRAGUS.

The MUSK ANTELOPE, *Neotragus moschatus*,—the *only* species— is a native of Abyssinia and Eastern Africa, where it bears the name of *Beni-Israel*, or "Children of Israel." It is one of the tiniest of antelopes, being hardly fourteen inches in height at the shoulder, and is so slightly made that it appears too fragile to live. Its legs are long, and not thicker than a lady's finger, the body is covered with fine long hairs, which are gray at the base, but a warm red at the top. A broad white stripe runs above and below the eyes. The hoofs and tear-bag are black. The male has a little pair of horns with ten to twelve rings, and pointed tips bent forward, and almost lost in the thick shock of hair in which they grow. Like some of its kindred, it lives in pairs, or in families consisting of the parents and their offspring.

It is found in the mountains as high as six thousand feet above the sea. It lives in the densest thickets, where the larger antelopes cannot enter, and it can pass through the narrowest clefts of the rock.

GENUS NANOTRAGUS.

The *nine* species of this genus comprise the smallest members of the family. They are all very much alike. The males have small, thin, upright horns, with a few rings or half-rings at the base. The head is round, the nose pointed, and the muzzle small.

The BLUE-BUCK, *Nanotragus Hemprichii*, is one of the most graceful of Ruminants. "The most practised eyes," writes Drayson, "are required to discover this buck in the bush, as its color is so similar to the gloom of the underwood, that if it did not shake the branches in its progress, it would be scarcely possible to see it. Long after the sportsman has become sufficiently acquainted with bush-craft to secure with certainty one or two red bucks during a day's stalking, he would still be unable to bag the little Blue-buck. Several times when I was with a Kaffir, who possessed eyes like those of an eagle, he would point, and with great excitement say, 'There goes a Blue-buck! there he is! there, there!' but it was of no use to me, I would strain my eyes and look to the spots pointed out, but could see no buck; and it was a considerable time before my sight became sufficiently quick to enable me to drop this little antelope with any certainty."

It is scarcely more than a foot in height, and about two feet long; and its color is a dark-blue, or mouse tint. It is found in Africa, south of the Sahara.

The sub-family CEPHALOPHINÆ contains *two* genera, one of which inhabits Africa, and the other the hilly parts of India.

GENUS CEPHALOPHUS.

The DUYKER-BOK, *Cephalophus mergens*, may be taken as the most typical of the *twenty-two* species. It derives its name of Duyker or Diver from the way in which it plunges into the bush. Drayson describes it as follows:

"On the borders of the bush, the antelope which is most commonly met is the Duyker, a solitary and very cunning animal. If the sportsman should happen to overtake this buck, it will lie still, watching him attentively, and will not move until it is aware that it is observed. It will then jump up and start off, making a series of sharp turns and dives, sometimes over bushes, and at others through them. When it perceives that it is observed, it will crouch in the long grass or behind a bush, as though it were going to lie down. This conduct is, however, nothing but a ruse for the purpose of concealing its retreat, as it will then crawl along under the foliage for several yards, and when it has gone to some distance in this sly manner, will again bound away. It is therefore very difficult to follow the course of a Duyker, as it makes so many sharp turns and leaps, that both 'spoorer' and dogs are frequently baffled.

"If the course of the buck can be watched, and the place discovered where it lies down after its erratic manœuvrings, it can be easily stalked by approaching it from the leeward side. One must, however, be a good shot to secure a Duyker with certainty, for the little creature is so tenacious of life that it will carry off a large charge of buckshot without any difficulty, and the irregular course which it then pursues requires great perfection and quickness in shooting with a single ball.

"The height of the Duyker-bok is about twenty-one inches at the shoulder, but the animal is somewhat higher at the croup, where it measures nearly twenty-three inches. It may be distinguished from the other species belonging to the large genus in which it is placed, by a ridge upon the front surface of the horns, which runs through the four or five central rings with which the horns are marked, but does not reach either to the tip or to the base. The general color of this animal is brown-yellow, fading into white on the abdomen and all the under parts, including the tail. The upper part of the tail is black, and there is a black streak running up the legs, and another on the nose."

The RHOODE-BOK, *Cephalophus natalensis*, is common in Natal. Its color is a deep reddish-brown; it stands above two feet high, and the horns are about three inches long, straight and pointed.

GENUS TETRACEROS.

The CHICKARA, *Tetraceros quadricornis*, is the best known of the two species of the genus.

In the scientific title of this very curious species of Antelope both words bear the same signification, namely, "four-horned." These singular animals are natives of India, where they are known also under the titles of CHOUSINGHA, or CHOUKA, the last word being derived from the native term *chonk*, a leap, which has been given to the animal in allusion to its habit of making lofty bounds.

The front pair of horns are very short, and are placed just above the eyes, the hinder pair being much longer, and occupying the usual position on the head. The females are hornless. The color of the Chousingha is a bright bay above, and gray-white below, a few sandy hairs being intermixed with the white. The length of the hinder pair of horns is rather more than three inches, while the front, or spurious horns as they are sometimes termed, are only three-quarters of an inch long. The height of the adult animal is about twenty inches.

The ALCEPHALINÆ form a sub-family divided into *two* genera, both inhabiting Africa, and Northeast to Syria.

GENUS ALCEPHALUS.

Of the *nine* species of this genus we will mention only the most important and typical, for all these antelopes are very much alike in habits and mode of life.

The HARTEBEEST, *Alcephalus caama*, may be easily known by the peculiar shape of the horns, which are lyrate at their commencement, thick and heavily knotted at the base, and then curve off suddenly nearly at a right angle. Its general color is a grayish-brown, diversified by a large, nearly triangular white spot on the haunches, a black streak on the face, another along the back, and a black-brown patch on the outer side of the limbs. It is a large animal, being about five feet high at the shoulder. Being of gregarious habits, it is found in little herds of ten or twelve in number, each herd being headed by an old male who has expelled all adult members of his own sex.

Not being very swift or agile, its movements are more clumsy than is generally the case with Antelopes. It is, however, very capable of running for considerable distances, and if brought to bay, becomes a very redoubtable foe, dropping on its knees, and charging forward with lightning rapidity. The Hartebeest is spread over a very large range of country, being found in the whole of the flat and wooded district between the Cape and the Tropic of Capricorn.

The SASSABY, *Alcephalus lunatus*, is reddish-brown, with a blackish-brown stripe down the middle of the face. It lives in herds of six or ten in the flat districts near the Tropic of Capricorn. It is a thirsty animal, and the hunter when he sees one of them knows that water is at no great distance from the spot where the game is.

GENUS CATOBLEPAS.

The *two* species of this genus are most remarkable animals. They seem a compound of horse, ox, and antelope. They have the head of an ox, the body and neck of a horse, and the cloven hoof of the antelope. Both sexes have horns.

The GNU, *Catoblepas gnu* (Plate XLVII), has a fine head, and peculiarly shaped horns which are bent downward, then upward, with a sharp curve. They live in large herds, and are curious in disposition, as well as very irritable in temper.

"They commence whisking their long white tails," says Cumming, "in a most eccentric manner; then, springing suddenly into the air, they begin pawing and capering, and pursue each other in circles at their utmost speed. Suddenly they all pull up together to overhaul the intruder, when some of the bulls will often commence fighting in the most violent manner, dropping on their knees at every shock; then, quickly wheeling about, they kick up their heels, whirl their tails with a fantastic flourish, and scour across the plain, enveloped in a cloud of dust." On account of these extraordinary manœuvres, the Gnu is called Wildebeest by the Dutch settlers.

The sub-family RUPRICAPRINÆ contains only *two* species, inhabiting the European Alps from the Pyrenees to the Caucasus. In appearance and habits they strongly resemble the goats. They are, however, true Antelopes, and may be distinguished by the shape of the horns, which rise straight from the top of the head for some inches, and then suddenly curve backward.

GENUS RUPRICAPRA.

The CHAMOIS, *Rupricapra tragus* (Plate XLVIII), is really a forest antelope, and wherever man spares them, they prefer the woods to the higher peaks of the mountains. At present they are most abundant in the Tyrol, where the chase of the Chamois is the favorite amusement of the

Emperor of Austria, and the princes of his house. It moves very swiftly over level ground, and is unsurpassed in traversing the Alpine rocks; the false hoofs of its hinder feet aid it greatly in descending the rocks. It is very wary, and possesses a keen scent; even an old footmark in the snow will startle it. It lives in small herds, which send out sentinels to watch while the rest are feeding, and give warning of a coming foe.

As the hind legs exceed the fore limbs in length, the Chamois is better fitted for the ascent of steep ground than for descending, and never exhibits its wonderful powers with such success as when it is leaping lightly and rapidly up the face of an apparently inaccessible rock, and taking advantage of every little projection to add impetus to its progress. Even when standing still, it is able to mount to a higher spot without leaping. It stands erect on its hind legs, places its fore-feet on some narrow shelf of rock, and by a sudden exertion, draws its whole body upon the ledge, where it stands secure.

The food of the Chamois consists of the various herbs which grow upon the mountains, and in the winter season it finds its nourishment on the buds of sundry trees, mostly of an aromatic nature, such as the fir, pine, and juniper. In consequence of this diet, the flesh assumes a rather powerful odor, which is decidedly repulsive to the palates of some persons, while others seem to appreciate the peculiar flavor, and to value it as highly as the modern gourmand appreciates the "gamey" flavor of long kept venison. The skin is largely employed in the manufacture of a certain leather, which is widely famous for its soft, though tough character. The color of the Chamois is yellowish-brown upon the greater portion of the body, the spinal line being marked with a black streak. In the winter months, the fur darkens and becomes blackish-brown. The face, cheeks, and throat are of a yellowish-white hue, diversified by a dark brownish-black band which passes from the corner of the mouth to the eyes, when it suddenly dilates and forms a nearly perfect ring round the eyes. The horns are jetty-black and highly polished, especially toward the tips, which are extremely sharp. There are several obscure rings on the basal portions, and their entire surface is marked with longitudinal lines.

Several varieties of the Chamois are recorded, but the distinctions between them lie only in the comparative length of the horns, and the hue of the coat. The full-grown Chamois is rather more than two feet in height, and the horns are from six to eight inches long.

The sub-family BUDORCINÆ comprises *one* genus of *two* species, which are found in Nepaul and Eastern Thibet. The sub-family NEMORHE-DINÆ embraces *two* genera of goat-like antelopes.

GENUS NEMORHEDUS.

The *nine* species of this genus range from the Eastern Himalayas to Northern China and Japan, and southward to Formosa and Sumatra. We mention only one species.

The GORAL, *Nemorhedus goral*, is the size of a goat; its horns are two feet long, thin and round, and placed close together; its ears are long and narrow; its coat is short, thick, of a gray or reddish-brown color, and is sprinkled on the sides with black or red. A white streak runs under the throat, and a ridge of hair on the spine is black. Both sexes have horns. It is as agile as the chamois, and possesses the same quick senses to descry danger. It seldom seeks the shades of the forest, but loves the rocks and rocky precipices.

GENUS APLOCERUS.

The *solitary* species of APLOCERUS lives in the Rocky Mountains and the Northern part of California. It is regarded by Brehm and other naturalists as a goat, rather than an antelope.

The MOUNTAIN GOAT, *Aplocerus Americanus*, is often confounded with the "Mountain Sheep." It stands two feet and a quarter high at the shoulder, and is about four feet in length. The horns are about eight inches in length; round at the root, and rising in a gentle curve upward, backward, and outward. The coat covering the whole body is of a uniform white color, and consists of long stiff hairs, with a fine, close under-wool. This thick coat makes the animal look much larger and more powerful than it really is. The hair forms a thick bush on the back of the head, and a mane on the neck, while the chin supports a long, abundant beard divided into regular locks; a collar of long hair goes round the neck, and extends itself over the shoulders and fore-arms.

According to Professor Baird, the Mountain Goat is most abundant in the mountains of Washington Territory. It lives in very high districts, and feeds on Alpine plants, the boughs of *pinus contorta*, and the like. During the summer it ascends to a height of fifteen thousand feet above the sea, and prefers to stay near the lower edge of the melting snow-fields; in winter it descends lower, but never leaves the mountain-peaks. In these deserts, where the human foot seldom treads, it moves, with the sure-footedness of its race, from rock to rock, or clambers up precipices which seem inaccessible. When alarmed, the troop, led by the males, rushes at full gallop to the edge of the most terrible precipices and plunges down, or leaps a chasm, one after another, treading exactly on the same spot, with the lightness of a winged creature, rather than like a quadruped. Endowed with keen powers of smelling and hearing, the Mountain Goat baffles in most cases the attempts of the hunter. The Indians occasionally hunt this animal, but without any eagerness. The flesh is of little value, as it is both very tough and has an odor which sickens even the Redskins. The pelt is valuable, and is carried in to the Hudson Bay Company's factories. Like other furs, its value varies with the fashion of the day. When monkey-skin muffs and collars were in vogue, the fleece of this antelope supplied material for an imitation.

CHAPTER XXX.

GOATS AND IBEXES.

THE GENUS CAPRA—THE GOATS—THE BEZOAR GOAT OR PASENG—THE CASHMERE GOAT—THE ANGORA GOAT—THE MAMBER GOAT—THE MARKHOR AND TAHR—THE EGYPTIAN GOAT—THE IBEXES—THE ALPINE IBEX—THE EYRENEAN IBEX—THE ARABIAN IBEX.

THE last sub-family of the BOVIDÆ is that named CAPRINÆ. It contains *two* genera, CAPRA and OVIBOS. The former very extensive genus has been often divided into numerous sub-genera, but it is more convenient to form only two divisions, "the Goats and Ibexes," and the "Sheep."

GENUS CAPRA.

The *twenty-two* species of *Capra* are equally divided between the Goats and the Sheep. The line of demarcation between the groups is by no means clearly drawn; the relationship between them is close, and no decisive characteristics can be described. In general, we may say that the Goats have erect horns, decidedly compressed, curved backward and outward, with a keel or ridge of horny substance in front. The males have a thick beard, and are notable for a very rank odor, which is not present in the male sheep.

THE GOATS.

The DOMESTIC GOAT (Plate XLIX) has shared the fate of our other domesticated animals—we cannot tell from what wild species it is derived. The Paseng, perhaps, has the best claim to be considered its ancestor. The goat was very early reduced to the service of man. During the Stone Age, judging by the Swiss-lake dwellings, it was more common than the sheep. In the oldest Egyptian monuments the goat appears, and herds of goats, as well as their milk and flesh, are repeatedly mentioned, together with the fact that the most ancient documents were written on goat-skin. In the Bible, the goat is frequently spoken of as supplying both flesh and milk, and its hair as furnishing raiment. One of the principal uses to which the skin of the animal was applied was the manufacture of leather, especially of leathern bottles for carrying water; or sacks, such as Joseph's brethren had, for conveying grain. In sacrifices the goat was in great requisition, and on the Great Day of Atonement it was the only animal that could be offered. Two were selected by lot, one for the Lord, the other for Azazel; the former was slain, and its blood sprinkled on the altar; the latter, the scapegoat, was driven into the wilderness.

Friendly as goats and sheep are, the flocks never mingle, not even when folded in the same enclosure. This instinctive separation of the animals led naturally to the simile, so frequently repeated of the just and the unjust, of the sheep and the goats. The goat, no doubt, gives much more trouble to the goatherd, than the sheep to the shepherd. The former is an erratic creature, climbing up the sides of the valleys, skipping and jumping, and venturing into places where man cannot set his foot. It is, too, more destructive than the sheep, and in Palestine has, by browsing on the young shoots, quite extirpated many species of trees.

In Palestine, at present, the most valuable herds are those of the Mohair or Angora Goat, or of the Syrian or Mamber Goat. From the coat of the former the costly coverings for the Tabernacle were made. Allusion to the long ears of the other variety is found in the Prophet Amos, ch. iii. ver. 12. The wild goat of Scripture was most probably a variety of the Ibex.

Goats, at present, are found everywhere. In some countries they are turned out to pasture in herds, and are watched by goatherds; in others, a few stray about near the houses of their owners. In all places they show by their habits that they are mountain-animals. They delight in clambering over rocks, stone walls, or anything that reminds them of their original home. They are very intelligent, and can easily be taught many tricks. In Spain they are employed as leaders to the flocks of sheep. In many countries of Europe a goat is kept in large stables to lead the horses from the stalls in case of fire, for nothing but the example of another animal will induce the horse to face the flame. In America, the Goat is an introduction by Europeans, and is abundant in

CASHMERE GOAT IBEX DOMESTIC GOAT
ANGORA GOAT

all parts, both north and south. Its maintenance costs little or nothing, and its milk is abundant, and can be made into a peculiarly-flavored cheese. The flesh of the young kid is delicate; its skin furnishes the best kind of leather, and its hair is fashioned into brushes or woven into cloth.

THE BEZOAR GOAT.

The BEZOAR GOAT or PASENG, *Capra ægagrus*, is rather larger than the Domestic Goat. It has very strong, large, curving horns, often nearly three feet in length, which have in front several protuberances. Both sexes have beards, and the coat consists of long, stiff, smooth hair, covering a short, fine wool. The color is a reddish-gray, passing into white on the abdomen.

The Paseng is found in Western and Central Asia, and extends to the islands of the Grecian Archipelago, abounding especially in Crete, where it frequents the loftier peaks. The herds usually consist of forty or fifty, and are dangerous, as they will attack a hunter, and, unless he is cautious, hurl him down the precipices. In the Caucasus they ascend up to the snow-line. In its mode of life it does not differ much from the chamois, climbing and leaping from cliff to cliff with equal agility and skill. The lynx and panther are its deadly foes in the Taurus range, but in all Western Asia it is eagerly hunted to obtain the Bezoar-stones. The Bezoars are balls which form in the intestines of many ruminants. The animal being partial to saline matters, gratifies its taste by licking pieces of rock containing saltpetre. Thus a variety of earthy and silicious particles are swallowed, which become agglutinated by the action of the stomach, and form curious pebble-like accretions. These are regarded as endowed with wonderful power as medicines, especially as safeguards against poison.

THE CASHMERE GOAT.

The CASHMERE GOAT *Capra laniger* (Plate XLIX), is remarkable for its soft, silky hair, from which the highly-prized shawls are manufactured.

This animal is a native of Thibet and the neighboring localities, but the Cashmere shawls are not manufactured in the same land which supplies the material. The fur of the Cashmere Goat is of two sorts—a soft, woolly under-coat of grayish hair, and a covering of long, silken hairs, that seem to defend the interior coat from the effects of winter. The woolly under-coat is the substance from which the Cashmere shawls are woven, and in order to make a single shawl, a yard-and-a-half square, at least ten goats are robbed of their natural covering. Beautiful as are these fabrics, they would be sold at a very much lower price but for the heavy and numerous taxes which are laid upon the material in all the stages of its manufacture, and after its completion upon the finished article. Indeed, the buyer of a Cashmere shawl is forced to pay at least a thousand per cent on the cost of his purchase.

Attempts have been made to domesticate this valuable animal in Europe, but without real success. It will unite with the Angora Goat and produce a mixed breed, from which may be procured very soft and fine wool, that is even longer and more plentiful than that of the pure Cashmere Goat. As a commercial speculation, however, the plan does not seem to have met with much success.

THE ANGORA GOAT.

The ANGORA GOAT, *Capra angorensis* (Plate XLIX), is the noblest of all the group. It is a large animal, with long peculiar horns, and magnificent hair. The horns in the male are compressed, not rounded, but with a sharp ridge; they slope almost horizontally backward, and make a wide double turn with the tips outward; in the female they are shorter, weaker, and round, and form only one curve. The face, ears, and lower joints of the legs are covered with short, smooth hair, the rest of the body is hidden with a thick long fleece, fine, soft, brilliant and silky, which curls in locks, and is chiefly composed of wool, with a few sparse hairs only appearing. The predominant color is a uniform dazzling white, although specimens with dark marks have been seen. This fleece falls off in handfuls in summer, but grows again very rapidly.

This goat derives its name from the town of Angora or Angola, in Asia Minor, where it was first seen by Europeans. The region they inhabit is dry and hot in summer, and very cold in winter, and a pure dry atmosphere is necessary for them. The animal is carefully washed and combed every month during the summer. The fleece is clipped in April, and packed for the market. Angora alone sends out 2,200,000 pounds, mostly shipped to England. The fineness of the wool decreases as the animal's age increases; hence the fleece of a one-year-old goat is the most precious; after their sixth year, the wool is unmerchantable. Angora goats have been introduced into various kingdoms of Europe, and preserve there all the fineness of their wool.

THE MAMBER GOAT.

The MAMBER GOAT, *Capra mambrica*, has long hair, but is distinguished from the previous goats by very long pendulous ears, which often exceed a foot in length. It is a large, high animal, with strong horns that describe a semicircle. The fleece is abundant, thick, and silky. This species is found in considerable numbers near Aleppo and Damascus, and thence has spread through a great part of Asia. The Kirghish Tartars dock the ears as they get in the way, when the creature is feeding.

THE MARKHOR AND TAHIR.

The MARKHOR, *Capra megaceros*, is found in Afghanistan, Thibet, Cashmere, and the Himalaya Mountains. As the specific title *mega-ceros* or "big-horn" indicates, its horns are of remarkable size. They grow, usually, to a length of forty inches, and have a semi-oval section with a prominent ridge at both sides: they stand close together, rise up straight, and make one and a half, or two spiral twists. In some bucks they perfectly resemble corkscrews. The coat is long on the shoulders and along the spine, giving the animal a kind of mane, and is more strikingly developed on the throat, chest, and chin, reaching often to the knee-joints. A grayish-brown is its prevailing color.

The TAHIR, or JEMLACK, *Capra Jemlaica*, is regarded by some as the representative of a sub-genus, *Hemitragus*. It is a handsome animal, inhabiting the loftiest mountains of India. It lives in herds, passes the day in the woods, and sallies out to feed in the evening.

THE EGYPTIAN GOAT.

The EGYPTIAN GOAT, *Capra Ægyptiaca*, is smaller than our common goat, and has thin, merely rudimentary horns, when horns are present. Usually, neither sex has these appendages. The color is reddish-brown. This animal is found everywhere domesticated in Lower Egypt and Nubia.

The DWARF GOAT, *Capra reversa*, stands about twenty inches high. It has short horns, curving slightly backward. The short, thick coat is dark, black and reddish-brown mixed together being the prevailing color, but it is sometimes marked with white patches. It is found in the countries between the White Nile and the Niger rivers in Africa, and probably extends into the heart of the continent.

THE IBEXES.

The IBEX inhabits the mountains of the Old World, and dwells on heights inaccessible to other large animals. Each species has only a narrow distribution. Europe contains at least three species, and the others are found in regions so widely separated as Siberia, Rocky Arabia, Abyssinia, and the Himalayas. The specific differences lie in the form of the horns.

THE ALPINE IBEX.

The IBEX, *Capra ibex* (Plate XLIX), is a stately creature nearly five feet in length and three feet in height. The body is compact, the head small but strongly arched at the brow, the legs powerful, the horns, which are common to both sexes, are large and strong, and curve in a semicircle backward. At the roots, where they are thickest, they stand close together, but gradually diverge and taper. Their section is nearly rectangular. The rings, which indicate the animal's age, form on the front of the horn strongly marked transverse ridges, being most clearly defined and most closely placed in the middle of the horn. The length of these horns is often upwards of a yard, and their weight nearly thirty pounds. The horns of the female are smaller than in the male, and round.

The Ibex was nearly exterminated some centuries ago, but for the last century has been carefully preserved in the Italian Alps. It no longer exists in the Tyrol or Switzerland, and for its preservation in the mountains between Piedmont and Savoy we must thank the late king of Italy, Victor Emmanuel, who took energetic measures to stop its destruction. At present, it is supposed, five hundred Ibexes exist in the hunting-grounds he possessed in the chain of Mont Blanc, in the communes of Cogne, Campigha, Ceresole, and Savaranche. Notices are put up, warning travelers not to shoot them, and the royal gamekeepers are ever on the alert. The Emperor of Austria has lately imported some to the Salzkammergut, where they are said to be increasing.

To hunt the Ibex successfully is as hard a matter as hunting the chamois, for the Ibex is to the full as wary and active an animal, and is sometimes apt to turn the tables on its pursuer, and assume an offensive deportment. Should the hunter approach too near the Ibex, the animal will, as if suddenly urged by the reckless courage of despair, dash boldly forward at its foe, and strike him from the precipitous rock over which he is forced to pass. The difficulty of the chase is further increased by the fact that the Ibex is a remarkably enduring animal, and is capable of abstaining from food or water for a considerable time.

It lives in little bands of five or ten in number, each troop being under the command of an old male, and preserving admirable order among themselves. Their sentinel is ever on the watch, and at the slightest suspicious sound, scent, or object, the warning whistle is blown, and the whole troop make instantly for the highest attainable point. Their instinct always leads them upward, an inborn "excelsior" being woven into their very natures, and as soon as they perceive danger, they invariably begin to mount toward the line of perpetual snow. The young of this animal are produced in April, and in a few hours after their birth they are strong enough to follow their parent.

The color of the Ibex is a reddish-brown in summer, and gray-brown in winter; a dark stripe passes along the spine and over the face, and the abdomen and interior faces of the limbs are washed with whitish gray. The Ibex is also known under the name of BOUQUETIN.

THE PYRENEAN IBEX.

The CABRAMONTES, *Capra Pyrenaica*, is found in the Sierra Nevada, Sierra Moreno, the Mountains of Toledo, the Pyrenees, and is especially abundant in the Sierra de Grados, which separates Old and New Castile. It is quite as large as the Alpine Ibex, but differs from it in the conformation of its horns. These, in the buck, stand very closely together—in fact, almost touching; they rise at first straight up for one-third of their length; they then spread out in a lyrate form, while the tips turn upward and toward each other; they are round in front, but form a short keel behind. The rings indicative of age are clearly visible, but do not form ridges, as in the Ibex. The horns of a buck eleven years old, were found to measure over two feet and a half in length. The color of the animal varies not only with age and season, but according to locality. The hair, beginning from the horns, and going down to the shoulder, forms a kind of mane nearly four inches long; the tail has a still longer tuft. In the Sierra de Grados, a dark-brown mixed with black is the prevailing color in summer; in winter, a brownish-black and gray hue predominate. In the Sierra Nevada the color is lighter, and the black less pronounced.

The Cabramontes usually lives in herds, divided according to the sex, often exceeding one hundred in number. The bucks, heedless of snow and cold, live in the highest part of the mountains, while the ewes seek the southern slopes. The herd is led by the oldest and strongest member. The leader advances ten or twelve yards, stops till the herd comes up, and then again advances in like manner. A herd, feeding, always appoints sentinels to give the alarm. A piping bleat gives the signal. The herd rushes away. Precipices, where man can see no possible foothold for any living creature, are scaled with easy rapidity and safety, not only by the old ones, but the youngest kids. The bucks are more watchful than the ewes, and take the additional precaution of having a rearguard; nor are they so timid, for, when disturbed, they do not at once take to flight, but leap on some rock and examine the intruder. The Spanish hunter has a hard task to bring down his game. He climbs by the wildest paths to the mountain ridge, then creeps on hands and knees to the edge of some precipice, where, after removing his hat, he lies flat down to look into the chasm below. If he sees a herd, he imitates their piping bleat, and by this device often attracts the bucks nearer to him. The flesh is highly prized, and the hide and horns have also their value.

Old writers used to relate that the horns of the Ibexes were of great service to the animal, for, when it leaped down a precipice, it alighted on its horns, and thus saved its skull. Of course this is mere fable.

THE ARABIAN IBEX.

The BEDEN, *Capra Syriaca*, is closely allied to the Ibex of the Alps. The differences between them lie in the horns, which have three angles instead of four. Its usual color is gray, becoming brownish in winter. The male has a black beard.

The Beden, or Wild Goat of Scripture, is still found in Palestine, in small herds of eight or ten. Its agility is extraordinary, and it flings itself with reckless accuracy from one craggy peak to another. Like the Ibex it is very wary, has very keen eyes and keen scent, and, like all gregarious animals, posts sentries to guard the herd. The flesh is excellent, and perhaps the reason why King David took up his abode at Engedi was the necessity of obtaining food for his followers, as it may be safely assumed that the Beden, which is still seen there, was more abundant in old days, before firearms rendered all wild animals afraid of the neighborhood of man. It is probable, too, that it was the Beden which Esau hunted with his quiver and his bow when he sought the savory meat his father loved.

CHAPTER XXXI.

THE SHEEP AND THE MUSK-OX.

THE AOUDAD—THE MOUFLON—THE ARGALI—THE KATSHKAR—THE BIG-HORN—ITS HABITS—
THE CRETAN SHEEP—THE SOUTHDOWN—THE LEICESTER—THE MERINO—
THE HIGHLAND SHEEP—THE GENUS OVIBOS—THE MUSK-OX OF NORTH AMERICA.

THE Sheep are distinguished from the Goats by the possession—as a rule—of the tear-bag, by the flat forehead, and the angular twisted horns, and by the absence of a beard. All wild sheep inhabit mountains of the Northern Hemisphere. Their proper home is Asia, but they extend as far as Africa and the Northern part of America. Every mountain-group in Asia has varieties peculiar to itself, while in Europe, Africa, and America, each have only one species. Many species and varieties are very close to each other, and are distinguished merely by the conformation of the horns. Sheep are like goats, children of the mountain, and live at heights where no other animal except the goat is found. The tame sheep is only a shadow of the wild one, and, unlike the goat, retains hardly a trace of its original qualities.

THE AOUDAD.

The AOUDAD, *Capra tragelaphus*, is a native of Northern Africa, where it is found only in the highest and almost inaccessible ridges of the Atlas range. It is a powerful and active animal, standing rather more than three feet at the shoulder. The horns are about two feet in length, and curve boldly backward. Its fleece consists of strong, hard, rough hair, and fine curling wool. The former turns on the back of the neck into a short mane, and is developed on the throat, breast, and fore-legs into a thick, long, bushy mass. The flesh of this wild sheep is highly prized by the Arabs, and resembles very much that of the deer. It lives a solitary life, and is never seen in herds.

The MOUFFLON, *Capra musimon*, is the only wild sheep found in Europe. It inhabits Sardinia and Corsica, frequenting the lofty peaks of those islands, but it is an error to confound with it the varieties found in the Balearic Islands and Spain, or in Greece. In olden days this variety of wild sheep was very abundant, as many as five hundred having been slain in a single chase; but at present, twenty or thirty head are the highest numbers killed, even when the sportsmen have all the needful appliances.

The Moufflon is one of the smallest of wild sheep, standing a little over two feet in height; the horns attain a length of two feet, the coat is pretty short and smooth, but in winter becomes very thick, and forms a kind of mane. Unlike the Aoudad, the Moufflon is found in tribes of fifty to a hundred, led by an old sturdy ram. Such a troop chooses for its dwelling some inaccessible height, and, like other social ruminants, throws out sentinels to give alarm. The movements of the Moufflon are lively, quick, and safe, but are deficient in endurance.

Tame Moufflons may be often seen in Corsica and Sardinia, but they are troublesome; the bucks, especially, losing all fear of man, and attacking him out of mere wantonness.

THE ARGALI.

The BEARDED ARGALI, *Capra argali* (Plate L), is the giant of the Sheep group, being nearly as large as a moderate sized ox. The horns of an adult male are nearly four feet in length, and measure nineteen inches in circumference at the base: they curve boldly downward till beneath the chin, then recurve and come to a point. The surface of these horns is covered with a set of deep grooves set closely together, and extending to the tips. Firmly set as these horns are, they are not unfrequently knocked off in the annual duels.

It is a mountain-loving animal, being found on the highest grounds of Southern Siberia and the mountains of Central Asia, and not fond of descending to the level ground.

Its power of limb and sureness of foot are truly marvelous, when the great size of the animal is taken into consideration. If disturbed while feeding in the valley, it makes at once for the rocks, and flies up their craggy surfaces with wonderful ease and rapidity. Living in such localities, it is liable to suffer great changes of temperature, and is sometimes wholly enveloped in the deep snow-drifts that are so common upon mountainous regions. In such cases they lie quietly under the snow, and continue respiration by means of a small breathing-hole through the snow. For these imprisoned Argalis the hunters eagerly search, as the animal is deprived of its fleet and powerful limbs, and is forced ignominiously to succumb to the foe, who impales him by driving his spear through the snow into the creature's body. Like others of the same group it is gregarious, and lives in small flocks.

THE KATSHKAR.

The KATSHKAR, *Capra polii*, is described by the old traveler, Marco Polo, as abundant in the elevated plateau of Pamere. It is as large as the Argali, measuring six feet in length, and four feet in height. The horns curve downward in a complete circle, and attain a length of nearly five feet. This sheep seems to be found in all the mountain table-lands of Asia, in the plateaus of Thian-Shan and North Thibet, and of the Aksai, where it ascends above the "timber-line." The herds consist of ten to fifteen, led by a buck whose snowy breast, long curved horns, and proud gait make him a noble object. This animal is hunted in a peculiar manner. The Cossacks and Kirghises go out in pairs to the chase. They are armed with long heavy muskets, which are fired from a rest. If the creature is not killed at the first fire, then the chase begins; one sportsman presses the animal closely, the other cuts off corners, and tries to conceal himself in places where the game must pass.

THE BIG HORN.

The BIG HORN, *Capra montana* (Plate L.), ranks next in size to the elk among the horned beasts of the Great West. It is a curious combination of the body of a deer and the head of a sheep; the horns are, as its common name indicates, of enormous size, and make a curve that is more than a complete circle. The head and horns often weigh sixty pounds. Its coat is thick with short grayish hair, changing in the fall into dun, and the hair becoming more than an inch long, and rather wiry. In winter the coat is increased by a layer of exceedingly fine wool which, though sometimes three inches long, never shows outside the hair, but lies curled up close to the skin.

ROCKY MOUNTAIN SHEEP
CRETAN SHEEP
DOMESTIC SHEEP
BEARDED ARGALI
HIGHLAND SHEEP
MERINO SHEEP

PLATE L   UNGULATA

The Big-horn is found in troops of twenty or thirty in number; they never quit the craggiest regions, but find their food upon the little knolls of green herbage that are sprinkled among the precipices, without being tempted by the verdure of the plains. They come down, however, from their rocky fastnesses to obtain water from the low-lying springs. They are very shy and suspicious, and, at the first appearance of a man, take flight. "What becomes of the Mountain Sheep," writes General Dodge, "when man invades his stronghold, it is impossible to say. Hundreds may be in a locality; man appears; a few, perhaps ten, are killed; the others disappear and leave no sign.

The Big-horn is an admirable climber, and runs up or down the faces of precipices where apparently no foothold exists. Their habits are those of other sheep. The lambs begin to be seen in June, when they are placed on some shelf of rock inaccessible to man or any beast of prey. The ewes and lambs, according to Richardson, form herds apart from the males. From the middle of August till November, the flesh of the Big-horn is in prime condition. According to General Dodge "it is impossible to describe it, but if one can imagine a saddle of most delicious 'Southdown' flavored with the richest and most gamey juices of the black-tail deer, he will form some idea of a feast of mountain-sheep in season, and properly cooked. Except in season, the mountain sheep is thin, tough, and the poorest food that the plains furnish to man."

THE FAT-TAILED SHEEP.

In several foreign breeds of the domestic sheep there is a curious tendency to the deposition of fat upon the hinder quarters. This propensity is not valued in our own country, where the sheep are almost invariably deprived of the greater portion of their tails by the hand of the shepherd, which in consequence are never developed. In some varieties, however, such as the steatopygous sheep of Tartary, the fat accumulates upon the hinder quarters in such enormous masses that the shape of the animal is completely altered. The fat of this portion of the body will sometimes weigh between thirty and forty pounds, and when melted down, will yield from twenty to thirty pounds of pure tallow. So inordinate is the growth of the fat, that the tail becomes almost obliterated, and is only perceptible externally as a little round fleshy button.

Some varieties present a different mode of producing fat, and deposit a large amount of fatty matter in the tail. Fat-tailed Sheep are found in many parts of the world, and are much valued on account of the peculiarity from which they derive their name. The Syrian variety is remarkable for the enormous dimensions of the tail, which in highly fattened and carefully tended specimens will weigh from seventy to eighty pounds. So large, indeed, are the tails, and so weighty are they, that the shepherds are forced to protect them from the ground by tying flat pieces of board to their under surface. Sometimes they add a pair of little wheels to the end which drags on the ground, in order to save the animal the trouble of drawing the bare board over the rough earth. The fat which is procured from the tail is highly valued, and is used in lieu of butter, as well as to "lard" meat that would otherwise be unpleasantly dry and tasteless. It is also melted down and poured into jars of preserved meat, for the purpose of excluding the air. These sheep are most carefully watched, and are generally fed by hand.

The Afghan Fat-tail has very fine silky wool, the Persian is remarkable for its contrasts of color, the body being white, the head a deep black. In all Northern and Central Africa, as well as in Arabia, these Fat-tailed Sheep are common.

THE CRETAN, OR WALLACHIAN SHEEP.

The CRETAN SHEEP (Plate L) is a native of Western Asia and the adjacent portions of Europe, and is very common in Crete, Wallachia, and Hungary. The horns of the Wallachian Sheep are strikingly like those of the Koodoo, or the Addas, their dimensions being proportionately large, and their form very similar. The first spiral turn is always the largest, and the horns are not precisely the same in every specimen. As a general rule, they rise boldly upward from the skull, being set almost perpendicularly upon the head; but in others, there is considerable variety in the formation of the spirals and the direction of the tips. In one specimen which was preserved in the gardens of the Zoological Society, the first spiral of the horns was curved downward, and their tips were directed toward the ground.

The fleece of this animal is composed of a soft woolly undercoat, covered with and protected by long drooping hairs. This wool is extremely fine in quality, and is employed in the manufacture of warm cloaks, which are largely used by the peasantry.

THE SOUTHDOWN.

The SOUTHDOWN (Plate L) is one of the short-wooled breeds, and is valuable, not only for its wool, but for the delicacy of its flesh. It has no horns. It derives its name from the South Downs, a range of chalk-hills in Sussex, Surrey, and Kent, which are covered with a short, sweet grass; but it is not confined to this region, but has been introduced wherever the soil and grass are suitable.

THE LEICESTER.

The Leicester and its varieties are long-wooled animals, which prefer low-lying pasturages to breezy downs.

The most celebrated breed of Leicester Sheep is that which is known as the Dishley breed, and which was developed by the persevering energies of a single individual against every possible discouragement. Mr. Bakewell, seeing that the whole practice of sheep-breeding was based on erroneous principles, struck out an entirely new plan, and followed it with admirable perseverance. The usual plan in breeding the old Leicester Sheep was to obtain a large body and a heavy fleece. Mr. Bakewell, however, thought that these overgrown animals could not be nearly so profitable to the farmer as a smaller and better proportioned breed; for the amount of wool and flesh which was gained by the larger animals would not compensate for the greater amount of food required to fatten them, and the additional year or eighteen months during which they had to be maintained.

His idea was, that three extra pounds of wool are not so valuable as ten or twelve pounds of meat, and that when the expense of keeping and feeding a sheep for eighteen months is taken into consideration, the balance is certainly on the wrong side. He therefore set himself to improve the flesh, letting the wool take care of itself at first, and then turned his attention to the fleece. It was found by experience that sheep with a heavy fleece fatten more slowly than those whose coat is moderately thick.

THE MERINO.

Originally, this animal is a native of Spain, a country which has been for many centuries celebrated for the quantity and quality of its wool.

The MERINO SHEEP (Plate L.), from which the fine Spanish wool is obtained, were greatly improved by an admixture with the Cotswold Sheep of England, some of which were sent to Spain in 1464, and the fleece was so improved by the crossing, that the famous English wool was surpassed by that which was supplied by Spain.

The Merino Sheep is but of little use except for its wool, as, although its mutton is sufficiently good when fattened, it consumes too much food, and occupies so much time in the process of ripening, that it is by no means a profitable animal. The Merino is larger in the limbs than the ordinary English Sheep, and the male is furnished with large spiral horns. The female is generally hornless, but sometimes possesses these appendages on a very small scale. It is liable to bear a black fleece, the sable hue continually making its appearance, even after long and careful crossing. By good management the black tint has been confined to the face and legs, but is ever liable to come out in spots or dashes in the wool. There is always a peculiar hue about the face of a Merino Sheep, not easy to describe, but readily to be recognized whenever seen.

In Spain, the Merinos are kept in vast flocks, and divided into two general heads, the Stationary and the Migratory. The former animals remain in the same locality during the whole of their lives, but the latter are accustomed to undertake regular annual migrations. The summer months they spend in the cool mountainous districts, but as soon as the weather begins to grow cold, the flocks pass into the warmer regions of Andalusia, where they remain until April. The flocks are sometimes ten thousand in number, and the organization by which they are managed is very complex and perfect. Over each great flock is set one experienced shepherd, who is called the "mayoral," and who exercises despotic sway over his subordinates. Fifty shepherds are placed under his orders, and are supplied with boys and intelligent dogs.

Under the guardianship of their shepherds, the Merino Sheep, which have spent the summer in the mountains, begin their downward journey about the month of September; and after a long and leisurely march, they arrive at the pasture-grounds, which are recognized instinctively by the sheep. In these pasturages the winter folds are prepared, and here are born the young Merinos, which generally enter the world in March, or the beginning of April. Toward the end of that month the sheep begin to be restless, and unless they are at once removed, will often decamp of their own accord. Sometimes a whole flock will thus escape, and, guided by some marvelous instinct, will make their way to their old quarters unharmed, except perchance by some prowling wolf, who takes advantage of the shepherd's absence.

THE HIGHLAND SHEEP.

The HIGHLAND SHEEP (Plate L) partakes in a great degree of the character of the wild animal. Pasturing together in enormous herds, and living upon vast ranges of bleak, hilly country, the light and active HIGHLAND SHEEP is a very intelligent and independent creature, quite distinct in character from the large, woolly, unintellectual animal that lives only in the fold, and is regularly supplied with its food by the careful hand of its guardian. It is very sensitive to atmospheric influences, and is so ready in obeying the directions of its own instinct, that a good shepherd when he first rises in the morning can generally tell where to find his sheep, merely by noticing the temperature, the direction of the wind, and the amount of moisture in the air and on the ground. As the Highland Sheep is liable to wander to considerable distances from its proper home, the shepherd is aided in his laborious task by several of those wonderful dogs whose virtues and powers have already been recorded in the course of this work.

GENUS OVIBOS.

The MUSK OX, *Ovibos moschatus* (Plate XLVI), the *only* species, is found in the Hudson Bay Territory, West Greenland, and other districts of Arctic America. It is a remarkable blending of the types of the ox and the sheep. Anatomy shows that it is more nearly allied to the latter. Richardson writes concerning it: " Notwithstanding the shortness of its legs, the Musk Ox runs fast, and climbs hills and rocks with great ease.

Its footmarks are very similar to those of the Caribou, but are rather longer and narrower. These animals assemble in herds of from twenty to thirty in August, and bring forth a calf about the latter end of May. If the hunters keep themselves concealed when they fire upon a herd, the poor creatures mistake the noise for thunder, and crowd together." The Musk Ox has a long-haired, woolly hide; the horns are remarkably broad at the base, where they approximate very closely; the first half of the horns is rough and light-colored, the tip smooth and black, and, after covering with a kind of long helmet the summit of the forehead, they turn down boldly behind the eye, and are again hooked upward at the tip.

The Musk Ox supplies the Esquimaux with a favorite article of food. Its flesh, as the name indicates, is marked by a strong musky odor, but except for a few weeks in the year, it is perfectly fit for food, and is fat and well-flavored.

ORDER VIII.
PROBOSCIDEA.

53. ELEPHANTIDÆ .... Elephants.

CHAPTER I.

ELEPHANTS IN GENERAL.

THE ORDER PROBOSCIDEA—DERIVATION OF NAME—THE FAMILY ELEPHANTIDÆ—FOSSIL ELEPHANTS—THE MAMMOTH—THE MASTODON—THE ELEPHANT—ITS TRUNK—ITS TUSKS—THE ELEPHANT IN HISTORY—IN THE EAST—IN ROME—IN MODERN TIMES—THE TWO SPECIES.

THE order PROBOSCIDEA contains the largest of terrestrial animals. The name is derived from the Greek word *proboscis*, "a trunk," and expresses the most conspicuous feature of the animals contained in it. It comprises only one family, the ELEPHANTIDÆ, or Elephants, those strange creatures which excite awe by their strength, and astonishment by their sagacity, and which form a clearly marked link between the creatures of the world we live in, and those which roamed over the surface of the globe in ages long before it assumed its present conditions of climate. A few words on the fossil remains of the extinct varieties of the elephant will form a fit prelude to our sketch of the survivors.

THE MAMMOTH.

The MAMMOTH, *Elephas primigenius*, has left its bones in abundance on the Arctic coasts of the continent of Asia and on the islands of New Siberia. In the latter, indeed, the soil seems formed of bones and tusks cemented together into a solid mass by sand and ice. When the thaws of summer loosen the sandy tundras near the rivers Obi, Yenisei, and Lena, heaps of huge tusks are revealed. Often the teeth are still fixed in the jaws, and huge ribs and thigh-bones have been found still covered with hide, and hair, and flesh, and still bloody. An old traveler, Ides, who in 1692 went through Siberia to China, writes in his account of his journey: "The natives call the beast that has left these remains the

Mammoth, and say it is enormously large, four or five yards high; that it has a long broad head and feet like a bear, that it lives and dwells under the earth, digging out a passage with its tusks. It seeks its food in the swamps, but dies when it touches sandy ground, perishing when it ascends into the outer air." The Russian explorer, Pallas, describes the heaps of bones he saw in his travels. In 1799, an entire carcass of a Mammoth was found in Siberia, near the mouth of the Lena. Unfortunately it was not till 1806 that it was scientifically examined by Professor Adams of Moscow. The Siberians had cut it up, and used its flesh as food for their dogs. The bears and other carnivorous animals had also consumed a great part of it, but a portion of the skin and one ear remained still untouched. He was able to distinguish the pupil of the eye, and the brain was also to be recognized. The skeleton was still entire, with the exception of one fore-foot. The neck was still clothed with a thick mane; and the skin was covered with blackish hairs and a sort of reddish wool in such abundance that what remained of it could only be carried with difficulty by ten men. Besides this, they collected more than thirty pounds weight of long and short hair that the White Bears had buried in the damp ground after they had devoured its flesh. The remains of this animal, which came to light when buried in the ice for probably many thousand years, are preserved in the Museum of the Academy of St. Petersburg.

This discovery excited considerable discussion among naturalists. The creatures had evidently lived in Siberia when the climate was less arctic. Whether their destruction was caused by a sudden change in the inclination of the earth's axis producing an arctic climate, or whether it arose from some vast deluge, may be still disputed.

THE MASTODON.

The MASTODON, *Elephas mastodon*, lived about the same time as the Mammoth. Ten or twelve species have been discovered in Europe, North and South America, and India. The United States are peculiarly rich in its remains; one species, the Gigantic Mastodon discovered in Ohio, has been, indeed, reconstructed. Barton relates that in 1761 the Indians had found five mammoth skeletons, near the heads of which, the finders declared, were "long noses with a mouth below them." Kalen also mentions a skeleton where the trunk could be distinguished. All the varieties resemble our present elephant. Some were larger, others smaller. Indian traditions called them the "Fathers of the Buffaloes," and relate that they lived at the same time as men of huge size, and were both destroyed by the thunderbolts of the Great Spirit. The Powhatan tribe in Virginia stated that the "Great Man" with his lightnings once smote the whole herd of these monsters because they were destroying the deer, bison, and other animals destined for the use of man. One large bull, however, caught the thunderbolts on his head and turned them aside, till at last he was struck in the flank. He fled, wounded, into the Great Lake, where he will live for ever.

European traditions, speaking of the bones of giants, refer to the remains of either the Mastodon or the Mammoth. The Spartans saw the bones of the hero Orestes in some bones twelve feet long, found in Thrace. A gigantic kneepan found near Salamis was attributed to Ajax, and other huge bones discovered in Sicily were confidently assigned to the one-eyed giant, Polyphemus.

THE ELEPHANT.

The most remarkable characteristic of the Elephant is its trunk. This organ, which is an organ at once of smell and touch, is an elongation of the nose, very remarkable for its mobility and sensibility. It contains an immense number of circular and longitudinal muscles, which enable it to turn in every direction, and to contract and elongate. It ends in a finger-like projection about five inches long, and in the ordinary actions of life is an instrument that performs all the functions of a hand. It seizes and picks up the smallest objects; it can uncork a bottle, or fire off a pistol. In the natural state, the Elephant makes use of it for conveying food to its mouth; for lifting heavy weights, and putting them on its back; for drinking, by filling it with water, and then letting the water pour down its throat. With this instrument it defends itself, and attacks others; it seizes its enemies, entwines them in its folds, squeezes them, crushes them, and tosses them into the air, or hurls them to the ground, afterwards to be trampled under its broad feet. The trunk is a conical tube, of an irregular form, very elongated, truncated and funnel-shaped at the end. The upper side is convex, and fluted along its breadth; the underside is flat, and furnished with two longitudinal rows of little eminences, which resemble the feet of silkworms.

The tusks are two enormously developed canine teeth projecting downward from the upper jaw, and they grow to an immense length and weight. The Elephant has one molar tooth on each side of the tusk-bearing jaw, composed of a number of bony plates covered with enamel. The section of the tusks shows a set of streaks radiating in curves from the centre to the circumference and forming lozenges where they intersect. The tusks are only renewed once, the molars repeatedly, the old ones being pushed forward by new ones coming from behind.

In order to support the enormous weight of the teeth, tusks, and proboscis, the head is required to be of very large dimensions, so as to afford support for the powerful muscles and tendons which are requisite for such a task. It is also needful that lightness should be combined with magnitude, and this double condition is very beautifully fulfilled. The skull of the Elephant, instead of being a mere bony shell round the brain, is enormously enlarged by the separation of its bony plates, the intervening space being filled with a vast number of honeycomb-like bony cells, their walls being hardly thicker than strong paper, and their hollows filled during the life of the animal with a kind of semi-liquid fat or oil. The brain lies in a comparatively small cavity within this cellular structure, and is therefore defended from the severe concussions which it would otherwise experience from the frequency with which the animal employs its head as a battering-ram.

In order to support the enormous weight which rests upon them, the legs are very stout, and are set perpendicularly, without that bend in the hinder leg which is found in most animals. It may seem strange, but it is nevertheless true, that localities which would be inaccessible to a horse are traversed by the Elephant with ease. In descending from a height, the animal performs a very curious series of manœuvres. Kneeling down, with its fore-feet stretched out in front and its hinder legs bent backward, as is their wont, the Elephant hitches one of its fore-feet upon some projection or in some crevice, and bearing firmly upon this support, lowers itself for a short distance. It then advances the other foot, secures it in like manner, and slides still farther, never losing its hold of one place of vantage until another is gained. Should no suitable projection be found, the Elephant scrapes a hole in the ground with its advanced foot, and makes use of this artificial depression in its descent,

If the declivity be very steep, the animal will not descend in a direct line, but makes an oblique track along the face of the hill. Although the description of this process occupies some time, the feat is performed with extreme rapidity.

THE ELEPHANT IN HISTORY.

The ancient Egyptians had seen both the African and Asiatic Elephant, having made the acquaintance of the latter during their campaigns in Assyria. On coins, the Greeks, who had encountered the Persians, always represented the Asiatic species, while the coins of the Romans represented the African species.

In India, the Elephant is called *Hathi*, "the beast with the hand," and the grandeur and state pomp of the mightiest Oriental kings, the enormity of whose magnificence sometimes reads like a fabulous wonder, seems almost inseparable from the early history of elephants. On all great occasions, and the assemblage of multitudes, the lofty and sagacious double forehead, with the quiet small eyes, enormous flaps of ears, and ever-varying attitude of "proboscis lithe," constitutes one of the most imposing figures of the majestic scene and its countless concourse. In the most ancient Sanscrit poems there are records of tame elephants in processions, a thousand years before the Christian era. We do not allude only to great state occasions, or to warlike processions, but even to religious ceremonies, since the elephant is found to occupy a post of extraordinary honor in the remotest records of the mythology of India. One of their most alarming deities rides upon his back; while the idol which is their symbol for wisdom and science, bears the form of a man with the head of an elephant. A few miles from the modern city of Kermanshah, the excavations of the rock display many finely carved figures, and the sides of some of the caves are covered with sculpture representing the hunting of wild boars along the banks of a river by men mounted on elephants, while others, in boats, are ready to attack the game when it takes to the water. The ancient Chinese represented the earth as borne upon the backs of eight elephants, whose heads were turned to the principal points of the compass. The same animal is a favorite figure of speech in their poetry. In Eastern architecture the elephant is likewise a very important personage at the gates of temples, on the walls of palaces, on the sides of tombs and pagodas, and in subterranean temples like those of Ellora. Even to the present time the Hindoos, on great occasions, select these creatures to bear the images of their gods, and we find them loaded with the most valuable ornaments in the mystic processions of Brahma and Vishnoo. The use of elephants is absolutely prohibited in the modern capital of Siam, excepting to personages of very high rank; and, in a portion of the Celestial Empire, the chief minister for the foreign department is expressly designated as "the Mandarin of Elephants."

The earliest account, which may be considered worthy to be regarded as history, of the employment of elephants as part of an army, is that which is given of the battle of Arbela (331 B.C., when Darius ranged fifteen of them in front of the centre of his grand line. They fell into the hands of the conqueror, Alexander, to whom a present was made of twelve more; but this great general was too wise to make use of them in his battle against Porus, as he had already perceived that they might prove very dangerous allies, if driven back mad with wounds and terror among the "serried ranks." Very soon, however, they were put to use, systematically; and gradually, by regular training, became very formidable. There were few wars in which the Romans were engaged, during the three hundred years that intervened between the time of Alexander the Great and Cæsar, in which these animals were not employed. Notwithstanding their military education, however, it must be admitted that the best fighting elephants not unfrequently caused their masters to lose the day by their insubordinate and disorderly conduct.

Of the tower which was fixed upon the back of the elephant, and filled with armed men, our impressions are chiefly derived from ancient medals and coins, pictures, bas-reliefs, and the writings of poets. In the Book of Maccabees it is said: "And upon the beasts there were strong towers of wood, which covered every one of them, and were girt fast unto them with devices; there were also upon every one two-and-thirty strong men, that fought upon them, beside the Indian that ruled him." This number is an exaggerated one. The usual number of men was four, beside the conductor.

The ancient armor of an elephant, in other respects, is highly interesting—a strange mixture of the terrible and grotesque. He was often baltcased with plates of metal, and wore a large breastplate, which was furnished with long sharp spikes, to render his charge into the ranks of the enemy more devastating; his tusks were fitted—in fact, elongated—with strong points of steel. Plumes of feathers, small flags, and bells, were also affixed to them. It is said that some of the most sagacious and skilful of the fighting elephants were taught the use of the sword, and, the handles being made suitable to the grasp of the trunk, they wielded enormous cimeters with extraordinary address. The Sultan Akbar had many of these sword-bearing elephants in his army.

According to Pliny, elephants were trained in Rome for the stage. He gives an account of a scene enacted by them, in which four of them carried a fifth in a litter, the latter representing an invalid. Others ranged themselves in a seated posture at a great banquet-table, and eat their food from large plates of gold and silver, with portentous gravity, that excessively delighted the spectators. Moreover, he and Suetonius both assure us that an elephant danced on the tight-rope! He walked up a slanting tight-rope from the bottom of the arena to the top of the amphitheatre; and on one great occasion a man was found daring enough, and confident enough in the performer's skill, to sit upon his back while he made the perilous ascent. There is no exaggeration in this statement. There was in Paris, in 1867, an elephant performing at the circus of the Boulevard du Prince-Eugène, which was called *L'Eléphant ascensioniste*; it had learned to balance its heavy mass on a tight-rope, like Blondin.

The Romans justly considered the Asiatic Elephant more intelligent and courageous than the African one. They introduced both in large numbers to their combats in the circus, and Pliny tells a touching story of the pathetic appeals which the poor creatures made to the spectators against the cruelty of their foes. After the fall of the Roman Empire, the first of these animals which was seen in Europe was sent to Charles the Great by the Caliph Haroun al Raschid. Pope Leo X received one as a present from the Sultan. It excited great curiosity, but soon died. At present both species are common in Zoological Gardens. But, naturally, the performing elephants we see are natives of Asia.

GENERAL HABITS.

Elephants of both kinds live in large forests, preferring those where water is most abundant. They are sometimes, however, found in Ceylon, at a height of six thousand feet above the sea. They are more nocturnal than diurnal animals. They live in herds, and the traveler who comes upon a herd without disturbing it, sees them feeding in the greatest peace and harmony. When they travel, no obstacles deter them. They climb steep hills, they swim lakes and streams, they easily force themselves through the thickest jungle, and form regular tracks as they march in Indian file. The old superstition that they could not lie down is unfounded; but may have arisen from the fact that they often sleep standing. The sight of the elephant is not highly developed, but its other senses are delicate, especially that of hearing. Every sportsman has learned that the snapping of a twig will start a whole herd. The intellectual faculties of the elephant resemble those of the higher mammals; he reflects before acting, he learns by experience, he is very susceptible of teaching, and remembers what he learns. They have the sense to avoid the neighborhood of trees during a thunder-storm;. a fact which proves a considerable reasoning power. Still, in his wild life the elephant is cunning rather than sagacious. The herds consist of various numbers, from ten up to hundreds. Kirk saw eight hundred together on the Zambesi river. Each herd consists of a single family, under the rule of a patriarch, who leads and guards the others, providing in every way for their safety. The elephant grows till it is about twenty, and probably lives till it is a hundred and twenty.

The Elephant is found in a wild state in Asia and in Africa. In the former continent it inhabits India, Burmah, and Siam, as well as the islands of Ceylon, Borneo, and Sumatra. In the latter it is a native of all the Interior whenever it is clothed with forests or with grass. Probably it never extended to the range of Mount Atlas. It was exterminated in the Cape of Good Hope at the end of the last century. In North as well as in South Africa it is receding before the advance of civilization.

CHAPTER II.

THE ASIATIC ELEPHANT.

THE ASIATIC ELEPHANT—ITS USE—MODE OF CAPTURE IN CEYLON—POINTS OF A GOOD ELEPHANT—WHITE ELEPHANTS—FUNERAL OF A WHITE ELEPHANT—THE DWARF ELEPHANT.

THE ASIATIC ELEPHANT, *Elephas Indicus* (Plate LI), is a powerful animal with a broad forehead, short neck, strong body, and pillar-like limbs. The head, more particularly, gives the overpowering impression which this great beast exercises on the spectator. It is high, short, and broad; the facial line is vertical, the cranium crowned by two elevated lateral bumps. Near the ears are two glands which, at certain times, secrete a most offensive fluid. The ears are of moderate size, four-sided, and ending in a point. The small, twinkling, but unpleasant eye is deep-set, and the eyelids are lined with thick, strong, black eyelashes. The centre of the eye is very small and round, the iris coffee-colored. The under lip is pointed, and usually hangs down. Between the eyes is the origin of the trunk, which gradually tapers to its extremity. The powerful tusks project from the upper jaw. The neck is short, the line of the spine is highest just behind the head, and slopes downward to the tail. The tail is placed high, and hangs down straight to the joint of the hind leg. The skin is covered with folds and wrinkles, and is almost destitute of hair. The few hairs on the back are of a dark-gray color. The size of the elephant is usually exaggerated. A very large male will measure from the end of the trunk to the end of the tail eight yards, and will stand about four yards high.

If it had not been for the presence of man upon the earth, the elephant would have been the lord of creation. But man has succeeded in reducing this monarch of the animal kingdom to his service. In India or Siam, when a troop of elephants has been discovered, the natives assemble and surround them, and drive them into a corral. At other times, a solitary male elephant which has been exiled from some troop is beguiled by a couple of tame females who caress him with their trunks while the men who had accompanied them are passing cords around his limbs. A well-trained elephant is of very great value in the Southeastern regions of Asia. Its strength is about five times that of the camel. In its wild state, the Indian Elephant is believed to attain to the age of two hundred years; but it rarely is so long-lived in a state of captivity. In war they are employed for carrying the sick, and camp equipage. The English in India harness them in their artillery trains. Moreover, the proprietors of large cultivated plains, in certain parts of India, have succeeded in making them draw plows. Never did a more monstrous beast of draught turn up the earth with a plowshare. A plowing elephant does the work of thirty oxen.

Without the presence of numerous elephants to grace it, no public *fête* in most parts of India is considered complete. It always figures in the suite of princes, and state processions.

It is especially useful for carrying sportsmen on its back in tiger hunting, and, if need be, for defending them against it, when this terrible animal turns to bay. The elephants which are used by the English army are usually captured by government officials in the island of Ceylon, where they are very numerous. The hunts are on a large scale, as well to rid the neighborhood of such dangerous visitors, as for supplying themselves with such docile workers. A large open space in the jungle is fenced round with strong bamboos, and a narrow entrance at one end is left open. Into this kraal or corral a large party of men drive the herd of wild elephants which are led by tame decoy elephants into the roads leading to the enclosure. When the herd is safely inside, the exit is barred, and the imprisoned animals make furious attempts to escape, charging the fence with desperate determination. When they are wearied out with their fruitless efforts, the best of the herd are selected for taming. Men, lasso in hand, creep quickly into the kraal. Each of them selects one of the largest and strongest of the group, and, having arranged the "lasso" for action, they apply a finger gently to the right heel of their beast, who, feeling the touch as though that of some insect, slowly raises the leg. The men, as the legs are lifted, place the running nooses beneath them, so that the elephants are quietly trapped, unknown to themselves, and with the utmost ease. The men now steal rapidly away with the ends of the ropes, and immediately make them fast to the ends of the nearest trees.

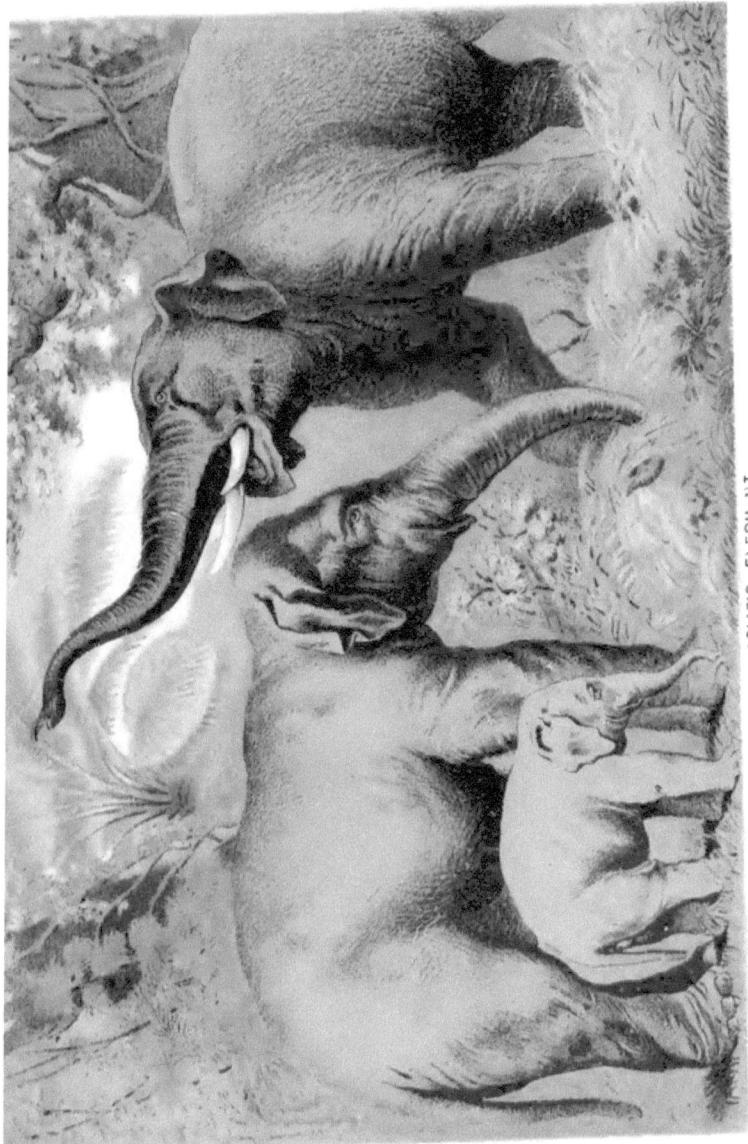

ASIATIC ELEPHANT

FEMALE AND YOUNG, MALE

But the work is not complete. It is necessary to secure the wild creatures more securely. Two tame elephants are now placed on active service. Walking slowly up to the nearest of the captured animals, they begin to urge him toward the tree to which he was fastened. At first the creature is stubborn; but a few taps on his great skull, and a mighty push on his carcass, send him a yard or two nearer his destination. As he proceeds, the man in charge of the rope gathers in the slack of it; and so matters go on between this party—a tap, a push, and a pull—until at length the three elephants are close to the trees. Two other villagers then come forward with a stout iron chain. The tame animals place themselves one on each side of their prisoner, pressing him between them so tightly as to prevent the possibility of his moving. In a minute or two the great chain is passed several times round the hind legs and the tree; and in this way the captive is left, helpless, and faint with struggling against his bonds.

The process of taming is soon accomplished. At the end of two or three months, the wild and unruly destroying monster of the jungle might be seen quietly and submissively piling logs of ebony in the government timber-yards, with a purpose-like intelligence little short of that of man.

A traveler writes: "These huge animals were generally employed in the commissariat timber-yard, or the civil engineer's department, either in removing and stowing logs and planks, or in rolling about heavy masses of stone for building purposes. I could not but admire the precision with which they performed their allotted task, unaided save by their own sagacity. They were one morning hard at work, though slowly, piling up a quantity of heavy pieces of ebony: the lower row of the pile had been already laid down, with mathematical precision, six logs side by side. These they had first rolled in from the adjoining wharf; and, when I rode up, they were engaged in bringing forward the next six for the second row in the pile. It was curious to observe those uncouth animals seize one of the heavy logs at each end; and, by means of their trunks, lift it up on the logs already placed, and then arrange it crosswise upon them with the most perfect skill. I waited while they thus placed the third row: feeling a curiosity to know how they would proceed when the timber had to be lifted to greater height. Some of the logs weighed nearly twenty hundred-weight. There was a short pause before the fourth row was touched; but the difficulty was no sooner perceived than it was overcome. The sagacious animals selected two straight pieces of timber, placed one end of each piece on the ground, with the other resting on the top of the pile, so as to form a sliding-way for the next logs; and, having seen that they were perfectly steady and in a straight line, the four-legged laborers rolled up the slope they had thus formed the six pieces of ebony for the fourth layer on the pile. Not the least amusing part of the performance was, the careful survey of the pile, made by one of the elephants, after placing each log, to ascertain if it were laid perfectly square with the rest."

A still more striking proof of sagacity exhibited by one of these commissariat elephants is told: "One evening, while riding in the vicinity of Kandy, my horse evinced some excitement at a noise which approached us in the thick jungle, and which consisted of a repetition of the ejaculation, *Urmph! urmph!* in a hoarse and dissatisfied tone. A turn in the forest explained the mystery, by bringing me face to face with a tame elephant, unaccompanied by any attendant. He was laboring painfully to carry a heavy beam of timber, which he balanced across his tusks, but the pathway being narrow, he was forced to bend his head to one side to permit it to pass endways; and the exertion and inconvenience combined, led him to utter the dissatisfied sounds which disturbed the composure of my horse.

"On seeing us halt, the elephant raised his head, reconnoitred us for a moment, then flung down the timber, and forced himself backward among the brushwood, so as to leave a passage, of which he expected us to avail ourselves. My horse still hesitated; the elephant observed, and impatiently thrust himself still deeper into the jungle, repeating his cry of *urmph!* but in a voice evidently meant to encourage us to come on. Still the horse trembled; and, anxious to observe the instinct of the two sagacious creatures, I forbore any interference; again the elephant wedged himself farther in among the trees, and waited impatiently for us to pass him, and after the horse had done so, tremblingly and timidly, I saw the wise creature stoop and take up his heavy burden, turn and balance it on his tusks, and resume his route, hoarsely snorting, as before, his discontented remonstrance."

The working elephant is not so valuable for certain kinds of work as has been supposed. In the unopened districts they are useful for carrying stores, and constructing bridges. But in more civilized districts, where the roads permit the use of horses and oxen, their services may be dispensed with. In Ceylon, an elephant costs six to seven shillings a day to feed, and he can work only four days a week. A stout horse, that works for five days, costs only two shillings and sixpence.

When the natives hunt the elephant merely for the sake of his ivory or his flesh, and do not care to take him alive, they achieve their object by stealing cautiously upon him as he dozes, and by gently tickling one of his hind-feet with a slight twig, they induce him to lift his foot from the ground. As soon as he does so, the hunters, who are furnished with a mallet and a sharp wooden spike about eight inches in length, drive the spike into his foot, and effectually lame him with a single blow. He is then quite at their disposal, and is easily dispatched. The flesh of the elephant is thought to be very poor indeed; but the heart, the tongue, the trunk, and the foot, are considered to be good eating.

The "points" of a good elephant are as important in India and Ceylon as those of a horse in Europe. In a native work upon the elephant, quoted by Sir E. Tennent, the points are given as follows: "The softness of the skin, the red color of the mouth and tongue, the forehead expanded and full, the ears large and rectangular, the trunk broad at the root, and blotched with pink in front, the eyes light and kindly, the cheeks large, the neck full, the back level, the chest square, the fore-legs short and convex in front, the hind quarters plump, five nails in each foot, all smooth, elastic, and round. An elephant with all these perfections will impart glory and magnificence to the king."

The herds in which the Indian elephants congregate are not of very great size, containing only from ten to twenty or thirty individuals, and consisting, as is generally thought by men of practical experience, of members of the same family. This opinion is strengthened by the fact that certain physical peculiarities, such as the shape of the trunk or the head, have been found in every member of the same herd. Sometimes these herds will associate with each other for a time, but at the smallest alarm each little flock assembles together independently of the others. It is remarkable that a whole herd has never been known to charge a foe simultaneously. The leader generally faces the enemy, while the remainder of the herd manœuvre in his rear; but that the entire herd should unite in a charge, is a circumstance never yet known to occur. The Asiatic elephant will permit the temporary society of other animals, and may be seen at a fountain, or feeding on an open space in close proximity to deer and wild buffaloes, neither animal displaying any aversion to, or fear of the other. In its general habits the elephant is restless and irritable, or rather "fidgety," never remaining quite still, but always in motion in some way or other.

The elephant is always guided by a mahoot, who sits astride its neck and directs its movements by his voice and a spiked hook. The persons who are carried on the elephant either sit in a howdah—a kind of carriage without wheels strapped on its back—or they sit upon a cushion. The latter mode is preferable, as the traveler can readily change his position. The motion is sometimes pleasant, sometimes fatiguing. At times the pace is so rapid that a man on horseback can with difficulty keep up. But this pace does not last long, and the animal usually travels about twenty-four miles a day.

Sometimes Albino elephants are discovered. The white color is really the result of hereditary disease. These specimens are highly prized in Siam; the whole country, indeed, has been called the land of the White Elephant.

The King of Siam once had no less than six of these wonders of the earth. They had apartments near the Palace, each having ten servants. Their food was fresh grass and sugar-cane, and bananas wreathed with flowers. Their dinner was set on a white table-cloth in a shady court near a marble fountain. Their tusks were ornamented with gold rings, their heads covered with a network of gold, and their backs supported a small embroidered cushion. The White Elephant is called "The Pure King," and "The Wonderful King." Whoever discovers a white elephant is rewarded by the Siamese kings, a grant of land is made to him, and his family, to the third generation, is exempt from taxes.

A white elephant died in his temple at Bangkok in Siam, in November, 1878, which was said to have been born in 1770. Each white elephant possesses its palace, a vessel of gold, and richly jewelled harness; several mandarins are attached to its service, and feed it with cakes and sugar-cane. The king is the only person to whom it bows the knee, and a similar salutation is returned by the prince. The one above mentioned had a magnificent funeral. A hundred priests officiated, the three surviving white elephants, preceded by trumpets, and followed by an immense concourse of people, accompanied the remains to the banks of the river, where the king and his nobles received them. A procession of thirty vessels formed the cortege to convoy the funeral car to the other bank for interment.

THE DWARF ELEPHANT.

There were lately landed in New York, by the ship Oxfordshire from Singapore, two strange creatures said to be captured about eight hundred miles from Singapore. They were described as being two dwarf elephants from the mountains of the Malay peninsula. They are the first elephants of their species ever brought to this country, and a gentleman who had visited all of the great Zoological Exhibitions in Europe declared that he had not seen anything like them. They are males, and their names are Prince and Sidney. Prince, the larger, is thirty-six inches tall. Sidney is only twenty-eight inches in height. They are supposed to be about five and seven years old, respectively. They are covered with a thick coat of bristly hair. Mr. Reiche calls it wool, and says that he shall style them the woolly elephants. He accounts for this growth by the fact that the elephants lived far up the mountains, where the climate is cold.

CHAPTER III.

THE AFRICAN ELEPHANT.

THE AFRICAN ELEPHANT—DIFFERENCE FROM THE INDIAN ELEPHANT—HUNTING THE ELEPHANT—
DELEGORGUE—GORDON CUMMING—THE ABYSSINIAN "HOCK-CUTTER"—CAPTIVE ELEPHANTS
—PLAY ELEPHANTS—ANECDOTE OF ELEPHANTS.

THE Elephant which is found in Africa differs considerably in appearance from the one just described. The head is rounder and narrower, the ears are enormous, the rims nearly meeting on the back of the head, and the tusk is much stronger.

The AFRICAN ELEPHANT, *Elephas Africanus* (Plate LII), is called by the Arabs *Fihl*. It is met with from the Cape of Good Hope as far north as Nubia and Cape Verd. It consequently exists in Mozambique, in Abyssinia, in Guinea, and in Senegal.

African Elephants live, like those of India, in troops more or less numerous. They are sometimes found alone; the Dutch call these *rodeurs*, rovers or prowlers. They were formerly much more common in the environs of the Cape of Good Hope than they are at present. Thunberg relates that a hunter told him that he had killed in these regions four or five a day, and that regularly. He added that the number of his victims had many a time amounted to twelve or thirteen, and even to twenty-two in one day. This may, perhaps, have been but a braggart's idle boast. Still they abound in the vast interior of Africa. In the present day they are not used by man for any warlike or domestic purpose, but are hunted for the sake of their ivory.

Delegorgue, a French traveler, has published, more recently, some curious accounts of their habits. Among these animals, gathered together in troops, there prevails a spirit of imitation which sometimes makes them all do exactly what the first has done. Delegorgue relates on this subject the following episode of one of his hunting excursions. A band of elephants was coming toward him and his two hunting com-

PLATE LII PROBOSCIDEA AND HYRACOIDEA

panions. He shot at the first of the troop; the elephant fell, sinking on its knees. A second elephant was then killed, and fell on its knees over the first. Another of the sportsmen then shot in his turn, and the elephant aimed at fell in the same manner over the two others. All the elephants fell thus on their knees, even to the very last of them, eleven in all, under the fire of the sportsmen.

Gordon Cumming gives thrilling accounts of his adventures with elephants. He sighted a herd, and selected two huge females. "They walked slowly past at about sixty yards, and the one I picked out was quietly feeding with two others on a thorny tree before me. My hand was now as steady as the rock on which it rested, so, taking a deliberate aim, I let fly at her head, a little behind the eye. She got it hard and sharp, just where I aimed, but it did not seem to affect her much. Uttering a loud cry, she wheeled about, when I gave her the second ball, close behind the shoulder. All the elephants uttered a strange rumbling noise, and made off in a line to the northward at a brisk ambling pace, their huge fan-like ears flapping in the ratio of their speed. I did not wait to load, but ran back to the hillock to obtain a view. Presently my men hove in sight, bringing the dogs; and when these came up, I waited some time before commencing the attack, that the dogs and horses might recover their wind. We then rode slowly toward the elephants, and had advanced within two hundred yards of them, when, the ground being open, they observed us, and made off in an easterly direction; but the wounded one immediately dropped astern, and next moment she was surrounded by the dogs, which, barking angrily, seemed to engross her attention. Having placed myself between her and the retreating troop, I dismounted, to fire within forty yards of her, in open ground. Colesberg, my horse, was extremely afraid of them, and gave me much trouble, jerking my arm when I tried to fire. At length I let fly; but, on endeavoring to regain my saddle, Colesberg declined to allow me to mount; and when I tried to lead him, and run for it, he only backed toward the wounded elephant. At this moment I heard another elephant close behind; and on looking about I beheld one of its friends with uplifted trunk, charging down upon me at top speed, shrilly trumpeting, and following an old black pointer named Schwart, that was perfectly deaf, and trotted along before the enraged elephant quite unaware of what was behind him. I felt certain that she would have either me or my horse. I, however, determined not to relinquish my steed, but to hold on by the bridle. Fortunately, however, the dogs took off the attention of the elephants; and just as they were upon me I managed to spring into the saddle, where I was safe. As I turned my back to mount, the elephants were so very near, that I really expected to feel one of their trunks lay hold of me. Returning to the charge, I was soon once more alongside, and, firing from the saddle, I sent another brace of bullets into the wounded elephant. Colesberg was extremely unsteady, and destroyed the correctness of my aim. The 'friend' now seemed resolved to do some mischief, and charged me furiously, pursuing me to a distance of several hundred yards. I therefore deemed it proper to give her a gentle hint to act less officiously, and accordingly, having loaded, I approached within thirty yards, and gave it her sharp, right and left, behind the shoulder; upon which she at once made off with drooping trunk, evidently with a mortal wound. Two more shots finished her; on receiving them she tossed her trunk up and down two or three times, and falling on her broadside against a thorny tree, which yielded like grass before her enormous weight, she uttered a deep hoarse cry and expired." For all elephant-shooting, leaden bullets are not to be relied on. Cumming used a two-ounce ball hardened with one-eighth lead. Shell-bullets also have been used with terrible effect.

Sir Samuel Baker describes the method of hunting pursued by the Aggageers or hock-cutters in Abyssinia. A party of seven, of which he was one, had started a fine bull elephant. It at once retreated, but when the hunters again drew near, it halted and faced them.

"'Be ready, and take care of the rocks!' said Taher Sherrif, as I rode forward by his side. Hardly had he uttered these words of caution, when the bull gave a vicious jerk with its head, and with a shrill scream charged down upon us with the greatest fury. Away we all went, helter-skelter, through the dry grass which whistled in my ears, over the hidden rocks, at full gallop, with the elephant tearing after us for about a hundred and eighty yards at a tremendous pace. I was not sorry when it gave up the hunt. We now quickly united and again followed the elephant, that had once more retreated. Then came the tug of war. Taher Sherrif came close to me and said, 'You had better shoot the elephant, as we shall have great difficulty in this rocky ground.' This I declined, as I wished to end the fight as it had been commenced, with the sword; and I proposed that he should endeavor to drive the animal to more favorable ground. 'Never mind,' replied Taher; 'Inshallah (please God), he shall not beat us.' He now advised me to keep as close to him as possible, and to look sharp for a charge.

"The elephant stood facing us like a statue; it did not move a muscle beyond a quick and restless action of the eyes that were watching all sides. Taher Sherrif and his brother now separated, and, passing at opposite sides of the elephant, joined each other about twenty yards behind it. In front were two of the hock-cutters; one of them, named Roder Sherrif, rode an experienced bay mare. She advanced close up to her antagonist until within eight or nine yards of his head. For an instant not a word was spoken. Then I saw the white of one of its eyes gleam. 'Look out, Roder! he's coming.' I exclaimed. With a shrill scream the elephant dashed upon him like an avalanche. Round went the mare as though upon a pivot, and away over rocks and stones, flying like a gazelle, with the monkey-like form of little Roder Sherrif leaning forward and looking over his left shoulder as the elephant rushed after him. For a moment I thought he must be caught. Had the mare stumbled, all were lost; but she gained in the race after a few quick, bounding strides, and Roder, still looking behind him, kept his distance so close to the elephant that its outstretched trunk was within a few feet of the mare's tail.

"Taher Sherrif and his brother Ibrahim swept down like falcons in the rear. In full speed, they dexterously avoided the trees until they arrived upon open ground, when they dashed up close to the hind-quarters of the furious elephant, which, maddened with the excitement, heeded nothing but Roder and his mare, now almost within its grasp. When close to the tail of the elephant, Taher Sheriff's sword flashed from its sheath, as, grasping his trusty blade, he leaped nimbly to the ground, while Ibrahim caught the reins of his horse.

"Two or three bounds on foot, with the sword clutched in both hands, and he was close behind the elephant. A bright glance shone like lightning as the sun struck upon the descending steel. This was followed by a dull crack, as the sword cut through skin and sinews, and settled deep in the bone about twelve inches above the foot. At the next stride the elephant halted dead short in the midst of its charge.

"Taher had jumped quickly on one side, and had vaulted into the saddle with his naked sword in hand. At the same moment Roder, who had led the chase, turned sharp round, and again faced the elephant as before. Stooping quickly from the saddle, he picked up from the ground a handful of dirt, which he threw into the face of the vicious-looking animal that once more attempted to rush upon him. It was impossible; the foot was dislocated, and turned up in front like an old shoe. In an instant Taher was once more on foot, and again the sharp sword slashed the remaining leg. The great bull elephant could not move! The first cut with the sword had utterly disabled it; the second was its death-blow."

ELEPHANTS IN CAPTIVITY.

It was a very long time believed that elephants would not breed except in a wild state. It is indisputable that the domesticated elephants in India have never produced any young. The first baby elephant seen in Europe arrived in England in 1851. The mother was a wild one, caught near Cawnpore, who gave birth to a female a few months after her capture. Two wild elephants just captured were put together by Corsa. Twenty-one months afterward a young one was born. It began to suck at once, turning back its trunk and taking the teat in its mouth. It grew rapidly, and at one year old measured nearly four feet in height. The suckling period lasts for two years. But in our own country a proof has been given of the unfounded character of the old supposition. A female elephant which had for years been traveling in one of our circuses, became the mother of the Baby Elephant which most of our readers have doubtless seen.

The BABY ELEPHANT (Plate LI), born while its mother was in close quarters in Philadelphia, in the winter of 1879, disposed forever of the assumption. We saw the little creature in Brooklyn at the opening of the present season, and found it to be as playful as a kitten. Its mother, however, did not exhibit any particular affection for it. She endured her offspring, nothing more. On one occasion she struck it with her trunk so violently that it was thrown a distance of quite twenty feet. It fell heavily on the ground, which, fortunately for its limbs and ribs, was soft, and covered with tan bark and straw. When the "baby" desired to feed, the mother would lie down with a sullen grunt, and permit it to take into its mouth one of her dugs. The milk is very rich, containing a high proportion of butter and sugar. It is said that the mother-elephant cares very little for her offspring, and if kept from it for forty or fifty hours, will never after recognize it.

The English novelist, Mr. Charles Reade, holds the opinion of all American showmen, that the elephant is only to be mastered by fear. He writes: "There is a fixed opinion among men that an elephant is a good, kind creature. The opinion is fed by the proprietors of elephants, who must nurse the notion or lose their customers, and so a set tale is always ready to clear the guilty and criminate the sufferer; and this tale is greedily swallowed by the public. You will hear and read many such tales in the papers before you die. Every such tale is a lie."

He then narrates the performances of a famous performing elephant, Mademoiselle Djek. She killed several of her keepers, but seemed to be devotedly attached to the most drunken and brutal of the lot. A comrade detected this keeper walking into her with a pitchfork, her trembling like a school-boy, with her head in a corner, and the blood streaming from her sides. The spectator next took an opportunity of trying the efficacy of this mode of treatment. "I took the steel rod, and introduced myself by driving it into Djek's ribs; she took it as a matter of course, and walked like a lamb. We marched on, the best of friends. About a mile out of the town she put out her trunk, and tried to curl it round me in a caressing way. I met this overture by driving the steel into her till the blood spurted out of her. If I had not, the siren would have killed me in the course of the next five minutes.

"Whenever she relaxed her speed I drove the steel into her. When the afternoon sun smiled gloriously on us, and the poor thing felt nature stir in her heart, and began to frisk in her awful, clumsy way, pounding the great globe, I drove the steel into her. If I had not, I should not be here to relate this narrative."

The fact that a creature can be cowed by cruelty is no proof that it is not amenable to kindness. Casanova, the great dealer in wild animals, bore witness how grateful the elephants he brought from Africa were to the European keepers who treated them kindly, while they could not bear the cruel natives.

Buffon relates the following trait: "A painter wished to make a drawing of the elephant of the menagerie of Versailles in an extraordinary attitude, which was with its trunk elevated in the air, and its mouth wide open. The painter's servant, to make it remain in this attitude, kept throwing fruit into its mouth, but oftener, by pretending to do so. The elephant, as if it knew that the painter's desire of making a drawing of it was the cause of its being thus annoyed, instead of revenging itself on the servant, addressed itself to the master, and discharged at him, through his trunk, a quantity of water, with which it spoiled the paper on which he was drawing."

In the London Zoological Garden the largest African elephant broke off his tusks. The stumps grew into the cheeks, causing intense pain. The intrepid superintendent undertook to perform a surgical operation and relieve the poor beast. Having prepared a gigantic hook-shaped lancet, he bandaged the creature's eyes, and proceeded to his task. It was an anxious moment, for there was absolutely nothing to prevent the animal killing his medical attendants upon the spot, and to rely upon the common sense and good nature of a creature weighing many tons and suffering from facial abscesses and neuralgia, argues, to say the least of it, the possession of considerable nerve. But Mr. Bartlett did not hesitate, and climbing up within reach of his patient, he lanced the swollen cheek. His courage was rewarded, for the beast at once perceived that the proceedings were for his good, and submitted quietly. The next morning, when they came to operate upon the other side, the elephant turned his cheek without being bidden, and endured the second incision without a groan.

Some elephants possess a taste for music. In 1813, the musicians of Paris met together and gave a concert to the male elephant, which was then in the Jardin des Plantes. The animal showed great pleasure at hearing sung *O ma tendre Musette!* But the air of *La charmante Gabrielle* pleased it so much that it beat time by making its trunk oscillate from right to left, and by rocking its enormous body from side to side. It even uttered a few sounds more or less in harmony with those produced by the orchestra.

An elephant was used in a spectacular play in Philadelphia. He was kept in a stable several blocks away, and taken to the theatre every evening at the proper point in the piece. One afternoon he took it into his head that the time had come to perform. Throwing his keeper aside, he burst into the street, overturned a wagon and several street-stands on his way to the theatre, smashed a door, and took his usual place on the stage. The absence of lights and audience seemed to convince him that he had made a mistake, and he suffered himself to be led back to the stable.

ORDER IX.

HYRACOIDEA.

54. HYRACIDÆ . . . . Rock Rabbits.

THE ROCK RABBITS.

THE ORDER HYRACOIDEA—THE GENUS HYRAX—ITS CHARACTERISTICS.

THE order HYRACOIDEA consists of only *one* family, HYRACIDÆ, which comprises only *one* genus. Naturalists are by no means agreed as to the classification of the animals comprised in this one order, family, or genus. Pallas regarded them as Rodents. Oken considered them nearest akin to the Marsupials. Huxley, however, who is followed by Wallace, raised them to the dignity of an independent order.

In the stony mountain ranges of Africa and Western Asia, the traveler sees in many spots a lively rabbit-like creature basking on a narrow ledge of rock. As he approaches, the animal takes fright, swiftly clambers up the precipice, vanishes in one of the numerous clefts, and then, safe and curious, turns and looks down with wondering eyes on the intruder. It is one of the Rock Rabbits.

GENUS HYRAX.

The Genus HYRAX contains *ten* species. The generic characteristics are the following. The body is long, the head large, the upper-lip cleft, the eyes small, the ear short, broad, and round, and almost hidden in fur, the neck short, the tail a mere rudiment, the legs moderately high and weak, the fore-feet divided into four, the hind-feet into three toes. A soft, thick coat covers the body and limbs.

The Rock Rabbits or Damans have been known from olden times. They are the "conies" of the Bible, "a weak people which had their houses in the rocks." Moses regarded them as unclean animals. At present, in Abyssinia neither Christian nor Mussulman will taste their flesh. In Arabia, however, the Bedouins eat it readily. The Hyrax, in its movements, seems to be a link between the clumsy ruminants and the lively rodents. When running on level ground they move their limbs deliberately and regularly. When alarmed, they fly to the rocks, and there display their wonderful agility. In disposition they display gentleness and incredible timidity. They love society, and are never seen alone. They are constant to the dwelling they have once selected, and are very voracious, the plants of their alpine homes supplying them with ample food. They bite off the grass with their teeth, and then move their jaws like animals chewing the cud. It is doubtful whether they actually ruminate, as the proper ruminants do.

THE ASHKOKO.

The ASHKOKO, *Hyrax Abyssinicus*, is from nine to twelve inches long. The coat consists of smooth fine hairs, and is of a dappled fawn-gray color. It is the characteristic animal of the stony mountain ranges, where troops may be seen lying on the rocks. They often are seen near the villages, where they have learned by experience that no one attempts to hurt them. But the sight of a European or the approach of a dog sets them off in full flight. A passing crow, or a swallow skimming the air near them, has the same effect. Yet they live in harmony with much more blood-thirsty animals, as, for instance, the ichneumon. They seldom leave their craggy homes. When the grass which grows in the ledges is eaten, they descend into the valleys, but a sentinel is then always posted on some prominent cliff, and a warning voice from him produces a speedy retreat. The power these creatures possess of running up smooth and perpendicular walls of rock is wonderful. Schweinfurth asserts that they can contract their feet so as to form a cavity in the centre of the sole, and thus adhere to the flat surface by atmospheric pressure alone.

The celebrated traveler Bruce was the first to give a detailed account of the Ashkoko. "It is found," he writes, "in Ethiopia in the caverns of the rocks, or under the great stones in the Mountain of the Sun behind the Queen's Palace at Koscam. It does not burrow or make holes like the rat or rabbit, for its feet are by no means adapted for digging. Several dozens of them may be seen sitting on the great stones at the mouth of caves, warming themselves in the sun, or enjoying the freshness of the summer evening. They do not stand upright upon their feet, but steal along as in fear, their belly being very close to the ground, advancing a few steps at a time, and then pausing."

This species is found plentifully on Mount Lebanon, and is sometimes called the Syrian Hyrax. The Arabs style it " Israel's Sheep," probably from its frequenting the rocks of Horeb and Mount Sinai, where the children of Israel made their forty years' wandering. It is known in Hebrew by the name Saphan, and is the animal erroneously called in the English version of the Bible the "coney."

There is no foundation for the species *Hyrax Hudsonicus* which was described by Pennant from a specimen in a museum.

THE KLIPDAS.

The KLIPDAS, *Hyrax capensis* (Plate LII) is sometimes called the Rock Badger. Kolbe, its first discoverer, and Buffon mention it as a Marmot. Blumenbach left it with the Rodentia. It is not much larger than a hare; the make is clumsy, rather long, low on the legs, with a thick head terminated by an obtuse muzzle. The fur is uniformly grayish-brown, with the inside of the ears white, and sometimes a blackish band is found on the back. In captivity these creatures soon become tamed, and are easily attached to their keeper. They are active and cleanly, and feed exclusively on vegetable substances, and are said to prepare a kind of nest or bed of dried leaves, grasses, and the like, in the cavities in which they reside.

THE TREE DAMAN.

The TREE DAMAN, *Hyrax arboreus*, is found in many of the forests of South Africa, and dwells in the hollows of decayed trees. It rather exceeds the size of the Klipdas, but resembles it in form, and in its manner of moving and sitting. In color it is a tawny-red, mottled with black upon the back, but the fur beneath is of a uniform dull white. The reddish color arises from the tips of most of the hairs being of that hue; the black variegations depend on a scanty intermixture of long hairs which are entirely of that color. The crown of the head has a predominance of black, the sides of the head are somewhat grayish. The ears are short and roundish, and beset outside with long, dusty-whitish hair. The teeth of this species differ a little from those of the Cape *Hyrax*, more particularly the incisors. The upper ones are more pointed, and the lower ones stand in pairs. "Little is known of its manners," writes Major Smith. "Almost the only observation that can be elicited from the farmers and inhabitants of the parts of the country in which it resides is, that it makes a great noise previous to the fall of rain."

The *Hyrax* or Daman, as the whole genus is sometimes called, is one of those links in the chain of nature's works which puzzle classifiers. Cuvier, the founder of modern scientific Zoology, placed these animals with the horse and the rhinoceros in the order Pachydermata. It is certain, indeed, that they resemble in their teeth and skeleton the huge and unwieldy rhinoceros.

ORDER X.

RODENTIA.

FAMILIES.

55. MURIDÆ - - - - - - - - - Rats.
56. SPALACIDÆ - - - - - - - - Mole Rats.
57. DIPODIDÆ - - - - - - - - Jerboas.
58. MYOXIDÆ - - - - - - - - Dormice.
59. SACCOMYIDÆ - - - - - - - Pouched Rats.
60. CASTORIDÆ - - - - - - - Beavers.
61. SCIURIDÆ - - - - - - - - Squirrels.
62. HAPLOODONTIDÆ - - - - - Sewellels.
63. CHINCHILLIDÆ - - - - - - Chinchillas.
64. OCTODONTIDÆ - - - - - - Octodons.
65. ECHIMYIDÆ - - - - - - - Spiny Rats.
66. CERCOLABIDÆ - - - - - - Tree Porcupines.
67. HYSTRICIDÆ - - - - - - - Porcupines.
68. CAVIIDÆ - - - - - - - - Cavies.
69. LAGOMYIDÆ - - - - - - - Pikas.
70. LEPORIDÆ - - - - - - - - Hares.

CHAPTER I.

RATS AND MICE.

THE ORDER RODENTIA—THE FAMILY MURIDÆ—RATS AND MICE—THE BLACK RAT—THE BROWN RAT—THE MOUSE—THE HARVEST MOUSE—THE BARBARY MOUSE—THE HAMSTER—THE MUSK RAT—THE WATER RAT—THE FIELD MOUSE—WILSON'S MEADOW MOUSE—LE CONTE'S MOUSE—THE COTTON RAT—THE LEMMING.

THE RODENTIA or Gnawers (from the Latin *rodere* "to gnaw"), constitute a well-defined order characterized by the possession of two long, curved, sharp-edged, rootless incisors in each jaw. The order contains a very considerable number of subdivisions, namely, *sixteen* families, some of which comprise over two hundred species. It may be inferred from this fact that the animals of the order differ very widely in all other respects than that which has given them their name. The gnawing teeth are necessarily placed in strong heavy jaws which are large in proportion to the head, for they not only require a stout support, but room for continual development. These teeth are worn continually by the friction they undergo, and necessitate a provision for their continual renewal; the base passes deeply into the jawbone, where a pulpy core supplies the material for constant growth. As those of the upper jaw meet those of the lower jaw at the tips, they are perpetually worn away by their action upon each other, and upon the hard food which they are formed for nibbling. The growth at the base and the wearing away of the tips thus balance each other. If one of the incisors is lost, the opposing one in the other jaw being no longer worn, grows till it projects like the tusk of an elephant. These incisors are chisel-shaped, and are provided in front with a layer of enamel hard as the hardest steel, while behind they are composed of a much softer material called ivory. This latter portion wears away much more rapidly than the front, and thus a sharp-cutting edge is preserved.

The RODENTIA are widely spread over the globe, and comprise nearly one-half of the mammalia. From the equator to the coldest latitudes they tenant rocks and mountains, plains and woods, and often devastate the cultivated domains of man.

RATS AND MICE.

The family MURIDÆ contains no less than *thirty-seven* genera, and *three hundred and thirty* species. The characteristics of the family may be summed up as follows: a pointed muzzle, a cleft upper-lip, round, deep-black eyes, and a long, naked tail covered with square scales, and bearing only a few stiff hairs. The fore-feet have four toes, the hinder ones five toes. The family is divided in common parlance into Rats and Mice. The former are stout and repulsive in appearance, the latter slender and graceful. The former are recent immigrants into Europe, the latter have, from the earliest periods, haunted the house.

GENUS MUS.

This genus is divided by some naturalists into *one hundred*, by others into *one hundred and twenty* species. We can find space to mention only the most marked, or the best known forms. Not a single indigenous species of the genus is found in either North or South America.

The BLACK RAT, *Mus rattus* (Plate LIII), is the Rat of fable. It was the Black Rat which, according to tradition, devoured the wicked Bishop Hatto, in his tower in the Rhine, which is still called the Mouse-tower, and which wrought such havoc in Hamelin till the charm of the Pied-piper freed the city.

In all probability the Black Rat came from Persia, where they still abound. In Europe, however, since the beginning of the last century, it has been nearly exterminated by the Brown Rat.

The BROWN RAT, *Mus decumanus* (Plate LIII), is larger than the Black Rat, and is of a brownish-gray color. It seems to have come from Central Asia. Pallas, in 1727, describes it as migrating from the vicinity of the Caspian, in consequence of an earthquake. It crossed the Volga in swarms, and proceeded steadily westward. In 1732 the same rat was brought by ship from the East Indies to England. It appeared in East

COMMON MICE
BARBARY MOUSE
COMMON RATS
HARVEST MICE
WATER RAT

PLATE LIII RODENTIA.

Prussia in 1750; in Paris in 1753; but did not reach Denmark and Switzerland till 1809. In 1760, it is said that the town of Jaik in Siberia was taken by storm by an innumerable army of rats. In 1755 the Brown Rat arrived in North America.

The Brown Rat is a fierce animal, and possesses great power of combination. A number of them have been known to kill a cat, and a large body would make short work of a man. It is exceedingly voracious, and has no scruple about devouring its companions if they are sick or wounded. They will enter a stable and nibble the horn of the horses' hoofs; they will creep among sleeping dogs and gnaw their feet. They have been known to attack children in the cradle. Yet they are not without their uses; they act as scavengers, and consume an immense quantity of refuse which otherwise would breed disease. They are cleanly in their personal habits, and spend much time in washing themselves. They have repeatedly been tamed by prisoners who have made their acquaintance in solitary confinement. They are subject to a very peculiar disease. The tails of several rats grow together and form what is called in Germany a "King-rat." In 1822, at Döllstedt, two miles from Gotha, two King-rats were captured; one of these groups consisted of twenty-eight, the other of fourteen rats; they were over a year old, and seemed in good health, but very hungry, as of course they had to live on the alms of other rats.

The MOUSE, *Mus musculus* (Plate LIII) is a pretty little creature with black bead-like eyes, velvety fur, and squirrel-like paws. Its back is a brownish-gray, its throat and abdomen a light-gray color. They are odd little creatures, full of curiosity, and easily tamed. White Mice (Plate LIII) are not uncommon; they are true albinos, and have red eyes. The common mouse is much more docile than the white variety. Mice make their nests in any quiet spot, and from any soft substance, and, like the rats, multiply rapidly.

The HARVEST MOUSE, *Mus minutus* (Plate LIII), is about five inches in length, including the tail. The abdomen is white, the back reddish-brown. It builds a pretty nest, which is thus described: "It was built upon a scaffolding of four of the rank grass-stems that are generally found on the sides of ditches, and was situated at some ten or eleven inches from the ground. In form it was globular, rather larger than a cricket-ball, and was quite empty, having probably been hardly completed when the remorseless scythe struck down the scaffolding and wasted all the elaborate labor of the poor little architect. The material of which it was composed was thin dry grass of nearly uniform substance, and its texture was remarkably loose, so that any object contained in it could be seen through the interstices as easily as if it had been placed in a lady's open-worked knitting basket. There was no vestige of aperture in any part of it, so that the method by which it was constructed seems quite enigmatical." This tiny creature is insectivorous to no small degree, and has a prehensile tail.

The BARBARY MOUSE, *Mus Barbarus* (Plate LIII), has a fur decorated with bold and elegant markings. It is larger than the common mouse, and lives in North and Central Africa. It is frequent in the Atlas range, and on the whole coast of Algeria. It burrows in the earth, and stores up provisions for the cold or wet season.

GENUS CRICETUS.

The *nine* species of this genus are found in Northern Asia and Europe, and in Egypt. They have thick bodies, short tails and legs. They build subterranean dwellings, where they accumulate food, and live.

The HAMSTER, *Cricetus frumentarius* (Plate LIV), is the best known member of the genus. It is of a grayish-fawn color on the back, deepening into black on the abdomen, and softening to yellow on the head. It has two large cheek pouches in which it carries off its plunder. It makes a very complicated burrow with two entrances, one perpendicular, the other sloping. It is torpid during a portion of the winter. It is an exceedingly destructive animal: a single one has been found with sixty pounds of corn in its burrow, and another with a hundredweight of beans. It is also very prolific, the female having several broods each year, with about ten in each brood. As soon as the young Hamsters are able to shift for themselves, an event which occurs in a wonderfully short time, they leave the maternal home, and dig separate burrows.

GENUS NEOTOMA.

The distinguishing marks by which this genus is separated from the *Arvicola* consist in the enamel of the teeth, which is so developed that a worn-down tooth of an old specimen exhibits insulated circles of enamel on the grinding surface. Of the *six* species, we describe two only.

The FLORIDA RAT, *Neotoma Floridiana*, has a light and agile form, bearing some resemblance to that of the squirrel; the eyes are large, the ears thin and prominent; the legs robust. The body and head are lead color, intermixed with yellowish black hair. The border of the abdomen is buff, the under surface of the body white. An adult specimen measures eight inches from nose to root of tail.

The Rat is found in all the States, but exhibits different habits in different circumstances. In Florida it burrows under stones and in rivers, in Georgia it prefers the woods. Sometimes it makes its nest in the fork of a tree; sometimes among vines near a sluggish stream. It forms, by piling up dry sticks, grasses, mud, and leaves, a structure impervious to rain, and inaccessible to the wild cat or racoon. The nests they build in trees are often of enormous size. Audubon saw some in Georgia ten to twenty feet from the ground which appeared larger than a cart-wheel, and contained a mass of leaves and sticks that would have more than filled a barrel. On the Missouri River, this rat lives in hollow trees; in the Rocky Mountains it nestles in clefts of the rock. In fact, it makes itself at home everywhere.

The ROCKY MOUNTAIN RAT, *Neotoma Drummondii*, is larger than the Florida Rat, and much more destructive. It gnaws everything left in its way, and is hated by the fur-traders, for it will cut their blankets to pieces while they are asleep, and gnaw through whole packs of furs in a single night.

It is very common near the Columbia River, in Oregon, and along the sides of the Rocky Mountains.

GENUS REITHRODON.

LE CONTE'S MOUSE, *Reithrodon Le Contei*, is about half the size of a full-grown mouse. The upper fore-teeth are deeply grooved, and yellow in color. The ears round, moderate in size, and nearly naked. The nails on the feet are long, but slightly hooked, and adapted to digging. The specimen described by Audubon came from the State of Georgia.

GENUS SIGMODON.

This genus was formed by Messrs. Say and Ord, and comprises *two* species which were previously placed among the *Arvicola*. The name of the genus is significative of its generic characteristic, namely, that the molar of the upper jaw has the form of the Greek letter S or Σ.

The COTTON RAT, *Sigmodon Hispidum*, is the most common wood-rat in the Southern States. It occasions, however, very little injury, as it prefers deserted old fields to cultivated grounds. It feeds on coarse grasses, and devours a considerable quantity of animal food, for it is decidedly carnivorous. Robins, partridges, or other birds that are wounded and drop among the long grass, are speedily devoured by them. They have been seen running about with crayfish in their mouths, and are especially fond of "fiddlers." They fight fiercely, so that it is almost impossible to keep more than one in a cage, for the strongest usually eats all the others. This species delights in sucking eggs. They come out at night from their holes which often run a distance of twenty or thirty yards under-ground, but so near the surface that they can be traced by the ridge in the ground. Each burrow contains only one family, but the various galleries often intersect. The Cotton Rat has obtained its name from the supposition that it makes its nest of cotton.

GENUS FIBER.

The *two* species of this genus inhabit the North American continent as far south as Mexico. They form a link between the Beavers and the Water Rats.

The MUSK RAT, *Fiber zibeticus*, may be described as a large water-rat with a short snout and a long tail. Near the tail is a gland which secretes a white oily fluid of a strong civet odor. The hind-feet are webbed; the incisor teeth are yellow, but the nails white. Audubon gives a very good account of its habits, which we quote:

"Musk Rats are very lively, playful animals when in their proper element, the water; and many of them may be occasionally seen disporting themselves on a calm night in some mill-pond or deep sequestered pool, crossing and recrossing in every direction, leaving long ripples in the water behind them, while others stand for a few moments on little hurdles or tufts of grass, or on stones or logs, on which they can get a footing above the water, or on the banks of the pond, and then plunge one after the other into the water. At times one is seen lying perfectly still on the surface of the pond or stream, with its body widely spread out, and as flat as can be. Suddenly it gives the water a smart slap with its tail, somewhat in the manner of the beaver, and disappears beneath the surface instantaneously, going down head foremost, and reminding one of the quickness and ease with which some species of ducks and grebes dive when shot at.

"At the distance of ten or twenty yards, the Musk Rat comes to the surface again, and perhaps joins its companions in their sports; at the same time others are feeding on the grassy banks, dragging off the roots of various kinds of plants, or digging underneath the edge of the bank. These animals seem to form a little community of social, playful creatures, who only require to be unmolested in order to be happy."

GENUS HESPEROMYS.

This genus, with its *ninety* species, replaces on the American continent the genus *Mus*.

The NEW JERSEY FIELD MOUSE, *Hesperomys campestris*, measures a little more than three inches in length. The fur is a leaden black, the head is large, the ears large, oval, blunt, and thinly covered with closely adpressed hair. The legs and feet are brown. The tail well covered with long hair.

The SONORA FIELD MOUSE, *Hesperomys Sonoriensis*, is slightly larger than the New Jersey species. The hair is slate color, mixed with brownish-gray; beneath, whitish, except on the throat. The head is elongated, the ears large, oval, and hairy. The feet are covered with short, whitish-brown hair. The tail of moderate length.

The TEXAN FIELD MOUSE, *Hesperomys Texana*, is smaller than the above described species, only measuring two inches in length. The head is large and blunt, the eyes prominent and dark-brown; ears large, erect, roundish, oval, blunt, sparsely covered outwardly with short, close-lying brown hairs, inwardly with gray. The hind-feet are furred, with the exception of the sole. The whiskers are long. The fur has rather a mottled appearance; beneath white, inclining to yellowish. The two colors are distinctly separated in a straight line. The feet are white, the hairs projecting over the nails. Its habitat is Western Texas.

GENUS ARVICOLA.

The *fifty* species of Arvicola are distributed over Europe, Northern Asia, and temperate North America.

The WATER RAT, *Arvicola amphibius* (Plate LIII), is found from the Atlantic to the Sea of Ochotsk, and from the White to the Mediterranean Sea, in plains, as well as in mountains.

The color of the Water Rat is a chestnut-brown, dashed with gray on the upper parts, and fading to gray below. The ears are so short that they are hardly perceptible above the fur. The incisor teeth are of a light yellow, and are very thick and strong. The tail is shorter than that of the common rat, hardly exceeding half the length of the head and body. The average length of a full-grown specimen is thirteen inches, the tail being about four inches and three-quarters long. It is not so prolific an animal as the Brown Rat, breeding only twice in the year, and producing from five to six young at a birth.

It is falsely accused of destroying fish; but it does immense harm by the tunnels it forms in embankments. Its food is entirely aquatic plants and roots.

The CAMPAGNOL, *Arvicola arvalis*, or Common Field Mouse of Europe, is a great foe to the European farmer. It eats his seed-corn, it plunders the ripening corn, it devours the contents of his ricks and barns. Its color is a ruddy brown, with gray on the chest. The tail is only one-third the length of the body.

WILSON'S MEADOW MOUSE, *Arvicola Pennsylvanica*, is the common Meadow Mouse of the Northern and Eastern States. In every meadow you may trace its small tortuous paths cut through the grass, leading to the root of a stump, or the borders of a ditch. If you dig up the nest, you find a family of five or ten. Their galleries do not run underground, but extend along the surface. The food of this species consists principally of roots and grasses; it will eat also bulbs, such as meadow-garlic, and red-lily. In very severe winters, when the ground is frozen, it sometimes attacks the bark of shrubs and fruit-trees, peeling it off, and destroying the tree. But Audubon thinks it does little harm to the farmer. It swims and dives well, but never visits outhouses or dwellings.

It is found in all the New England States, and is abundant in New York. It is common near Philadelphia, and has been traced southward to the northern boundary of North Carolina, and northward to Hudson's Bay.

GENUS MYODES.

The *four* species of the genus are compactly built, short-tailed animals. The head is large, the under lip deeply cleft, the ear small, and all the feet have five toes.

The LEMMING, *Myodes lemmus* (Plate LIV), is the most mysterious animal in all Scandinavia.

At uncertain and distant intervals of time, many of the northern parts of Europe, such as Lapland, Norway, and Sweden, are subjected to a strange invasion. Hundreds of little, dark, mouse-like animals sweep over the land, like clouds of locusts suddenly changed into quadrupeds, coming from some unknown home, and going no one knows whither. These creatures are the LEMMINGS, and their sudden appearances are so entirely mysterious, that the Norwegians look upon them as having been rained from the clouds upon the earth.

The Lemming feeds upon various vegetable substances, such as grass, reeds, and lichens, being often forced to seek the last-named plant beneath the snow, and to make occasional air-shafts to the surface. Even when engaged in their ordinary pursuits, and not excited by the migratorial instinct, they are obstinately savage creatures. Mr. Metcalfe describes them as swarming in the forest, sitting two or three on every stump, and biting the dogs' noses as they came to investigate the character of the irritable little animals. If they happened to be in a pathway, they would not turn aside to permit a passenger to move by them, but boldly disputed the right of way, and uttered defiance in little sharp, squeaking barks.

The color of the Lemming is dark brownish-black, mixed irregularly with a tawny hue upon the back, and fading into yellowish-white upon the abdomen. Its length is not quite six inches, the tail being only half an inch long.

CHAPTER II.

MOLE RATS, POUCHED RATS, AND BEAVERS.

THE MOLE RAT—THE JERBOA—THE ALACTAGA—THE CAPE LEAPING HARE—THE HUDSON BAY JUMPING MOUSE—THE FAT DORMOUSE—THE COMMON DORMOUSE—THE POUCHED RATS—THE BEAVERS—THE AMERICAN BEAVER—THE EUROPEAN BEAVER.

THE family Spalacidæ or Mole Rats have a straggling distribution over the continents of the Old World. They are misshaped, subterranean rodents, with all the unpleasant characteristics of the mole. They inhabit dry sandy plains, and make burrows which run for long distances. They do not form societies, but live solitary in their dens; they avoid the light, and seldom come aboveground; they dig with great rapidity, often in a perpendicular direction. The family comprises seven genera.

GENUS SPALAX.

The Mole Rat. *Spalax typhlus* (Plate LIV), the *only* species, is known also by its Russian name, Slapush. The head is pointed and large, the neck short, immovable, and as thick as the tailless body; the legs are short, with large paws armed with strong claws. The eyes are two small, round black specks, which lie under the fur-covered skin; so that even if they were susceptible to light, the brightest rays of the sun could not reach them. The ears are very large, and the animal's powers of hearing are very acute. The fur is thick, close, and smooth, but some stiff bristles grow on the cheeks. The color is generally a yellowish-brown, the chin and paws being, however, of a dirty-white hue. The total length is about ten inches.

The Mole Rat abounds in the country adjacent to the Don and the Volga, and is especially numerous in the Ukraine. It frequents cultivated and fertile districts, where countless heaps mark its subterranean

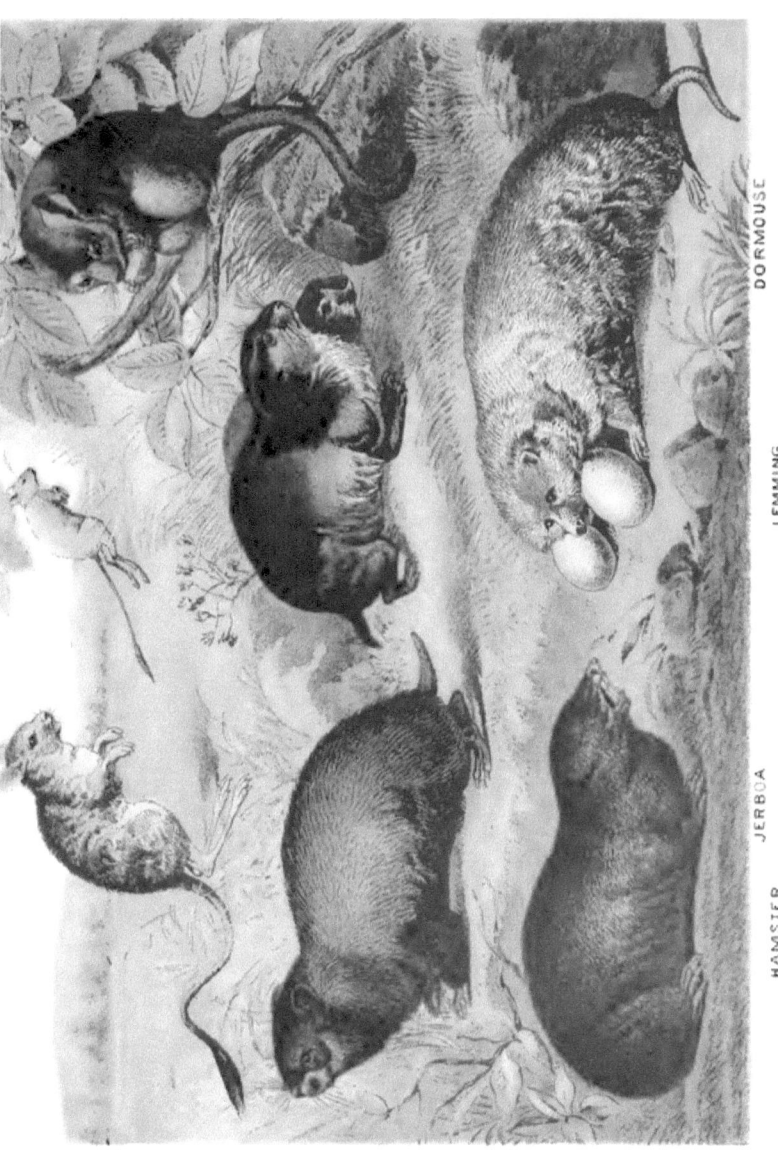

HAMSTER JERBOA LEMMING DORMOUSE
MOLE RAT FUR COUNTRY POUCHED RAT
PLATE LIV RODENTIA

chambers and passages. The tunnels formed by this burrower are very complicated, and can be traced through moist and sloppy marshes, as well as on the steep hill-sides. The peasants have a notion that if any one will seize a Slapush in his bare hands, permit the animal to bite him, and then squeeze it to death, he will have the power of curing goitre by touch.

The family DIPODIDÆ, containing *three* genera, consists of the Jumping Mice or JERBOAS. They are found in Arabia, Egypt, and Abyssinia, but extend to the Caspian Sea and eastward to India, as well as over a large part of Africa. Isolated forms occur in North America.

GENUS DIPUS.

The JERBOA, *Dipus Ægyptus* (Plate LIV), is hardly larger than an ordinary rat. The color of its fur is a light dun, washed with yellow, the abdomen being nearly white. The tail is of very great length, is cylindrical in shape, and tufted at its extremity with stiff black hairs. It is abundant in North Africa, and lives in society, forming large warrens in the regions it inhabits. The Jerboa is admirably adapted for digging in hard and burning ground, as its feet are guarded by a thick covering of stiff, bristly hair. The body is about six, the tail eight inches in length.

The ALACTAGA, *Dipus alactaga*, or the Jumping Rabbit of Siberia, has a body about seven inches, and a tail ten inches in length. The head is pretty, with lively, prominent eyes, long narrow ears—as long as the head itself, and eight rows of long black hairs on each upper lip. The color of the fur is reddish yellow. On the approach of cold weather, it falls asleep, after having carefully stopped up the entrances to its burrow, and remains torpid from September to the end of April.

GENUS PEDETES.

The CAPE LEAPING HARE, *Pedetes capensis*, is a solitary and isolated species found in South Africa from the Cape of Good Hope to Mozambique and Angola. The Dutch Boers have named it the SPRING HAAS.

With the exception of shorter ears, and the elongated hinder limbs, the Spring Haas is not unlike our common hare. The fur is of a dark fawn, or reddish-brown, perceptibly tinged with yellow on the upper parts, and fading into grayish-white beneath. In texture it is very similar to that of the hare. The tail is about as long as the body, and is heavily covered with rather stiff hairs, which at the extremity are of a deep black hue.

GENUS JACULUS.

The *single* species is found from Nova Scotia and Canada, southward to Pennsylvania, and westward to California and British Columbia.

The JUMPING MOUSE, *Jaculus Hudsonius*, has the general form of its Old World kindred, but its tail is more mouse-like. This tail is thin, round, and very long, tapering down to a fine point, and sparsely sprinkled with short hairs. The fur is a dark liver color, with a blackish tinge on the flanks. The body measures about three inches, the tail about five inches in length. It lives near the forests, remains concealed by day, but comes out in bands by night. Its burrows are about twenty inches below the surface, but in cold weather they are sunk still lower. At the beginning of winter the creature forms a hollow ball of clay, rolls itself up in it, and lies in complete torpidity till springs returns. In summer it is very active, and hops around on its hind legs. Audubon thinks that there is no animal so agile. He had two young ones in captivity, and placed a foot deep of earth in the cage; the mice made a burrow in it, and kept themselves closely hidden as long as it was light, and concealed by night whatever was placed in the cage.

The family MYOXIDÆ is a small one containing only a *single* genus of *twelve* species. In form and habit the Dormice approximate to the squirrels; the head is narrow, the eyes pretty large, the ears large and bare; the fur is rich and soft, the tail is bushy, and the hairs on it grow in a double row. These small rodents are found over all the Northern regions of the Old World, and over most parts of Africa.

GENUS MYOXUS.

The FAT DORMOUSE, *Myoxus glis*, is found in Southern and Eastern Europe. They live in the forests, concealing themselves by day in hollow trees, or even in the deserted burrows of the hamster, but coming out at nightfall to feed. When autumn approaches, they store up their winter supplies; they make in some deep hole a nest of soft moss, roll themselves up with several of their companions, and long before the cold weather comes, fall asleep.

The Fat Dormouse is six inches long in the body, with a tail five inches long. Its soft, thick fur is of a uniform ashy-gray, the tail is bushy, the hairs falling in two rows, of a brownish-gray color, with a white stripe on its under surface. It is the smallest of all hibernating animals.

The COMMON DORMOUSE, *Myoxus avellanus*, or *Muscardinus avellanarius* (Plate LIV), is found all through Central Europe from Sweden and England to the borders of Turkey.

The total length of this pretty little animal is rather more than five inches, the tail being two inches and a half long. The color of its fur is a light reddish-brown upon the back, yellowish-white upon the abdomen, and white on the throat. The tail is thickly covered with hair, which is arranged in a double row throughout its length, and forms a slight tuft at the extremity. The head is rather large in proportion to the body, the ears are large and broad, and the eye full, black, and slightly prominent.

Like the rest of its tribe, it lays in a stock of provisions in the fall, and becomes exceedingly fat toward the end of autumn. This hoard is not gathered into the nest, but is hidden away close to the spot where the nest is placed. Comparatively little of the store is eaten during the winter, but it is of very great service in the earlier part of the spring, when the Dormouse is awake and lively, and there are as yet no fresh fruits on which it could feed.

THE POUCHED RATS.

The family SACCOMYIDÆ or Pouched Rats, are almost wholly confined to the Rocky Mountains and the elevated plains of Central North America. It is divided into *eight* genera and *thirty-three* species. They are remarkable for the possession of long and deep cheek-pouches opening from without, and lined with short hair. The mouth is, as it were, double, the outer portion being very wide and hairy, behind which is the small inner mouth.

GENUS DIPODOMYS.

The *five* species of this genus are found in Mexico, California, up to the Columbia River, with *one* specimen in South Carolina. They have slender bodies, long hind-legs, and a long tail with a brush at the extremity.

The POUCHED JERBOA, *Dipodomys Philippi*, is a native of California, where it lives in the most desert-like districts. It is probable that it can live without water for a long time, or is satisfied with the dewdrops which form on some plants. Audubon writes that in Western Mexico and California they were so tame that they could have been caught by the hand without difficulty. "This species hops about kangaroo-fashion, and jumps pretty far at a leap. Where the men encamped they came smelling about the legs of the mules, as if they were old friends. These animals appear to prefer the sides of stony hills, which afford them secure places to hide in, and they can easily convey their food in their cheek-pouches to their nests."

GENUS GEOMYS.

The FUR COUNTRY POUCHED RAT or MULO, *Geomys bursarius* (Plate LIV), is the most remarkable of the *five* species of the genus.

It is of a pale-gray color, but the exterior of the pouches, the abdomen, and the tail are covered with white hair. The teeth are yellow, and are marked with deep longitudinal grooves. The central claw of the fore feet is very strong, and measures nearly half an inch in length.

It is rather gregarious in its habits, associating together in moderately large bands, and undermining the ground in all directions. Its burrow is not very deep, but runs for a considerable distance in a horizontal direction, and along its course occasional hillocks are thrown up, by means of which it may be traced from the surface. A more detailed account is found in the following extract from Audubon:

"Having observed some freshly thrown up mounds in M. Choeteau's garden, several servants were called and set to work to dig out the animals, if practicable, alive; and we soon dug up several galleries worked by the Muloes, in different directions.

"One of the main galleries was about a foot beneath the surface of the ground, except when it passed under the walks, in which places it was sunk rather lower. We turned up this entire gallery, which led across a large garden-bed and two walks into another bed, where we discovered that several fine plants had been killed by these animals eating off their roots just beneath the surface of the ground. The burrow ended near these plants under a large rose-bush. We then dug out another principal burrow, but its terminus was among the roots of a large peach-tree, some of the bark of which had been eaten off by these animals. We could not capture any of them at the time, owing to the ramification of their galleries having escaped our notice while following the main burrows. On carefully examining the ground, we discovered that several galleries existed that appeared to run entirely out of the garden into the open fields and woods beyond, so that we were obliged to give up the chase."

The SALAMANDER, *Geomys pinetus*, is also called the Southern Goffer or Georgian Hamster. It does not remain underground during the winter, but continues its digging throughout the year. It frequents the high pine barren regions from the middle of Georgia and Alabama to the south point of Florida.

We have refrained from using the term "Goffer" in describing these pouched rats. Goffer in the Southern States is often applied to a land tortoise, and even in the North is used with great laxity for describing various kinds of both rats and squirrels.

THE BEAVERS.

The family CASTORIDÆ consists of only *one* genus—CASTOR. The animals embraced in the genus have from the earliest times attracted the attention of observers. Aristotle classes them with the Sea-otters. Pliny speaks of the tenacity of the Beaver's bite, of its felling trees, and of its possessing a fish-like tail. Solinus affirms that Beavers are found only in the waters of the Black Sea. Olaus Magnus adds that they are found in the Rhine, the Danube, and many rivers of Scandinavia. It seems, from these accounts, that the Beaver was once widely diffused, but has been ruthlessly exterminated in most parts of the Old World. Yet it exists in considerable numbers in Central and Northern Siberia, and in spite of continuous persecution still is found in the northern districts of America.

GENUS CASTOR.

The genus CASTOR comprises *two* species: the American, ranging over the whole of North America from Labrador to New Mexico, and the European, which appears to be confined to the temperate regions of Europe and Asia from France to the Amoor, over which extensive region it doubtless ranged in prehistoric times.

THE AMERICAN BEAVER.

The AMERICAN BEAVER, *Castor Canadensis* (Plate LV.), is distinguished from its European congener by its more arched profile, its narrower head, and its darker coat. It is a social animal, and displays a singular mixture of reason and instinct. The societies in which it lives vary considerably in numbers; they dwell near small clear rivers and creeks, or close to large springs, rarely taking up their abode on the banks of lakes. In such situations it executes works which would do credit to any engineer. The object of these works is to keep the water always above the level of the entrance-tunnel which leads to its house. The dams constructed for this purpose are formed of tree-branches, mud, and stones; they are ten or twelve feet thick at the bottom, and about two feet at the top; in length they vary, some attaining the dimensions of two or three hundred feet. When the stream to be stopped is gentle, the dam is carried straight from bank to bank; when it is impetuous, the dam is curved upward against the stream. The logs forming the dam are laid horizontally, and kept down by stones; they are about three feet in length, and vary extremely in thickness. Generally, they are about six or seven inches in diameter, but they have been known to measure no less than eighteen inches in diameter. An almost incredible number of these logs are required for the completion of one dam, as may be supposed from the fact that a single dam will sometimes be three hundred yards in length, ten or twelve feet thick at the bottom, and of a height varying according to the depth of water. Before employing the logs in this structure, the Beavers take care to separate the bark, which they carry away, and lay up for a winter store of food.

Near the dams are built the Beaver-houses, or "lodges," as they are termed; edifices as remarkable in their way as that which has just been mentioned. They are entirely composed of branches, moss, and mud, and will accommodate five or six Beavers together. The form of an ordinarily sized Beaver's lodge is circular, and its cavity is about seven feet in diameter by three feet in height. The walls of this structure are extremely thick, so that the external measurement of the same lodges will be fifteen or twenty feet in diameter, and seven or eight feet in height. The roofs are all finished off with a thin layer of mud, laid on with marvelous smoothness, and carefully renewed every year. As this compost of mud, moss, and branches is congealed into a solid mass by the severe frost of a North American winter, it forms a very sufficient defence against the attacks of the Beaver's great enemy, the wolverene, and cannot readily be broken through, even with the help of iron tools. The precise manner in which the Beavers perform their various tasks is not easy to discern, as the animals work only in the dark. The females, however, it is known, are the architects and builders, the males merely carriers and laborers.

Around the lodges the Beavers excavate a rather large ditch, too deep to be entirely frozen, and into this ditch the various lodges open, so that the inhabitants can pass in or out without hindrance. This precaution is the more necessary, as they are poor pedestrians, and never travel by land as long as they can swim by water. Each lodge is inhabited by a small number of Beavers, whose beds are arranged against the wall, each bed being separate, and the centre of the chamber being left unoccupied.

In order to secure a store of winter food, the Beavers take a vast number of small logs, and carefully fasten them under water in the close vicinity of their lodges. When a Beaver feels hungry, he dives to the store-heap, drags out a suitable log, carries it to a sheltered and dry spot, nibbles the bark away, and then either permits the stripped log to float down the stream, or applies it to the dam.

The Beaver does not possess a pleasing appearance. Its thickset shape, its large head, small eyes, cloven upper lip, which shows its powerful incisors, its long and wide tail, flattened like a spatula and covered with scales, combine to give it an awkward appearance. Its hind feet are larger than the fore, and are fully webbed. Owing to the deep separation of the fingers, and the existence of certain fleshy tubercles placed on the lower face of the extremities, they fulfil to some extent the functions of thumbs, those in front more especially. The muzzle is prolonged a little way beyond the jaws, and the nostrils are remarkably mobile. The ears are also movable; they do not show much, and the animal has the faculty of placing them close to its head when it dives, so as to prevent water entering the auditory passage. Its coat is well adapted to the requirements of an aquatic life, and is composed of a fine thick woolly substance, which lies close upon the skin, and is impervious to water.

The color of the long shining hairs which cover the back of the Beaver is a light chestnut, and the fine wool that lies next to the skin is a soft grayish-brown. The total length of the animal is about three feet and a half; the flat, paddle-shaped, scale-covered tail being about one foot in length. The flesh of the Beaver is eaten by the trappers, who compare it to flabby pork. The tail is something like beef marrow, when properly cooked, but it is too rich and oily to suit the taste of most persons. The female Beaver produces about three or four young at a litter, and the little creatures are born with open eyes.

The Beaver secretes a curious odoriferous substance called *castoreum* or "barkstone." According to Audubon, "If two Beaver lodges are tolerably near each other, the inhabitants of the one lodge, which we will call lodge A, go to a little distance for the purpose of ridding themselves of the superabundant castoreum. The Beavers of lodge B, smelling the castoreum, go to the same spot, and cover the odoriferous substance with a thick layer of earth and leaves. They then place their own castoreum upon the heap, and return home. The inhabitants of lodge A then go through precisely the same process, until they have raised a mound some four or five feet in height."

THE EUROPEAN BEAVER.

The EUROPEAN BEAVER, *Castor fiber*, lives usually under conditions which do not require it to build a house. Hence in France it has abandoned a communal life, and takes refuge in the rocky crevices that overhang streams. In place of being a builder, it has become a miner. On the Rhine the Beavers frequent uninhabited islands; their burrows communicate with the water by a long gallery opening beneath the surface. These excavations are sometimes of considerable size, one having been seen that measured fifty feet in length, and was divided into several compartments.

Colonies of Constructive Beavers are yet to be found in Europe. This fact was noted in 1787 by a German observer, not far from Magdeburg, on one of the affluents of the Elbe. A number had collected in this place, and had built huts in every respect similar to those of American Beavers. Such colonies are, as may be imagined, excessively rare, and excite the greatest amount of curiosity.

An attempt to rear the Beaver in a domestic state was undertaken by Exinger, of Vienna, on the banks of a large pond situated in the vicinity of Modin, Poland. The Beavers belonged to those which burrow in the

BEAVERS

PLATE LV RODENTIA

ground. This observer was able to study them for six years. They were very timid, and scarcely ever left their retreat until nightfall. At the approach of winter, Exinger had the willow and poplar trees cut down, and laid them on the bank of the pond, the trunks in the water. In the first cold days the Beavers dragged these trees to the bottom of the pond, and ranged them side by side, weaving them in such a manner as to form a solid and resisting mass. When the winter was prolonged, Exinger broke the ice and introduced some fresh trunks of trees, so as to furnish an additional supply of provisions for the prisoners.

Dr. Sace, in alluding to this example, remarks that there is here an excellent means of utilizing the immense marshes of the East and North of Europe, in favoring the settlement and multiplication of Beavers. In America the Beaver, in settled districts, is looked on as a nuisance, by backing the water till it overflowed arable land.

In 1868 some American Beavers were sent to the Jardin des Plantes in Paris. They were placed in a large wooden box, with a door opening upon a pond. They at once set to work to improve their home, and protect themselves from the weather. For this purpose they turned up the turf of their little yard and carried it on to the roof of their hut. In a word, they executed a special work which was not in accordance with their habits. They took great pains to keep the floor of their house scrupulously clean.

A Beaver was brought to England in the year 1825, very young, very small, and very woolly. The building instinct showed itself early. Before it had been a week in its new quarters, as soon as it was let out of its cage, and materials placed in its way, it went to work. It had soon learned to know its name, and when called by it—"Binny"—it answered with a low plaintive cry, and came to its owner. Its strength, when it was half grown, was great. It would drag along a large sweeping-brush, or a warming-pan, grasping the handle with its teeth, so that it came over its shoulders, and advancing with the load in an oblique direction till it arrived at the point where it wished to place it.

The long and large materials were always taken first, and two of the longest were generally laid crosswise, with one of the ends of each touching the wall, and the other ends projecting out into the room. The area formed by the crossed brushes and the wall he would fill up with hand-brushes, rush-baskets, books, boots, sticks, clothes, dried turf, or anything portable. As the work grew high, he supported himself on his tail, which propped him up admirably; and he would often, after laying on one of his building materials, sit up over against it, appearing to consider his work, or, as the country people say, "judge it." After this pause he would sometimes change the position of the material "judged," and sometimes he would leave it in its place.

After he had piled up his materials in one part of the room (for he generally chose the same place), he proceeded to wall up the space between the feet of a chest of drawers which stood at a little distance from it, high enough on its legs to make the bottom a roof for him; using for this purpose dried turf and sticks, which he laid very even, and filling up the interstices with bits of coal, hay, cloth, or anything he could pick up. This last place he seemed to appropriate for his dwelling; the former work seemed to be intended for a dam. When he had walled up the space between the feet of the chest of drawers, he proceeded to carry in sticks, clothes, hay, cotton wool, etc., and to make a nest. When he had done this to his satisfaction, he would sit up under the drawers and comb himself with the nails of his hind feet.

Binny generally carried small and light articles between his right fore leg and his chin, walking on the other three legs; and large masses, which he could not grasp readily with his teeth, he pushed forward, leaning against them with his right fore paw and his chin. He never carried anything on his tail, which he liked to dip in water, but he was not fond of plunging in the whole of his body. If his tail was kept moist, he never cared to drink; but if it was kept dry it became hot, and the animal appeared distressed, and would drink a great deal.

Binny must have been captured too young to have seen any of the building operations of his parents, or their co-mates; but his instinct impelled him to go to work under the most unfavorable circumstances, and he busied himself as earnestly in constructing a dam, in a room up three pair of stairs in London, as if he had been laying his foundation in a stream or lake in Upper Canada.

CHAPTER III.

THE SQUIRRELS AND MARMOTS.

THE FAMILY SCIURIDÆ—THE EUROPEAN SQUIRREL—THE JAVANESE SQUIRREL—THE HARE SQUIRREL—THE BLACK SQUIRREL—THE GRAY SQUIRREL—THE NORTHERN GRAY SQUIRREL—THE RED SQUIRREL—THE LONG-HAIRED SQUIRREL—THE FLYING SQUIRREL—THE AMERICAN FLYING SQUIRREL—THE TAGUAN—THE CHIPMUCK—THE LEOPARD MARMOT—THE MARMOT—THE BABAC—THE WOODCHUCK—THE PRAIRIE DOG.

THE family SCIURIDÆ comprehends the Squirrels, the Marmots, and the Prairie Dogs. It is very widely spread over the earth, being especially abundant in the North Temperate Zone, but having no representatives in the West Indian Islands, Madagascar, or Australia. It contains *eight* genera.

GENUS SCIURUS.

The *one hundred* species of this genus occupy the area of the whole family wherever woods and forests occur. They are pretty little animals, elegant in form, and rapid in movement. They are readily recognized by their long tails, raised like a plume above their heads, and by their abundant, clean, and glossy fur. The woods are their natural home, and we see them passing incessantly from branch to branch, from tree to tree, with wonderful agility, and untiring restlessness. Their usual food consists of nuts, but at times they suck birds' eggs.

THE SQUIRRELS.

The EUROPEAN SQUIRREL, *Sciurus Europeus*, is usually of a ruddy-brown upon the back, and a grayish-white on the under portions of the body. It is, however, very variable in its color. It has a summer and winter coat, the latter being always much lighter in its hue. In Siberia it becomes of a fine grayish-slate color. These graceful, fascinating Rodents live in couples.

The JELERANG, *Sciurus Javanicus*, is one of the handsomest of the Squirrels. It is a native of Java, part of India, and Cochin China. Its total length is about two feet, the tail and body being equal to each other in measurement. In color it is one of the most variable of animals, so that it has been more than once described under different names. The usual color of the Jelerang is a dark brownish-black on the back, the top of the head yellowish, and the sides and abdomen golden yellow.

The HARE SQUIRREL, *Sciurus leporinus* (Plate LVI), is one of the most beautiful of our native squirrels, being remarkable for its splendid tail, with its rich white border. The color of the body resembles that of the common English hare, becoming somewhat lighter on the sides. It is a native of California, and smaller than the Gray Squirrel, to which it bears a great similarity.

The BLACK SQUIRREL, *Sciurus niger* (Plate LVI), is a native of many parts of North America. The whole of its fur, with very slight and variable exceptions, is of a deep-black color; even the abdomen, which in most animals is lighter than the back, displays the same inky hue. The total length of the animal is about two feet ten inches, including a tail thirteen inches long. It is vanishing before the inroads of the Gray Squirrel, and seems to be a timid creature, flying in terror from the anger of the Red Squirrel. When undisturbed, it is an active and lively animal, and is remarkable for a curious habit of suddenly ceasing its play, and running to some water to refresh itself. After drinking, which it does by putting its nose and mouth in the water, it carefully washes its face.

The GRAY SQUIRREL, *Sciurus Carolinensis*, has, on the back, for three fourths of its length, fur of a dark lead color, succeeded by a slight indication of black edges with yellowish-brown in some of the hairs, giving it on the surface a dark grayish-yellow tint. The feet are light-gray, three-fourths of the tail is a yellowish-brown, the remainder black, edged with white, the lower surface of the body white. It differs in many respects from the Northern Gray Squirrel; its bark is more shrill, and instead of mounting the tree when alarmed, it plays around the trunk. It is less wild than the Northern species, and haunts swampy places, or trees overhanging rivers, and is constantly found in the cypress swamps. It is abundant in Florida, Georgia, and the Carolinas, but does not extend northward.

The NORTHERN GRAY SQUIRREL, *Sciurus leucotis*, is the most active species in the Atlantic States. It sallies forth with the sun, and for four or five hours industriously searches for food. After a siesta at noon, it resumes its labors. It is not, in Audubon's opinion, a very provident animal, and lays in but a small stock of food in its nest, and does not gather and bury any winter stores. In fact it is, in a cold climate, in a state of partial torpidity requiring little food. This Squirrel is, unfortunately, fond of green corn and young wheat, and hence is regarded with hatred by the farmer. In Pennsylvania, an ancient law offered three pence a head for every squirrel destroyed, and in one year, 1749, eight thousand pounds sterling were spent in such payment, a sum representing the death of six hundred and forty thousand squirrels.

This Squirrel is styled by Audubon the Migratory Squirrel, from the long migrations it occasionally performs. At such periods they congregate in various districts of the northwest, and turn their steps in an eastern direction. "Onward they come," writes Audubon, "devouring on their way everything that is suited to their taste, laying waste the corn and wheat-fields of the farmer, and as their numbers are thinned by the gun, the dog, and the club, others fall in and fill the ranks. It has often been inquired how they cross rivers like the Hudson and the Ohio. It has even been asserted that they carry to the shore a piece of bark, seat themselves upon it, and hoist their tails as a sail. Unfortunately, the story is not true." The same naturalist saw a migration which crossed the Hudson. He says they swam clumsily.

The RED SQUIRREL or CHICKAREE, *Sciurus Hudsonius* (Plate LVI), is the most common species around New York, and throughout the Eastern States. It is fearless to a great degree of the presence of man, and in its quick graceful motions from branch to branch, reminds one of a bird. It is always neat and cleanly in its coat, industrious, and provident. The Chickaree obtained its name from its noisy chattering note which it repeats at frequent intervals. Unlike the Gray Squirrel, it exhibits the greatest sprightliness amid the snows and frosts of our Northern regions, and consequently consumes in winter as great a quantity of food as at any other season. It wisely makes ample provision, and the quantity of nuts it often lays up is almost incredible. It is too cautious to trust a single hoard, but forms several. Shell-barks, butter-nuts, and the like, it obtains by gnawing off the portion of the branch on which the nuts hang. After having thrown down a considerable quantity, it drags them into a heap, covers them with leaves till the thick outer covering falls off or opens, and then carries off the nuts more conveniently. But even if these stores of nuts fail, the Chickaree can live on the cones of the pine and fir-tree. In the southern part of New York, and in more Southern States, it is satisfied with a hollow tree for its winter residence, but in Northern New York, Massachusetts, Canada, and further north, it digs deep burrows in the earth. It can swim and dive moderately well.

The LONG-EARED SQUIRREL, *Sciurus macrotus* (Plate LVI), is a remarkable species found in Borneo. Its title is not due to the length of the ears, but to the very long hair-tufts with which these organs are decorated. This fringe of hair is about two inches in length, of a glossy blackish-brown color, and stiff in texture. The color of the back and exterior of the limbs is a rich chestnut-brown, which fades into paler fawn along the flanks, and is marked by a single dark longitudinal stripe, extending from the fore to the hinder limbs. This dark band is narrow at each end, but of some width in the centre. The inside of the limbs is a pale chestnut, and the paws are jetty black. The tail is remarkably bushy, reminding the spectator of a fox's "brush," and is generally of the same color as the back, but grizzled with yellowish-white hairs, which are thickly sown among those of the darker hue.

The two following genera comprise the Flying Squirrels. Their common character consists in their being provided with wing-like membranes, extending along the flanks between the fore and hind legs. These membranes are covered with hair, and form regular parachutes, which enable them to sustain themselves in the air longer than the rest of the family, and thus perform extraordinary leaps.

GENUS SCIUROPTERUS.

This genus is divided into *sixteen* to *nineteen* species, and comprises the Flat-tailed Flying Squirrels, which range from Lapland and Finland to North China and Japan, and southward through India and Ceylon, to Malacca and Java; while in North America they occur from Labrador to British Columbia, and south to Minnesota and California.

The FLYING SQUIRREL, *Sciuropterus volucella*, may be taken as a type of the genus. It is called also the Assapan. It is the smallest of all the squirrels, measuring, tail and all, only about nine inches in length. Its

HARE SQUIRREL
BLACK SQUIRREL
TAGUAN FLYING SQUIRREL
LONG EARED SQUIRREL
RED SQUIRREL
GROUND SQUIRREL

PLATE LVI. RODENTIA.

fur is delicate and soft, brownish gray on the back, lighter on the sides of the neck, yellowish-white on the whole underside. The paws are silver-white, the flying membrane is edged with black and white, and the bright eyes are of a black-brown color.

The Flying Squirrel is a harmless and very gentle species, becoming tame in a few hours. After a few days it will take up its residence in some crevice, or under the eaves, and will remain there for years. They are gregarious, and live in considerable communities. There is nothing resembling flying in their movements, they merely descend from a high position by a gliding course, alight on a tree, ascend it at once, and again descend with their membrane expanded. They do not build nests, like the true squirrels, but confine themselves to a hollow in a branch.

The LAUTAGA, *Sciuropterus Sibericus*, inhabits the northern portion of Eastern Europe, but is most common, as its name indicates, in Siberia. It is about the same size as our native Flying Squirrel. The thick, silky fur is of a grayish-brown in the summer season; it grows longer in winter, and its color changes to a silver-gray.

GENUS PTEROMYS.

The twelve species of this genus comprise the Round-tailed Flying Squirrels. It is a more southern form than the preceding genus, and is confined to the wooded regions of India from the Western Himalayas to Java and Borneo, with species in Formosa and Japan.

The TAGUAN, *Pteromys petaurista* (Plate LVI), is a native of India, where it is tolerably common. It is rather a large species, as its total length is nearly three feet, the tail occupying about one foot eight inches, measured to the extremity of the long hairs with which it is so thickly clothed. The general color of this animal is a clear chestnut, deepening into brown on the back, and becoming more ruddy on the sides. The little pointed ears are covered with short and soft fur of a delicate brown, and the tail is heavily clad with bushy hairs, grayish-black on the basal portions of that member, and sooty-black towards the extremity. The parachute membrane is delicately thin, scarcely thicker than ordinary writing-paper, when it is stretched to its utmost, and is covered with hair on both its surfaces, the fur of the upper side being chestnut, and that of the lower surface nearly white. A stripe of grayish-black hairs marks the edge of the membrane, and the entire abdomen of the animal, together with the throat and the breast, is covered with beautiful silvery grayish-white fur.

THE MARMOTS.

From Squirrels which flit through the air with the graceful gliding of a bird, we pass now to the genera composing creatures which burrow in the ground, or pass a more or less subterranean life.

GENUS TAMIAS.

The *five* species of the genus are chiefly North American, ranging from Mexico to Puget Sound on the West, and from Virginia to Montreal on the East coast. One species is found over all Northern Asia. The animals of this genus differ from the true squirrels in several important particulars. They have a sharp convex nose, adapted to digging in the earth; they have longer heads, and their ears are set further back than those of squirrels; they have a more slender body, and shorter extremities. They have check pouches, and round narrow tails.

The CHIPMUCK, *Tamias lysteri* (Plate LVI), is also called the Chipping Squirrel or Ground Squirrel. It is found in most parts of the United States, and its beautiful markings attract the attention of all who see it. It is among quadrupeds, what the wren is among birds, lively, busy, agile, and graceful. Its clacking resembles the chip, chip, chip of a young chicken. "We fancy we see one of these sprightly Chipping Squirrels," writes Audubon, "as he runs before us with the speed of a bird; skewering along a log or fence with his chops distended by the nuts he has gathered, he makes no pause till he reaches the entrance of his subterranean storehouse. Now he stands upright, and his chattering cry is heard, but at the first step we make toward him he disappears. As we remove stone after stone from the aperture leading to his deep burrow, and cut through the tangled roots, we hear his angry querulous tones. We are within a few inches of him, and can see his large dark eyes. But at this moment out he rushes, and ere we can grab him has passed us, and finds security in some other hiding-place." The same writer caught one of these animals in Louisiana which had sixteen chinquapin nuts in its cheek-pouches, and saw another with a table-spoonful and a half of trefoil in the same receptacles. It is a very prudent creature, continuing to add to its winter store till the frost stops its work. Dr. Wright watched one collecting its store. It took a hickory nut and pushed it into one of its pouches with both paws, then it stuffed two into the other cheek, and finally took another nut in its teeth. It always carried four nuts at a time. The stock they thus accumulate is very large; often a peck of corn or nuts has been found in spring in the burrows of one of these Squirrels.

The Chipmuck has a slender body, and is beautifully marked. The anterior portion of the back is hoary-gray, the hinder portion reddish-fawn. A dark line runs along the back, a broad yellowish-white line runs from the shoulders to the thighs, bordered on each side with black. The fur on the body is a beautiful downy-white.

The BURUNDUK, *Tamias striatus*, is smaller than the common squirrel, but stouter built. The head is long, the eyes large, the ears short and small, the limbs strong, the soles of the feet bare, the tail is only slightly bushy. The color of the short, rough, thick-lying fur is, on the head, neck, and sides, yellowish; on the back there are five black stripes; the two lowest inclosing a yellowish band between them. The abdomen is grayish-white; the tail black above, yellowish below; the whiskers are black. The Burunduk is a native of Siberia.

GENUS SPERMOPHILUS.

The *twenty-six* species of this genus are distributed from the Arctic Ocean to Mexico on the west coast of this continent, but they do not pass east of Lake Michigan and the Lower Mississippi. In the Old World they extend from Siberia through South Russia to the Amoor and Kamschatka, being most abundant in the desert plains of Tartary and Mongolia. These pretty creatures have a long head, a slender body, ears hidden in their fur, a short tail, only bushy at the extremity, five toes on the hind feet, and large cheek-pouches.

The LEOPARD MARMOT, *Spermophilus Hoodii* (Plate LVII), is about the same size as the Chipmuck, and is remarkable for the brilliant and conspicuous manner in which its fur is diversified with contrasting hues. Along the back are drawn eight pale yellowish-brown bands, and nine dark-brown bands of greater width. The five upper bands are marked with pale spots. The coloring is slightly variable, both in distribution and depth of tint, for in some specimens the dark bands are paler than in others, while in several specimens the pale spots have a tendency to merge altogether and form bands. The average length of this creature is nearly eleven inches, the tail slightly exceeding four inches in length. The cheek-pouches are moderate in dimensions. It is an inhabitant of Northern America.

The burrow of the Leopard Marmot is generally driven perpendicularly into the ground, to the depth of four or nearly five feet; but on the plains of the Upper Missouri, where the soil is sandy, the burrow is almost horizontal, and lies barely a foot below the surface. Its bite is very severe, and it snaps fiercely at all who try to capture it.

The LINE-TAILED SPERMOPHILE, *Spermophilus grammurus*, has a wide bushy tail, and large, pointed ears. The length of the animal is about twelve inches; the color is grayish, the tail is white and black, in alternating longitudinal bands, two black, and three white. It is found in Colorado.

GENUS ARCTOMYS.

The *eight* species of the genus are found in the northern parts of North America as far as Virginia and Nebraska to the Rocky Mountains and British Columbia, but not in California. In the Eastern Hemisphere they extend from the Swiss Alps to Lake Baikal and Kamschatka, and as far south as the Himalayas. The generic characteristics are strong claws, fit for burrowing, a tail of moderate length, short limbs, and a hare-lip.

The MARMOT, *Arctomys marmota* (Plate LVII), is about the size of an ordinary rabbit, and not very unlike that animal in color. The general tint of the fur is grayish-yellow upon the back and flanks, deepening into black-gray on the top of the head, and into black on the extremity of the tail. It is common in all the mountainous districts of Northern Europe, and lives on the high peaks of the Swiss and Savoy Alps, in the vicinity of the glaciers. It forms small societies, composed of two or three families, and digs out burrows on the slopes exposed to the sun.

Marmots have a summer and winter residence—a town and country mansion. In summer they betake themselves to the highest part of the mountain, where they devote themselves to breeding and rearing their young, the number of which varies from two to four, and who remain with the parents until the following summer. When autumn arrives, they descend to the region of pasturage, and dig out a new burrow for their winter home, which is always deeper than the summer retreat. It is then they make hay—cutting grass, turning and drying it, which when cured, they carry into the chamber appointed for its reception.

The BABAC, *Arctomys babak* (Plate LVII), extends from Southern Poland and Galicia to the Amoor. It inhabits the steppes and stony plateaux, avoiding sandy spots and woodlands. Its thick fur is of a ruddy-yellow on the upper side of the body, rather darker on the head; the tip of the tail is almost black, the throat and chest grayish-white.

The Babacs live in settlements containing a very large population, which in the summer present an animated sight. At sunrise young and old leave their burrows, and eagerly lick up the dew; then they eat the leaves and blades of grass, and seem to sport merrily on the hillocks they have thrown up. But they vanish at once when any danger comes in sight. In June they begin to collect their winter provisions, and when the autumn is felt gather their hay-harvest. In the first half of September they retire into their winter abode, stop the entrance to the burrow with stones, sand, and grass, and pass into a state of semi-torpidity.

The WOODCHUCK or POUCHED MARMOT, *Arctomys monax* (Plate LVII), has a thick body and very short legs. The head is short and conical, the ears short, rounded, and thickly covered with hair, eyes moderate, whiskers numerous. The fore-feet have four toes and a rudimentary thumb, the hind-feet have five toes. The name Woodchuck properly belongs to this animal, although it is often applied to the Pecan.

The Woodchuck is subject to many variations in the color of its fur. Generally the back is grizzly or hoary, the cheeks light-gray, whiskers black; head, nose, feet, and tail dark-brown; eyes black; the whole under-surface a reddish-orange. It is fond of sitting in an erect posture, sitting on its rump, and letting its fore-legs hang loosely. It keeps in this position when feeding, inclining the head and forepart of the body forward and sideways. If pursued they run very fast for eight or ten yards, then stop and squat down; then start again. When walking leisurely, they place their feet flat upon the ground. They sleep during the greater part of the day, but leave their burrows early in the morning and in the evening. They climb trees very awkwardly. They become torpid in the winter, and remain underground till the grass has sprung up.

GENUS CYNOMYS.

The two species inhabit the plains east of the Rocky Mountains from the Upper Missouri to the Red River and the Rio Grande. Their burrows are numerous on the prairies, and the manner in which they perch themselves on little mounds and gaze on intruders is noticed by all travelers.

The PRAIRIE DOG, *Cynomys ludovicianus* (Plate LVII), is found in vast numbers wherever the soil is favorable for its subterranean works, and where there is vegetation sufficient to support it. Its color is reddish-brown on the back, mixed with gray and black. The abdomen and throat are grayish-white; the extreme half of the tail is covered with a brush of deep blackish-brown hair. The cheek-pouches are rather small, and the incisor teeth are large and protruding from the mouth. The length of the animal rather exceeds sixteen inches, the tail being a little more than three inches long. The cheek-pouches are about three-quarters of an inch in depth, and are half that measurement in diameter.

As long as no danger is apprehended, the little animals are all in lively motion, sitting upon their mounds, or hurrying from one tunnel to another as eagerly as if they were transacting the most important business. Suddenly a sharp yelp is heard, and the peaceful scene is in a moment transformed into a whirl of indistinguishable confusion.

As it is so wary an animal it is with difficulty approached or shot, and even when severely wounded it is not readily secured, owing to its wonderful tenacity of life. A bullet that would instantly drop a deer has, comparatively, no immediate effect upon the Prairie Dog, which is capable of reaching its burrow, even though mortally wounded in such a manner as would cause the instantaneous death of many a larger animal.

The mode by which this animal enters the burrow is very comical. It does not creep or run into the entrance, but makes a jump in the air, turning a partial somersault, flourishing its hind legs and whisking its tail in the most ludicrous manner, and disappearing as if by magic.

The burrows of the Prairie Dog are generally made at an angle of forty degrees, and after being sunk for some little distance run horizontally, or even rise toward the surface of the earth. It is well known that these burrows are not only inhabited by the legitimate owners and excavators, but are shared by the burrowing owl and the rattlesnake.

POUCHED MARMOT
PRAIRIE DOG
LEOPARD SPERMOPHILE
COMMON MARMOT
BABAC

PLATE LVII RODENTIA

According to popular belief, the three creatures live very harmoniously together; but careful observations have shown that the snake and the owl are interlopers, living in the burrows because the poor owners cannot turn them out, and finding an easy subsistence on the young Prairie Dogs. A rattlesnake has been killed near a burrow, and when the reptile was dissected, a Prairie Dog was found in its stomach.

From the most recent accounts, it appears that the Prairie Dog does not hibernate, but that it is as fresh and lively during winter as in the heat of summer.

Prairie Dogs who live far away from any river or stream obtain drink from wells dug by themselves that have concealed openings. No matter what the depth, the dogs will keep digging until they reach water. A frontiersman named Leach, formerly of Mercer county, Pennsylvania, says he knows of one such well two hundred feet deep, and having a circular staircase leading down to the water. Every time a dog wants a drink she descends this staircase, which, considering the distance, is no mean task.

GENUS ANOMALURUS.

The *five* species of this African genus consist of animals which resemble Flying Squirrels, but differ from all other members of the family in internal structure. They are found only in the Island of Fernando Po and West Africa.

CHAPTER IV.

THE SEWELLELS, PORCUPINES, AND CAVIES.

THE FAMILY HAPLOODONTIDÆ—THE FAMILY CHINCHILLIDÆ—THE CHINCHILLAS AND VISCACHAS—THE OCTODONTIDÆ—THE HUTIA CONGA—THE DEGU—THE TUKOTUKO—THE CUNDI—THE COYPU—THE GROUND PIG—THE CANADIAN PORCUPINE—THE BRAZILIAN PORCUPINE—THE COMMON PORCUPINE—THE TUFTED-TAILED PORCUPINE—THE AGOUTI—THE MOSEY PACA—THE CAVIIDÆ—THE GUINEA PIG—THE MARA—THE PACAS.

THE family HAPLOODONTIDÆ has been constituted by Professor Lilljeborg to comprise some curious rat-like animals having affinities with both the Beavers and Marmots, and inhabiting the West Coast of America from the southern part of British Columbia to the mountains of California. It contains only one genus.

GENUS HAPLOODON.

According to Wallace there are *two* species of this genus, their generic characteristics are a stout, heavy form, a cylindrical body, short legs, flat head, bushy tail, and very long whiskers. The eyes are small, the under-fur is woolly, the walk is plantigrade.

The SEWELLEL, *Haploodon rufus*, was first discovered during the expedition of Clark and Lewis to the Columbia River in 1805. It lives in communities like the prairie dog, sitting upon mounds at its burrows, and whistling as the latter do in the early morning. It is about as large as the musk rat; its color is brown mixed with black, with grayer tints on the abdomen. Little is known of its habits, but it seems to be ascertained that it is not a hibernating animal. The name is derived from the Nisqually language. The Indians hunt them for their skins, which they sew together and dress with the fur on. It is said to cut bundles of herbs and place them out to dry for future use. One species is found north of the Columbia River, the other in California.

The family CHINCHILLIDÆ is confined to the Alpine zones of the

Andes from the boundary of Ecuador and Peru to Southern Chili, and over the pampas to the Rio Negro and the Uruguay. It comprises *three* genera.

GENUS CHINCHILLA.

The genus is usually divided into *two* species, according to the difference of their fur. They are "rabbits with long tails," a link, in fact, between the rabbit and the mouse. Their color is a light gray, with white and dark brown markings. They are found in the Andes of Peru and Chili, south of nine degrees S. lat.; at about eight thousand to twelve thousand feet above the sea level.

The CHINCHILLA, *Chinchilla lanigir* (Plate LVIII), lives chiefly among the higher mountainous districts, where its thick silken fur is of infinite service in protecting it from the cold. It is a burrowing animal, banding together in great numbers in certain favored localities. The food of the Chinchilla is exclusively of a vegetable nature, and consists chiefly of various bulbous roots, which it disinters by means of its powerful fossorial paws. While feeding, it sits upon its hinder-feet, and conveys the food to its mouth with its fore-feet, which it uses with singular adroitness. It is an exceedingly cleanly animal, as might be supposed from the beautiful delicacy of its fur.

As far as is known, the Chinchilla is not a very intelligent animal, seeming to be hardly superior to the guinea-pig in intellect, and appearing scarcely to recognize even the hand that supplies it with food. They are of a gentle nature and easily tamed, and the Chilenos are fond of keeping them in their houses.

GENUS LAGIDIUM.

The members of this genus have longer ears, and more bushy tail than those of the preceding genus. As yet only *three* species are known, and they inhabit the loftiest plateaus of the Cordilleras from eleven thousand to sixteen thousand feet. They have four toes, while the Chinchillas have only four on their hind-feet.

The ALPINE VISCACHA, *Lagidium Cuvieri*, has a thick snout, long hind-legs, and a long tail. The fur is very soft and long, and of an ash-gray color which, on the sides, approaches to yellow. The upper surface of the tail is covered with long hair of a brownish-black color. The whiskers are remarkably long, and reach to the shoulders.

GENUS LAGOSTOMUS.

This genus, which contains only *one* species, has a close resemblance to the Chinchilla. They have four toes on the front feet, and three on the hind; the latter armed with long claws.

The VISCACHA, *Lagostomus trichodactylus* (Plate LVIII), is the Viscacha of the Pampas. Their habitat is the vast plains of South America, or the basin of La Plata river. They live in communities, and hollow out very deep burrows. Grasses and vegetables constitute the chief part of their food. Their usual posture is that generally assumed by rabbits; and they use their feet to convey their food into their mouths. Their movements are very active, and they are excessively wary and difficult to approach. They are hunted for the sake of their fur. Their burrows are dug in common, and inhabited in common, and are provided with countless exits and entrances: often forty or fifty tunnels have been found, and the burrow is divided into as many chambers as there are families. During the day they live underground, at sundown they steal out one by one, and in the twilight a numerous company assembles. The horsemen of the Pampas and the condor are deadly enemies to the Viscachas. These persecuted creatures display an almost human affection for each other: if one is wounded when outside the burrow, its companions carry it off into the safest recess. The Indians believe that if a burrow is stopped up, the inhabitants of the neighborhood assemble and dig out their buried kinsfolk.

The family OCTODONTIDÆ includes a number of curious and obscure rat-like animals, mostly confined to the mountains and open plains of South America, but having a few stragglers in other parts of the world. Of the *eight* genera, two are peculiar to the West India islands; and two to Africa.

GENUS CAPROMYS.

The generic characteristics of the *three* species are a short, thick body with powerful hindquarters; a short, thick neck; a long, broad head; broad, almost hairless ears; large eyes; strong legs, with five toes on the hind, and four on the front feet. The fur is abundant, smooth, and glistening.

The HUTIA CONGA, *Capromys pilorides*, is described by Oviedo in

1525 as a rabbit-like animal found in San Domingo which constituted the chief food of the natives. It measures about a foot and a half in length, and stands about eight inches high. The color of its coat is yellowish-gray and brown, with a reddish tinge on the shoulders; the paws are black, and a longitudinal stripe on the abdomen is gray. This animal lives either in trees, or in the thickest bushwood, and is only visible by night. The disproportionate development of the hind-quarters renders it awkward on land, but it climbs skilfully, making use of its tail. On the ground it sits up like a rabbit, and makes short leaps. In many parts of Cuba it is still hunted for the sake of its flesh. Its food consists of fruits, leaves, and bark. When kept in confinement, it shows a preference for strong-smelling plants, such as mint and thyme, which other rodents do not like.

GENUS OCTODON.

This genus contains *three* species. The animals contained in it have a thick compact body, a short thick neck, a large head, a tail bushy at the end, and long hind-legs. All the feet have five toes.

The DEGU, *Octodon Cummingii*, is of a brownish-gray color on the back, and a grayish-brown below; the root of the tail is nearly white; the inside of the ears is white, the tail black at the extremity. The Degu is one of the commonest animals in Central Chili, being found by hundreds in every fence or thicket; it runs fearlessly about the high roads, and boldly enters the gardens and fields. It seldom attempts to climb the trees, but at the first sign of danger rushes with upright tail into one of the many entrances into its burrow. It does not hibernate, but collects provisions, like the hibernating animals. It soon becomes tame in captivity.

GENUS CTENOMYS.

The *six* species of this genus are found in the Pampas in the Campos of Brazil, in Bolivia, and in Terra del Fuego. Their small eyes and ears indicate their underground habits. Their limbs are short, the coat is smooth. They are found in astonishing numbers in the plantless desert of the plateaus of the Cordilleras, where the soil is perforated like a sieve with their burrows.

The TUKOTUKO, *Ctenomys Magellanicus*, is betrayed to the traveler by the peculiar grunting sounds which, in regular succession, are heard proceeding from the earth. The animal that utters them is about the size of the hamster, of a brownish-gray color on the body, and with white feet and tail. The Tukotuko was first described by Darwin, who discovered it at the eastern entrance to the Straits of Magellan. Wide, dry, sandy, and barren plains are its home. Here it burrows, like a mole for a long distance, during the night. By day it rests. Its motion on the surface of the ground is clumsy, it cannot leap over the slightest obstacle, and when out of its hole, can be easily captured. In all probability it does not hibernate.

GENUS CTENODACTYLUS.

We take this genus as the representative of the African branch of the family. The head is thick and blunt, the ears round and short, the eyes moderate in size, the whiskers uncommonly long, the limbs strong, the hind ones longer than the fore ones; the tail is a mere stump, but is covered with long bristles.

The GUNDY, *Ctenodactylus massoni*, is found in the wildly romantic valleys of the Djebel Aures, and in the southern elevations of Algeria, near the Sahara. In the winter months it is seen at midday on the rocks, but always high enough to be safe from surprise. With its head directed to the valley, and lying close to the rock, it looks like a part of it. It is everywhere common, living in clefts, and under stones, and is remarkable for its agility. The slightest noise makes it hop back into its hole, which usually defies all the sportsman's efforts. The best time to watch it is the morning, as at sunrise it begins to descend into the valley in search of food. When it reaches the fields, it sits up like a rabbit, and gnaws the stalks of growing corn. Traps and nooses of hair are employed to catch it, and the Arab children amuse themselves by setting them. The flesh is very like that of a chicken. The velvety fur is used to make purses.

The family ECHIMGIDÆ or Spring Rats, is best known as the COYPU or large beaver-like water-rat of Peru and Chili. It is divided into *ten* genera, two of which inhabit South Africa, but all the rest are confined to the continent of South America east of the Andes.

VISCACHA COMMON PORCUPINE CANADIAN PORCUPINE
CHINCHILLA   COYPU

GENUS MYOPOTAMUS.

The COYPU, *Myopotamus coypu* (Plate LVIII), the *only* species, greatly resembles the beaver. It is about the same size, and has palmated feet, but its tail is cylindrical and scaly. It is common in Chili and La Plata, and is also, though more rarely, found in Brazil and the other states of South America, where the natives incessantly persecute it on account of its valuable fur. As it remains in its burrow during the day, it is hunted at night with dogs. Some time ago the exportation of Coypu skins was carried on very extensively.

The color of the Coypu is a light reddish-brown. It is agile and lively, and very amusing in its behavior. If anything is thrown into its trough, it takes it up in its fore-feet, and washes it with the skill of a laundress. It lives on the banks of streams and lakes where the water-plants are large enough to hide it. It dives badly, but swims well. It walks slowly, owing to its shortness of limb, and only ventures on land when passing from one stream to another.

GENUS AULACODES.

The GROUND PIG, *Aulacodes swinderianus*, is the *only* species of this African genus. It is one of the links between the beavers and the porcupines, and has a considerable affinity with the latter animals.

It is found in many parts of Southern Africa, as well as on the coast of Guinea, where it is not at all uncommon. The hair of this animal is rather peculiar, and approximates closely to the quill-hairs of the true porcupines, being either flat and grooved above, or developed into flexile spines. The tail is but sparsely covered with hair, and is rather short in proportion to the size of its owner. The hinder feet are only furnished with four toes, armed with large, rounded, and rather blunt claws. The ears are short and rounded.

The Porcupines are divided into two families. The Tree Porcupines, or CERCOLABIDÆ, a group of *three* genera entirely confined to America, where they range from the northern limit of trees on the Mackenzie River to the southern forests of Paraguay. They are absent from the Southern United States. The True Porcupines, or HYSTRICIDÆ, also have *three* genera, and are confined to the Eastern Hemisphere.

GENUS ERETHIZON.

The *three* species are distributed as follows: One, throughout Canada and as far south as Northern Pennsylvania and west to the Mississippi; the second from California to Alaska and west to the Missouri; the third in the northwest part of South America.

The CANADIAN PORCUPINE, *Erethizon dorsatum* (Plate LVIII), is also called the Urson. Its chief food consists of living bark, which it strips from the branches as cleanly as if it had been furnished with a sharp knife. When it begins to feed, it ascends the tree, commences at the highest branches, and eats its way regularly downward.

In the work of Messrs. Audubon and Bachman is a very amusing little story of the manner in which the tame Urson above mentioned repelled an attack made upon it by a fierce dog.

"A large, ferocious, and exceedingly troublesome mastiff belonging to the neighborhood, had been in the habit of digging a hole under the fence, and entering our garden. Early one morning we saw him making a dash at some object in the corner of the fence, which proved to be our Porcupine, which had, during the night, made its escape from the cage.

"The dog seemed regardless of all its threats, and probably supposing it to be an animal not more formidable than a cat, sprang upon it with open mouth. The Porcupine seemed to swell up in an instant to nearly double its size, and as the dog pounced upon it, it dealt him such a sidewise blow with its tail, as to cause the mastiff to relinquish his hold instantly, and set up a loud howl in an agony of pain. His mouth, tongue, and nose were full of Porcupine quills. He could not close his jaws, but hurried, open-mouthed, off the premises."

GENUS CERCOLABES.

The *twelve* species of this genus range from Mexico to Paraguay on the east side of the Andes. They are distinguished by long tails, usually prehensile, and short quills.

The BRAZILIAN PORCUPINE or COENDOO, *Cercolabes prehensilis* (Plate LIX), as might be presumed, from the prehensile tail and the peculiarly armed claws, is of arboreal habits, finding its food among the lofty branches of trees. On the level ground it is slow and awkward, but it climbs with great ease, seldom using its tail, except as an aid in descent.

The total length of the Coendoo is about three feet six inches, of which the tail composes one foot six inches. Its nose is thick and blunt, like that of the common porcupine, and the face is furnished with very long whisker-hairs of a deep black. The numerous spines which cover the body are parti-colored, being black in the centre and white at each extremity. Their length is rather more than two inches on the back, an inch and a half on the fore-legs, and not quite an inch on the hinder-limbs. A number of short quills are also set upon the basal half of the tail, the remainder of that organ being furnished with scales, and tapering to its extremity. The color of the scales is black. The entire under surface of the tail is covered with similar scales, among which are interspersed a number of bright chestnut hairs. The abdomen, breast, and inner face of the limbs are clothed with dense, brown, coarse hairs. It is a nocturnal animal; sleeping by day, and feeding by night.

GENUS HYSTRIX.

Southern Europe, Africa, all India, Ceylon, and South China are the homes of the *five* species of HYSTRIX. They are all remarkable for the coat of spines or quills with which they are defended.

The COMMON PORCUPINE, *Hystrix cristata* (Plate LVIII), is found in Europe, Africa, and India. It is over two feet in length. The spines or quills vary considerably in length, the longest quills being flexible, and not capable of doing much harm to an opponent. Beneath these is a plentiful supply of shorter spines, from five to ten inches in length, which are the really effective weapons of this imposing array. Their hold on the skin is very slight, so that when they have been struck into a foe, they remain fixed in the wound, and unless immediately removed work sad woe to the sufferer. For the quill is so constructed, that it gradually bores its way into the flesh, burrowing deeper at every movement, and sometimes even causing death.

The Porcupine is a nocturnal animal, and is therefore not often seen even in the localities which it most prefers. It is said not to require the presence of water, and its food is entirely of a vegetable nature. This animal takes up its abode in deep burrows which it excavates, and in which it is supposed to undergo a partial hibernation.

GENUS ATHERURA.

The Brush Tailed Porcupines, divided into *five* species, are found in West Africa, India, Siam, Borneo, and Sumatra. They are small, with short naked ears, and a long tail, partly covered with scales, and ending in a tuft of most extraordinary formation.

The TUFTED TAILED PORCUPINE, *Atherura Africana* (Plate LIX), is a comparatively slender animal, about two feet in length.

The quills which cover the body are very short in proportion to the size of the animal, and are flattened like so many blades of grass. The tail is scaly throughout a considerable part of its length, but at the tip is garnished with a tuft of most extraordinary-looking objects, which look very like narrow, irregular strips of parchment. The coloring of the quills is diversified, but as a general rule they are black toward the extremity and white toward the base.

The Agoutis and Cavies are placed by Mr. Wallace in the family CAVIDÆ. There is a striking external resemblance between them, they inhabit the same regions of South America, and with one exception are all found east of the Andes. They are divided into *six* genera.

GENUS DASYPROCTA.

The AGOUTI, *Dasyprocta aguti* (Plate LIX), is the type of the *nine* species of the genus. It ranges from Mexico to Paraguay, one species inhabiting the small West India Islands of St. Vincent, Lucia, and Grenada. The habits of all the species are very similar.

The Agoutis are natives of South America and the West Indies. Woods spreading over hills and mountains are the localities where they generally take up their abode; and the clefts of rocks, or the hollows in trees, serve for their retreats. If ready-made places of shelter are not procurable, they dig burrows. They are nocturnal in their habits, and feed principally on roots and fruit. But when in captivity they are omnivorous, and manifest an unbearable voracity, for they gnaw everything they can get at. They are hunted by dogs, but although tolerably swift are unable to sustain a long chase. When running, they resemble the common hare, and, like that animal, are apt to overbalance themselves when running down hill. They can be easily tamed, but display little affection.

GENUS CÆLOGENYS.

The members of this genus are found from Guatemala to Peru. There is perhaps only *one* species in the genus, the so-called second species found in Eastern Peru being doubtful. They are remarkable for the extraordinary development of a portion of the skull. The cheek-bone is enormously enlarged into a mass of bone, concave and rough on the exterior; the lower edge of this bony development descends beneath the lower jawbone. Connected with this curious structure is a cheek-pouch with a very small opening, for which no use has been discovered. There are also two large cheek-pouches which open into the mouth. The name *Cælogenys* means hollow-cheeked.

The SOOTY PACA, *Cælogenys paca* (Plate LIX), is a pretty animal, the rows of white spots which decorate its sides standing out in pleasing contrast to the rich black-brown hue with which the remainder of the fur is tinged. The throat and abdomen are white, and the lowermost of the four rows of white spots is often nearly merged into the white fur of the under portions of the body. The coloring is rather variable in different individuals. The paws are light flesh-color, and the large full eyes are dark-brown. The total length of this animal is about two feet. It is an active animal, in spite of its clumsy looks, and not only runs with considerable speed, but is a good swimmer, and can jump well.

The favorite localities of the Pacas are in wooded districts, in marshy grounds, or near the banks of rivers. Their domiciles are excavated in the ground, but are at no great depth, and are remarkable for the admirable state of cleanliness in which they are preserved by the inhabitants.

GENUS HYDROCHÆRUS.

The *only* species inhabits the banks of rivers from Guayana to La Plata. Its size, its harsh coarse hair, its hoof-like toes and clumsy bearing are so swinish, that it seems to be a kind of wild hog. From this external resemblance the name *Hydrochærus* or Water Hog is derived.

The CAPYBARA, *Hydrochærus capybara* (Plate LIX), is the largest of all the living rodent animals, rather exceeding three feet in total length, and being so bulkily made that when it walks its abdomen nearly touches the ground. The muzzle of this animal is heavy and blunt, the eyes are set high in the head, and are moderate in size, the tail is wanting, and the toes are partially connected together by a development of the skin. The color of the Capybara is rather undeterminate, owing to the manner in which the hairs are marked with black and yellow, so that the general idea which its coat presents is a dingy, blackish-gray, with a tinge of yellow. The hairs are rather long, and fall heavily over the body. The incisor teeth are of enormous dimensions, and the molars are very curiously formed, presenting some analogy to those of the elephant.

It is a water-loving animal, and not only swims well, but is a good diver. It is gregarious, being generally found in small herds upon the banks of the streams.

GENUS CAVIA.

Of the *nine* species into which the genus is divided, one is found west of the Andes in Peru, the others are distributed from Brazil to the Straits of Magellan. The common name Guinea Pig is quite a misnomer, as they are not pigs, and are not found in Guinea.

The GUINEA PIG, *Cavia cobaya* (Plate LIX), is so well known, as hardly to require description. In their native wilds these pretty creatures lead a nocturnal life, and either dig out burrows for themselves, or find a retreat among the herbage. In captivity they eat bread, roots, vegetables, and grasses. It is often thought that they never drink, but this is a mistake. The color is very variable, but is generally composed of white, red, and black in patches. Their domestication dates back to a very distant period. This fact may at least be inferred from their being marked by large black and yellow patches on a white ground, a peculiarity of color which they presented even before their introduction into Europe in the middle of the sixteenth century.

GENUS DOLECHOTIS.

The PATAGONIAN CAVY or MARA, *Dolechotis Patagonicus*, is an animal which is remarkably swift for a short distance, but is so easily fatigued that it can be run down by a man on horseback. It is more tameable than the agouti, and is often kept in a state of domestication. It is generally found in couples, a male and his mate occupying the same "form." It does not seem to burrow, nor to keep so close to its retreat as the agouti, but is fond of crouching in a form like our common hare. It is about thirty inches in length, and about nineteen inches high at the crupper, which is the most elevated part of the animal. At the shoulder

BRAZILLIAN PORCUPINE
SOOTY PACA
CAPYBARA
GUINEA PIG
TUFTED TAILED PORCUPINE
AGOUTI

PLATE LIX RODENTIA

it hardly exceeds sixteen inches. The fur of this animal is soft and warm, and from the contrasting colors of black, white, and golden brown, presents a very handsome appearance.

The family LAGOMYIDÆ consists of small Alpine and desert animals which range from the south of the Ural Mountains to Cashmere and the Himalayas, at heights of eleven thousand to fourteen thousand feet, and northward to the Polar regions. In America they are confined to the Rocky Mountains from about forty-two to sixty degrees north latitude.

GENUS LAGOMYS.

This *solitary* genus is divided into *eleven* species. The animals comprised in it have the dentition of the hare, are nearly tailless, and have the form of the *Anicola*. Hence the generic title which means Hare-Mouse.

The NORTH AMERICAN PIKA, *Lagomys princeps*, is often called the "Little Chief Hare." It is about seven inches in length, of a grayish-brown color, varied with black and yellowish-brown. These animals range along the summits of the Rocky Mountains near the forest line. They do not wander far from their homes, and are very timid, retreating at the slightest alarm. They are not hares in any of their habits, but sit up like marmots or prairie dogs.

The ALPINE PIKA, *Lagomys Alpinus*, has been described by Pallas, Radde, and Przewalski. It is found on all the mountain heights of Inner Asia, and occurs even in Kamschatka. It prefers woody regions, and avoids the bare high steppes. In the latter it is replaced by a species named OTOGONO, *Lagomys otogono*. This species lives in clefts in the rocks or in burrows that it digs. Usually large numbers of these subterranean dwellings are found together, so that when one Pika is discovered hundreds or thousands are not far off. During the winter they do not come above-ground; in summer, too, when the weather is bright, they keep concealed, but on a dull day they are active. They are active and industrious creatures, and collect large quantities of hay which they shelter from the rain by a covering of leaves. They make runs under the snow from their burrows to their haystacks, and have the precaution of supplying each run with an air-hole. They have many enemies. Every bird or beast of prey attacks them. In hard winter the Mongols feed their horses with the fodder the poor Otogono have collected.

CHAPTER V.

HARES AND RABBITS.

THE FAMILY LEPORIDÆ—THE AMERICAN HARES—THE POLAR HARE—THE NORTHERN HARE—THE WOOD HARE—THE BRAZILIAN HARE—THE JACKASS RABBIT—THE AFRICAN HARES—THE SAND HARE—THE EUROPEAN HARE—THE COMMON HARE—THE ALPINE HARE—THE RABBIT—THE WILD RABBIT—THE DOMESTIC RABBIT.

THE family LEPORIDÆ consists of only *one* genus divided into *thirty-five* to *forty* species. Only one species occurs in South America, while one or two North American species extend into Mexico and Guatemala. In the Old World they extend from Iceland to Japan, and as far north as Greenland and the Arctic Ocean.

GENUS LEPUS.

The Hares and Rabbits are the only rodents that have more than two front teeth. They have a long body, with high hinder limbs, a long skull, with large ears and eyes, five toes on the front, four on the hind-feet, very mobile deeply-cleft lips, a thick fleecy coat, and a short, erect tail. All the species live on soft, juicy plants, but yet eat anything else that they can get. Their motions are peculiar. When walking slowly they seem clumsy and awkward, but in full speed they are very fleet, and can in the wildest race make turns of wonderful quickness.

THE HARE.

The POLAR HARE, *Lepus timidus*, inhabits the whole Arctic Circle. Like other Polar animals, it changes its color according to the season. In winter it is pure white, excepting the tips of the ears, which are black. In summer it is of a light-yellowish brown-gray, with black ears. Its length is about two feet, its weight eight to ten pounds.

The NORTHERN HARE, *Lepus Americanus* (Plate LX), presents itself in four different varieties. In winter it is white, the tips of the ears having a narrow black border, in summer it is of a pale yellowish-brown. The species in some of its varieties is found as far south as New Mexico; it occurs in most parts of the State of New York, and exhibits more of the habits of the English Hare than any of our native species. In the Eastern and Middle States the color undergoes very slight change.

The WOOD HARE, *Lepus sylvaticus*, is also called the Gray Rabbit. It is found throughout the United States and Canada. It is very timid, and does not burrow, but makes a "form." It is very like the MARSH HARE, *Lepus palustris*, but the latter has a shorter tail and broader ears, and is of a yellowish-brown color. The Marsh Hare is common in Florida and Mexico in bushy swamps and the borders of creeks.

TROWBRIDGE'S HARE, *Lepus Trowbridgii*, is the smallest of our hares; the tail is rudimentary, the ears are as long as the head. The color is yellowish-brown. It is found east of the Cascade Range, and on the coast from Fort Crook to Cape St. Lucas.

The JACKASS RABBIT, *Lepus callotis* or *Texianus* (Plate LX), has very large ears, more than one-third longer than the head. A dark-brown stripe is seen on the top of the neck, and a black stripe from the rump extends to the root of the tail, and along its upper surface to the tip. The upper surface of the body is mottled deep buff and black, the throat and belly white, the under side of the neck dull rufous.

It received its name from the troops in the Mexican War, who found it very good eating, and it formed an important article of provisions for J. W. Audubon during his travels in Mexico. It inhabits the southern parts of New Mexico, the western parts of Texas, and the elevated lands westward of the *tierra caliente* or coast lands of Mexico. On the Pacific coast it is replaced by the California Hare.

Till it was better known, this hare was described as enormously large, and equal in size to a fox. A fine large specimen, however, only measures about one foot nine inches in length.

AFRICAN HARES.

The species are distinguished by their diminutive size and extremely long ears. They are of a sandy color, and live in the deserts or their neighborhood.

The SAND HARE, *Lepus Æthiopicus*, is found in the mountains and coast-lands of Abyssinia. As the inhabitants, both Christian and Mohammedan, observe the precept of the Law of Moses which enumerates the Hare among the unclean beasts not fit for food, the Sand Hare has acquired no fear of man, and, in place of running away, merely walks to some bush. The European sportsman, who gives chase, sees the animal as he approaches again move slowly to some other bush. Even the sound of the gun does not quicken its pace for any length of time. If pursued by dogs, however, it alters its tactics, and displays great fleetness, agility, and endurance.

The COMMON HARE, *Lepus vulgaris* (Plate LX), is distinguished from the rabbit by the redder hue of its fur, the great proportionate length of its black-tipped ears, which are nearly an inch longer than the head; by its very long hind-legs, and its large and prominent eyes. The color of the Common Hare is grayish-brown on the upper portions of the body, mixed with a dash of yellow; the abdomen is white, and the neck and breast are yellowish-white. The tail is black on the upper surface and white underneath, so that when the creature runs it exhibits the white tail at every leap. Sometimes the color of the Hare deepens into black, and there are many examples of albino specimens of this animal.

The Hare does not live in burrows, like the rabbit, but only makes a slight depression in the ground, in which she lies so flatly pressed to the earth that she can hardly be distinguished from the soil and dried herbage among which she has taken up her abode. In countries where the snow lies deep in winter, the Hare lies very comfortably under the white mantle which envelops the earth, in a little cave of her own construction. She does not attempt to leave her "form" as the snow falls heavily around her, but only presses it backward and forward by the movement of her body, so as to leave a small space between herself and the snow. By degrees the feathery flakes are formed into a kind of domed chamber, with the exception of a little round hole which serves as a ventilating aperture. This air-hole is often the means of her destruction as well as of her safety, for the scent which issues from the aperture betrays her presence.

The Hare is by no means a timid animal. It fights desperately with its own species, and in defence of its young it will attack even man. In England the hare is shot, and hunted either with greyhounds or a pack of hunters. Its long and powerful limbs enable it to make prodigious bounds, and it has been known to leap over a wall eight feet high.

It is a wonderfully cunning animal, and is said by many who have closely studied its habits to surpass the fox in ready ingenuity. Sometimes it will run forward for a considerable distance, and then, after returning for a few hundred yards on the same track, will make a great leap at right angles to its former course, and lie quietly hidden while the hounds run past its spot of concealment. It then jumps back again to its track, and steals quietly out of sight in one direction, while the hounds are going in the other.

The Hare also displays great ingenuity in running over the kind of soil that will best suit the formation of her feet, and has been known on more than one occasion to break the line of scent most effectually by leaping into some stream or lake, and swimming for a considerable distance before she takes to the land again.

In pursuing the Hare with greyhounds, the dogs are held in couples in a leash, and are let loose when a hare is seen. This is the method adopted in coursing matches. In this sport a judge rides after the dogs, and makes note of their performance; the number of turns, and the like, which each of the contestants successfully achieve, and not the mere seizing of the hare, decide the victory.

The Hare has been often tamed. The poet Cowper amused his solitude with his tame hares, and celebrated them in his verse. Dr. Franklin had a hare that used to sit between a cat and a greyhound before the fire, and lived on the best terms with them, and they have been taught various tricks, such as beating drums, firing pistols, and dancing.

The SNOW HARE, *Lepus alpinus*, is very similar to the American Hare. It is more lively and daring than the common species. The head is shorter, rounder, and more arched; the ears are smaller, the eyes are dark-brown. In winter the Snow Hare is pure white, except the tips of the ears; but in summer is a uniform grayish-brown. The IRISH HARE, *Lepus Hibernicus*, closely resembles the Snow Hare, but does not change its color. In the high north, the Hares remain white all the year through.

THE RABBIT.

The RABBIT or CONEY, *Lepus cuniculus* (Plate LX), is distinguished from the Hare by its smaller dimensions, its grayer color, and its shorter ears, as well as by its habits. It has been extensively acclimatized in America. It exists in great numbers in Sable Island, Nova Scotia, and on Rabbit Key, near Key West. It is found everywhere in Europe, and is supposed to have spread northward from Africa.

The Rabbit is one of the most familiar of British quadrupeds, having taken firm possession of the soil into which it has been imported, and multiplied to so great an extent that its numbers can hardly be kept within proper bounds without annual and wholesale massacres. As it is more tameable than the hare, it has long been ranked among the chief of domestic pets, and has been so modified by careful management that it has developed itself into many permanent varieties, which would be considered as different species by one who saw them for the first time. The little brown, short-furred wild Rabbit of the warren bears hardly less resemblance to the long-haired, silken-furred, Angola variety, than the Angola to the pure Lop-eared variety with its enormously lengthened ears and its heavy dewlap.

In its wild state the Rabbit is an intelligent and amusing creature, full of odd little tricks, and given to playing the most ludicrous antics as it gambols about the warren in all the unrestrained joyousness of habitual freedom. No one can form any true conception of the Rabbit nature until he has observed the little creatures in their native home; and when he has once done so, he will seize the earliest opportunity of resuming his acquaintance with the droll little creatures.

The female Rabbit is exceedingly prolific, and has seven or eight litters a year, with from four to eight in each. Some days before bringing forth, the Rabbit excavates a chamber which is specially destined for its progeny. This burrow, which is straight or crooked, as the case may be, invariably terminates in a circular apartment, furnished with a bed of dry herbage, which again is covered with a layer of down, that the mother has torn from the lower portion of her body. On this bed the young are deposited.

TEXAS or JACKASS RABBIT
NORTHERN HARE SUMMER COAT
DOMESTIC RABBIT
COMMON WILD RABBIT
NORTHERN HARE WINTER COAT
COMMON HARE
PLATE LX RODENTIA

ORDER XI.

EDENTATA.

FAMILIES.

71. BRADYPODIDÆ - - - - - Sloths.
72. MANIDIDÆ - - - - - - - Scaly Ant Eaters.
73. DASYPODIDÆ - - - - - - Armadillos.
74. ORYCTEROPODIDÆ - - - - Ant Bears.
75. MYRMECOPHAGIDÆ - - - - Ant Eaters.

EDENTATA

CHAPTER I.

THE SLOTHS AND ARMADILLOS.

THE EDENTATA—THE SLOTHS—THE TWO-TOED SLOTH—THE AI OR THREE-TOED SLOTH—THE SPOTTED SLOTH—THE SCALY ANT-EATERS—THE CHALAGOS—THE PANGOLINS—THE TATOUHOU—THE GIANT ARMADILLO—THE TATOUAY—THE ARMADILLO—THE MAIS—THE PICHIEGO.

THE animals constituting the present Order are the scanty survivors of numerous families which, in earlier periods of the world's history, assumed vast proportions. Many of the fossil species whose bones are dug out of the bowels of the earth, were as large as the ox, the rhinoceros, and the elephant. Of the species now existing, none exceed three feet in length. These singular animals are almost confined to South America, two families are scantily represented in Africa, and one of these extends over all the East.

The name, EDENTATA, means "toothless"; all the order, however, are not completely devoid of teeth, although this is the case with several species; in all, the incisors are wanting, and there is an empty space in front of their jaws. The teeth, in those that have any, are all alike, and have only one fang. The order is divided into *five* families.

THE SLOTHS.

The family BRADYPODIDÆ is formed by *three* genera of arboreal animals, confined to the great forests of South America from Guatemala to Brazil and Eastern Bolivia. The Sloths, from their climbing habits, were for a long time classed among the monkeys. On the ground they move with extreme slowness, and this inactivity is the source of the common English name. But if we follow with our eye their motions on a tree, in the midst of those conditions of existence which are natural to them, they leave on our mind a very different impression. They embrace the branches with their strong arms, and bury in the bark the enormous claws which terminate their four limbs. As the last joint of their toes is movable, they can bend them to a certain extent, and thus convert their claws into powerful hooks, which enable them to hang on trees. Hidden in the densest foliage, they browse at their ease on all that surrounds them; or, firmly fixed by three of their legs, they avail themselves of the fourth to gather the fruit and convey it to their mouths. They appear sleepy during the day, their eyes not being fitted for sunlight. Their stomach, like that of the ruminants, is divided into four compartments, but it is not known whether they chew the cud. They have no tail, or visible ear. Their fur is harsh, long, and abundant.

GENUS CHOLOEPUS.

The *two* species are sloths with two toes on the fore-limbs, and the sexes are alike in appearance. They are found in the virgin-forests from Costa Rica to Brazil.

The TWO-TOED SLOTH or UNAU, *Choloepus didactylus* (Plate LXI), measures about thirty inches in length. The hair on the forehead and neck is of an olive-green gray color; on the body, olive-gray; on the breast, olive-brown. The muzzle is hairless and flesh-colored; the soles of the feet are also flesh-colored, while the nails are bluish-gray. They pass their whole lives suspended with their backs downward from the branches of trees, hooking their curved talons over it. Their food consists of leaves, buds, and young shoots.

GENUS BRADYPUS.

The Sloths forming the *two* species of this genus possess three toes on the fore-limbs, and extend from the Amazon River to Rio de Janeiro.

The AI or THREE-TOED SLOTH, *Bradypus tridactylus* (Plate LXI), is rather smaller than the Unau. The color of this animal is rather variable, but is generally of a brownish-gray, slightly variegated by differently tinted hairs, and the head and face being darker than the body and limbs. The hair has a curious hay-like aspect, being coarse, flat, and harsh toward the extremity, although it is very fine toward the root. The cry of this creature is low and plaintive, and is thought to resemble the sound Ai. The head is short and round, the eyes sunk in the head, and nose large and moist.

GENUS ARCTOPITHECUS.

The *eight* species of this genus have three toes on the fore-limbs, and the males have a colored patch on the backs. They extend from Costa Rica to Brazil and Eastern Bolivia.

The SPOTTED SLOTH, *Arctopithecus flaccidus*, differs from the other Sloths only by possessing a curious black spot on the back which looks like a hole in the trunk of a tree, and is found only in the males. The fur of most of the sloths has a greenish tinge very like that of the "vegetable horsehair" which clothes many of the trees in Central America, and this probably conceals them from their enemies, the harpy-eagles.

THE SCALY ANT-EATERS.

The family MANIDIDÆ or Scaly Ant-eaters are the only Edentata found out of America. They extend in Africa to the West Coast and the Cape, and in Asia from the Himalayas to Ceylon, Borneo, and Java, as well as to South China.

GENUS MANIS.

The *solitary* genus comprises *eight* species. In all, the upper part of the body is covered with large flat scales of horn, which overlap each other like the plates in a fir-cone, and which resemble the scales of a fish more than anything else. The body and tail are long, the head small, the legs short, the claws long and fit for digging. The horny scales do not occur on the under-side of the body; they are sharp at the edges, and uncommonly hard. Between these scales are a few thin hairs. The snout has no scales, but is protected by a horny skin.

The PHATAGIN or LONG-TAILED MANIS, *Manis longicaudatus* (Plate LXI), is a native of West Africa. The body is two feet, the tail three feet in length. When attacked, it rolls itself up into a ball. The negroes kill it with clubs, sell its hide to Europeans, and eat the flesh, which is white and tender. The tongue of the Phatagin is long, and is either inserted into the holes of the ants on which it lives, or laid across their path.

The PANGOLIN or SHORT-TAILED MANIS, *Manis brevicaudatus*, is an Eastern species. It has five toes. Sir Emerson Tennent gives the following short account of it: "Of the Edentates, the only example in Ceylon is the Scaly Ant-eater, called by the Singalese, Caballaya, but usually known by its Malay name of Pengolin, a word indicative of its faculty of 'rolling itself up' into a compact ball, by bending its head toward its stomach, arching its back into a circle, and securing all by a powerful hold of its mail-covered tail. When at liberty, they burrow in the dry ground to a depth of seven or eight feet, where they reside in pairs, and produce annually two or three young."

THE ARMADILLOS.

The family DASYPODIDÆ ranges from South Texas to the plains of Patagonia. It is divided into *six* genera, which are all very much alike. The legs are short, the body compact, the tail of medium length, the horny plates form rings around the body; the other horny integuments on the head and tail are composed of a number of small plates joined together, and forming patterns which differ in the various species. The whole of the animal, even to the long and tapering tail, is covered with these horny scales, with the exception of the upper part of the legs.

GENUS TATARIA.

The TATOUHON, *Tataria septemcinctus*, is a native of Guinea, Brazil, and Paraguay, and is about thirty inches in length. Its color is a very dark brown-black, whence it is often called the BLACK TATU.

According to Audubon: "The Armadillo is not a fighting character, and, in general, does not evince any disposition to resent an attack. In fact, one of them, when teased by a pet parrot, struck out with its claws only till pressed by the bird, when it drew in its head and feet, and, secure in its tough shell, yielded, without seeming to care much about it, to its noisy and mischievous tormentor, until the parrot left it to seek some less apathetic and more vulnerable object."

GENUS PRIMODONTES.

The GIANT ARMADILLO, *Primodontes gigas*, measures more than four feet six inches in length, the head and body being rather more than three feet long. The teeth are very remarkable, there being from sixteen to eighteen small molars on each side of the jaws. The tail is about seventeen inches long, and tapers gradually to a point from the base, at which spot it is nearly ten inches in circumference. This member is covered with regularly graduating horny rings, and when dried and hollowed, is used as a trumpet by the Botocudos. The Tatou is found in Brazil and Surinam.

GENUS XENURUS.

The TATOUAY, *Xenurus unicinctus*, is mostly remarkable for the undefended state of its tail, which is devoid of the bony rings that encircle the same member in the other Armadillos, and is only supplied with a coating of brown hair. For about three inches of the extremity the under-side of the tail is not even furnished with hair, but is quite naked, with the exception of a few rounded scales.

GENUS DASYPUS.

The ARMADILLO, *Dasypus sexcinctus* (Plate LXI), is about two feet long, and is protected by a coat of six rings of armor-plates. It lives in burrows about two yards long, which, at the entrance, only admit the creature's body, but enlarge in the interior, where it can turn round. Here it passes most of its time. As it is a nocturnal animal, its eyes are more fitted for the dark than for the bright glare of sunlight, which dazzles the creature, and sadly bewilders it. If it should be detected on the surface of the ground, and its retreat intercepted before it can regain its hole, the Armadillo rolls itself up, and tucking its head under its chest, draws in its legs and awaits the result. When taken in hand, it kicks so violently with its powerful legs, that it can inflict severe lacerations with the digging claws.

GENUS TOLYPEUTES.

The APAR, *Tolypeutes tricinctus*, is remarkable for the solid manner in which its armor is attached. It is sometimes called the MATAKO or the BOLITA, from its rolling itself into a ball. It measures about eighteen inches in length. The three rings which characterize it gradually narrow from the back to the sides. The rest of the body is hidden by a coat of rough irregular-shaped plates, each of which is composed of a number of small irregular pieces. The color of the animal is a plumbous gray, the skin between the rings is white, but dark on the lower side of the body, where some very strong plates cover the upper parts of the limbs.

GENUS CHLAMYDOPHORUS.

The PICHIAGO, *Chlamydophorus truncatus*, is a native of Chili.

The top of the head, the back, and the hind-quarters are covered with a shelly plate, which runs unbroken to the haunches, over which it dips suddenly, looking as if the creature had been chopped short by the blow of a hatchet, and a piece of shell stuck on the cut extremity. The remainder of the body is covered with long silken hair, very like that of the mole in its soft texture. It is a very little creature, scarcely surpassing the common mole in dimensions, and living, like that animal, almost entirely below the surface of the earth. Its feet are formed for burrowing, and are most powerful instruments for that purpose, though they are not well fitted for rapid progress over the ground.

Its food consists, as far as is known, of worms, and other subterraneous creatures, in addition to those which it may catch in its nocturnal expeditions into the open air. As is the case with the mole and other subterranean animals, the eyes are of minute dimensions, and are hidden under the soft and profuse fur of the face.

CHAPTER II.

THE AARD VARK AND ANT-EATERS.

THE AARD VARK OF THE CAPE—THE GREAT ANT-EATER OR TAMANOIR—THE TAMANDUA—
THE LITTLE ANT-EATER.

THE family ORYCTEROPIDÆ consists of only *one* genus. The animals belonging to it have the general form of an Ant-eater, but the bristly skin and obtuse snout of a pig.

GENUS ORYCTEROPUS.

The AARD VARK, *Orycteropus capensis*, is the best known of the *two* species.

It is short-legged; its claws are thick, sharp, and almost like hoofs. Its skin is hard, and covered with scanty and rough hair; its head, which is very long and tapering, is terminated by a kind of snout. Its mouth is furnished with molar teeth of a very peculiar structure. If a horizontal section is made of one of these teeth, it presents the appearance of a piece of cane.

The Aard Vark measures rather more than three feet in length, not including the tail, which is about a foot and a half long. Its height is eighteen inches. It lives in burrows, which it hollows out with great rapidity. When its head and fore-feet are buried in the ground, it maintains its position with so much obstinacy that the strongest man is unable to draw it out. Its food consists of ants, or rather *termites*, insects which are commonly designated by the name of white ants, on account of their resemblance to very large specimens of the race. It is well known that these termites live enclosed in a mound of earth in the form of a dome. The Aard Vark, squatting down by the side of it, scratches till an entrance is effected through the walls, and immediately legions of the insects rush out to defend their habitation. Without losing a moment, the quadruped darts out its tongue, which is covered with a viscous humor, into the midst of the restless crowd, and then draws it back covered with the victims.

This exclusive description of food communicates to the flesh a strongly acidulated taste; nevertheless the Hottentots and the colonists at the Cape of Good Hope are partial to it, and hunt these animals. A slight blow on the head with a stick is sufficient to kill it. The Aard Vark is represented in Northeastern Africa by the species *Orycteropus Ethiopicus*, and it is probable that a third species occurs in Senegal.

The family MYRMECOPHAGIDÆ comprises *three* genera of true Ant-eaters. They have little resemblance to the Aard Vark. The body is more elongated, the head and snout more pointed, and the tail is much longer. The hinder legs are weaker than the fore ones. The mouth is very small, the tongue long, thin, and round like a worm. The eyes and ears are small. There is not a trace of teeth.

GENUS MYRMECOPHAGUS.

The GREAT ANT-EATER, *Myrmecophagus jubatus* (Plate LXI), the *only* species, is also called the ANT BEAR or TAMANOIR. It possesses a wonderfully elongated and narrow head, and is thickly covered with long, coarse hair, which on the tail forms a heavy plume. The color of this animal is brown, washed with gray on the head and face, and interspersed with pure white hairs on the head, body, and tail. The throat is black, and a long triangular black mark arises from the throat, and passes obliquely over the shoulders. There are four toes on the fore-feet, and five on the hinder. In total length it measures between six and seven feet, the tail being about two feet six inches long.

The claws of the fore-feet are extremely long and curved, and the animal folds them back upon a thick, tough pad which is placed in the palm, and seems to render the exertion of walking less difficult. As, however, the Ant Bear is forced to walk upon the outer edge of its fore-feet, its progress is a peculiarly awkward one, and cannot be kept up for any long time. Its mode of feeding is similar to that of the Aard Vark. It extends its long extensile tongue, which is covered with a glutinous secretion to which the ants adhere, and which prevents them from making their escape during the short period of time that elapses between the

SLOTH AI
GREAT ANT-EATER LITTLE ANT-EATER
ARMADILLO PHATAGIN

PLATE LXI EDENTATA

moment when they are first touched and that in which they are drawn into the mouth. It makes no burrows, but contents itself with the shade of its own plumy tail when it retires to rest. While sleeping, it likes a rough bundle of hay thrown on the ground. The eye is very sly and cunning in its expression. It is a native of Paraguay, Brazil, and Guinea.

It is nocturnal, solitary, and listless in its habits, and delights in damp forests and marshy savannahs, in which its insect food is most abundant. The female only produces a single young one at a time, which she constantly carries on her back. In the gardens of the Zoological Society of London, which was in possession of two specimens, they were fed on bread soaked in milk, and eggs; but it became certain that they had also a taste for blood, as they were one day noticed sucking the flesh of a rabbit which had been given them.

GENUS TAMANDUA.

The *two* species of four-toed Ant-eaters extend from Guatemala and Ecuador to Paraguay. They possess one of the characteristics which are peculiar to American monkeys, that of grasping branches firmly by the tail, a portion of which is bare of hair underneath.

The TAMANDUA, *Tamandua tetradactyla*, has four toes on the hind, five on the fore-feet. It inhabits the same districts as the Great Ant-eater, but extends into Peru. Its total length is about forty inches, its average height about a foot or fourteen inches. The head is not so elongated, nor the muzzle so prominent as in the Great Ant-eater; the neck is thick, the ears oval, and the hair short over the whole body. Its color, too, is much lighter than that of the Tamanoir, and a black stripe passes over each shoulder. It is much more active than the preceding species, and is an excellent climber of trees, which it ascends in search of the insects on which it feeds. The tail is long and tapering.

A live specimen has been carried to London, where it soon came to know its keeper, leaping upon his shoulder, and poking its worm-like tongue into every fold of his clothing, as well as into his ears, nose and eyes. When strangers approached, it ran to the bars of its cage, and put its tongue out to lick their hands; its food was milk, sweet biscuits, and chopped meat.

GENUS CYCLOTHURUS.

The two species of Two-toed Ant-eaters are found in South America from Honduras and Costa Rica to Brazil.

The LITTLE ANT-EATER, *Cyclothurus didactylus* (Plate LXII), is a truly curious animal, possessing many of the habits of the two preceding animals, together with several customs of its own. The head of this creature is comparatively short; its body is covered with fine silken fur, and its entire length does not exceed twenty or twenty-one inches. The tail is well furred, excepting three inches of the under surface at the extremity, which is employed as the prehensile portion of that member, and is capable of sustaining the weight of the body as it swings from a branch. On looking at the skeleton, a most curious structure presents itself. On a side view, the cavity of the chest is completely hidden by the ribs, which are greatly flattened, and overlap each other so that on a hasty glance the ribs appear to be formed of one solid piece of bone. There are only two claws on the fore-feet and four on the hinder limbs. Its silky coat is usually fox-red, but darker specimens have been seen.

The Little Ant-eater is a native of tropical America, and is always to be found on trees. It possesses many squirrel-like customs, using its fore-claws with great dexterity, and hooking the smaller insects out of the bark crevices in which they have taken refuge. While thus employed it sits upon its hind limbs, supporting itself with its prehensile tail. The claws are compressed, curved, and very sharp, and the animal can strike smart blows with them. It is a bold little creature, attacking the nests of wasps, putting its little paw into the combs, and dragging the grubs from their cells.

Like its larger relations, it is nocturnal in its habits, and sleeps during the day with its tail safely twisted round the branch on which it sits.

ORDER XII.

MARSUPIALIA.

FAMILIES.

76. DIDELPHIDÆ . . . . Opossums.
77. DASYURIDÆ . . . . Native Cats.
78. MYRMECOBIIDÆ . . . . . Native Ant-eater.
79. PERAMELIDÆ . . . . . . Bandicoots.
80. MACROPODIDÆ . . . . . Kangaroos.
81. PHALANGISTIDÆ . . . . . Phalangers.
82. PHASCOLOMYIDÆ . . . . Wombats.

CHAPTER I.

THE OPOSSUMS AND BANDICOOTS.

THE MARSUPIALS—THE TRUE OPOSSUMS—THE VIRGINIAN OPOSSUM—MERRIAN'S OPOSSUM—THE CRAB-EATING OPOSSUM—THE YAPOCK—THE POUCHED MOUSE—THE TASMANIAN DEVIL—THE NATIVE CAT—THE ZEBRA WOLF—THE NATIVE ANT-EATER—THE STRIPED KANGAROO—THE CHŒROPUS.

THE name MARSUPIALIA is derived from the Latin word *marsupium* "a purse," and indicates the characteristic peculiarity of the whole Order. The pouch varies considerably in the various species. Usually it is tolerably large, and it contains the *mammæ* or teats. The young, when born, are exceedingly minute, and are transferred by the mother into the pouch, when they instinctively attach themselves to the teats. By degrees they loosen their hold, and put their little heads out of the living cradle. In a few weeks more they leave the pouch entirely, but their mother is always ready to receive them again into their cradle.

The Order is divided into *seven* families, six of which are Australian, and one American.

The family DIDELPHIDÆ or TRUE OPOSSUMS range through all the wooded districts of South America from Texas to the River Platt, with one species extending to the Hudson River, and west to the Missouri. They are most numerous in the great forest region of Brazil, but the number of species is very uncertain.

GENUS DIDELPHYS.

With the exception of the Yapock, and a small rat-like animal, HYRACODON, found in Ecuador, all the members of the family belong to this genus, and form *twenty* species.

The OPOSSUM, *Didelphys Virginiana* (Plate LXII) is about the size of a large cat. They have the thumbs opposable and nailless, the tail generally bare and prehensile for its terminal half, or more. The mouth is provided with fifty teeth. They sally out at twilight or at night to feed on small quadrupeds, birds, eggs, insects, molluscs, and even fruits or young vegetable shoots, from which they suck the sap. The females are remarkably prolific; they have from ten to fifteen young at a litter, and nurse their progeny with tender solicitude. The Opossum being plantigrade, walks and runs awkwardly, but climbs well.

MERRIAN'S OPOSSUM, *Didelphys dorsigera* (Plate LXII), has no true pouch, the place being indicated only by a fold of skin. The mother carries her young on her back, where they cling tightly to her fur with their feet, and twine their tails around hers. It is a very small animal, measuring when full-grown only six inches from the nose to the root of the tail. The fur is very short, and of a pale grayish-brown on the upper portions. Round the eye is a deep brown mark, the forehead, cheeks, limbs, and feet are of a yellowish-white, inclining to gray. Its native country is Surinam.

The CRAB-EATING OPOSSUM, *Didelphys cancrivora*, is not so large an animal as the Virginia Opossum, being only thirty or thirty-one inches in total length, the head and body measuring sixteen inches, and the tail fifteen. It can also be distinguished by the darker hue of its fur, the attenuated head, and the uniformly colored ears, which are generally black, but are sometimes of a yellowish tint. It is peculiarly fitted for dwelling in trees, where it swings with its prehensile tail.

GENUS CHEIRONECTES.

The YAPOCK, *Cheironectes yapock* (Plate LXII), is the *only* species of the genus. It may be considered as the Marsupial representative of the Otter.

The general hue of the fur is a pale fawn-gray, with a very *watery* look about it, and set closely upon the skin. Four dark bands of sooty-black are drawn across the body in a peculiar, but extremely variable manner. The hind-feet are furnished with a membranous web. The claws are small and weak, and the thumb-joint is not opposable to the others. The ears are moderate in size, sharp, and pointed, and the head

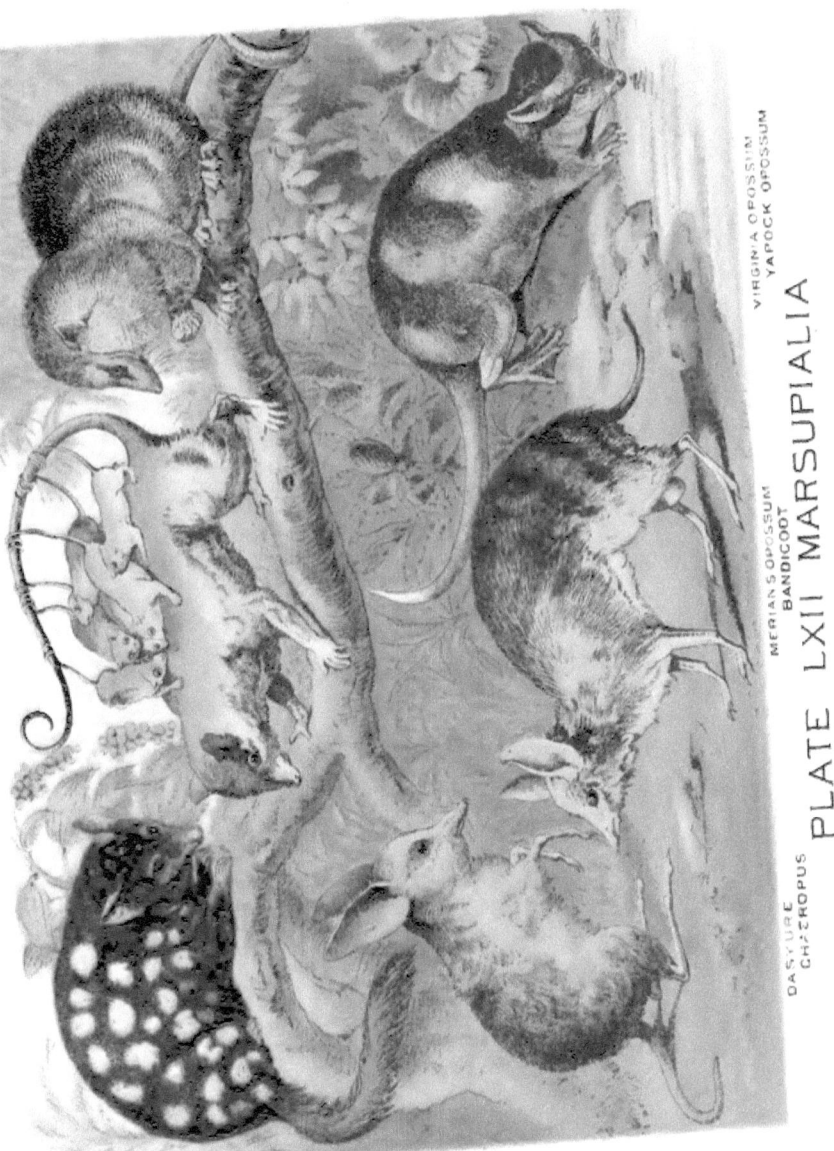

PLATE LXII MARSUPIALIA

VIRGINIA OPOSSUM
YAPOCK OPOSSUM
MERIAN'S OPOSSUM
BANDICOOT
DASYURE
CHŒROPUS

tapers rapidly toward the nose. The Yapock has large cheek-pouches, extending far backward along the sides of the mouth, and capable of containing a large supply of food.

The family DASYURIDÆ forms a group of Carnivorous and Insectivorous Marsupials ranging from the size of a wolf to that of a mouse. They are found in Australia, Tasmania, and New Guinea, and are placed in *ten* genera. They all possess sharp-edged and pointed teeth.

GENUS ANTECHINUS.

The *twelve* species are distributed over all Australia and Tasmania, and are the most common of Australian quadrupeds. They are all small, hardly exceeding the size of a common mouse.

The YELLOW-FOOTED POUCHED MOUSE, *Antechinus flavipes*, is a very pretty little creature. The face, head, and shoulders are dark-gray, the sides of the body a warm chestnut. The chin, throat, and abdomen are white, the tail black, and often tufted. They are arboreal animals, and run up and down a perpendicular trunk with singular activity.

GENUS SARCOPHILUS.

The TASMANIAN DEVIL, *Sarcophilus ursinus* (Plate LXIII), is the *only* species. Its fur is of a deep dead-black, without any glossiness, and there is a conspicuous white mark across the breast. The body is heavy and thick-set, the muzzle blunt, the mouth wide. Its length, exclusive of the tail, is about twenty inches. It is a creature of unparalleled ferocity and stupidity, and lives in a state of passionate and purposeless anger. It is incapable of fear, and more than a match for most dogs. It is now nearly exterminated, as it plays sad havoc in the poultry-yard. It burrows in the ground.

GENUS DASYURUS.

The DASYURE or NATIVE CAT, *Dasyurus viverrinus* (Plate LXII) is of a dark brown, almost black color, diversified with many spots of white scattered at random over the whole of the body. The tail is moderately long, but not prehensile, and is thickly covered with hair. They make their abodes in the hollows of decayed trees. This is the best known of the *four* species.

GENUS THYLOCINUS.

The ZEBRA WOLF, *Thylocinus cynocephalus*, the only species, is the largest and most destructive of the Carnivorous Marsupials. They are also called "Native Tigers" or "Native Hyænas," and being destructive to sheep, have been extensively destroyed by the farmers. The general tint of the fur is a grayish-brown, washed with yellow. It is about equal to the jackal in size, and if attacked will fight in the most desperate manner.

The family MYRMECOBIIDÆ consists of only one genus and one species, the Native Ant-eater, a small, bushy-tailed, squirrel-like animal, found in the South and West of Australia. It is remarkable for possessing no fewer than fifty-two teeth. The pouch is rudimentary.

GENUS MYRMECOBIUS.

The NATIVE ANT-EATER, *Myrmecobius fasciatus* (Plate LXIII) is a beautiful little animal. The general color is a bright fawn on the shoulders, which deepens into blackish-brown from the shoulders to the tail, the fur of the hinder portions being nearly black. Across the back are drawn six or seven white bands, broad on the back, and tapering off towards their extremities. The under parts of the body are of a yellowish-white. The tail is thickly covered with long, bushy hair, and has a grizzled aspect. Its long tongue is nearly as thick as a common black-lead pencil, and is capable of protrusion to some distance.

The number of its young averages from five to eight. Its usual habitation is in the trunk of a fallen tree, or in a hollow in the ground. It is a native of the borders of the Swan River.

The family PERAMELIDÆ are small Insectivorous Marsupials, approximating in form to the kangaroo. They range over the whole of Australia. They have a rat-like aspect, and their gait is a mixture of running and jumping. The pouch opens backward.

GENUS PERAMELES.

The STRIPED BANDICOOT, *Perameles fasciata* (Plate LXII), is of a blackish-yellow color. Over the hinder quarters are drawn some boldly marked black lines, which, when viewed from behind, form a singular and rather pleasing pattern, the dark stripes being made more conspicuous by bands of whitish-yellow. These marks continue as far as the root of the tail, and a single, narrow dark line runs along the whole upper side of the tail, which is of the same color as the body. The fur is rather light upon the head, and the under parts of the body, together with the feet, are white, slightly tinged with gray.

The LONG-NOSED BANDICOOT, *Perameles nasuta*, is not unlike the preceding animal in form. The face, head, and body are of a brown tint, penciled with black on the upper portions, and the sides are of a pale brown. The edge of the upper lip is white, as are also the under portions of the body, and the fore-legs and feet. The fur is very harsh to the touch. The total length of this animal is about twenty-one inches, the tail being five inches in length.

The genus contains eight species, one of which, *Perameles Gunii*, is peculiar to New Guinea.

GENUS CHŒROPUS.

The *solitary* species of this genus is a very pretty little animal with something of the appearance of the mouse-deer. The name, *Chœropus* or "Swine-footed," has been given to it from the fact that the fore-feet have only two toes of equal length, with hoof-like claws, which leave a track very much like that of the hog.

The CHŒROPUS, *Chœropus castanotis* (Plate LXII), is about equal in size to a small rabbit, and the soft, woolly fur is much of the same color as that of the common wild rabbit.

It is an inhabitant of New South Wales. Its speed is considerable, and its usual haunts are among the masses of dense scrub foliage that cover so vast an extent of ground in its native country. Its nest is similar to that of the Bandicoot, being made of dried grass and leaves rather artistically put together, the grass, however, predominating.

CHAPTER II.

THE KANGAROOS, PHALANGERS, AND WOMBATS.

THE KANGAROO—THE WOOLLY KANGAROO—THE WALLABEE—THE ROCK KANGAROO—THE TREE KANGAROO—THE KANGAROO HARE—THE JERBOA KANGAROO—THE POTOROO—THE KOALA—THE SOOTY PHALANGIST—THE VULPINE PHALANGIST—THE CUSCUS—THE TAGUAN—THE GREAT FLYING PHALANGER—THE SUGAR SQUIRREL—OPOSSUM MOUSE—THE WOMBAT.

THE most prominent characteristic of the family MACROPODIDÆ is the relative disproportion of their anterior and posterior members. While the former are short and weak, the latter are singularly long, thick, and strong. Thence the name of *Macropodidæ* or "long-footed." The tail is long and powerful, and constitutes a sort of fifth member, destined to facilitate in the Kangaroos that mode of progression which is peculiar to them.

GENUS MACROPUS.

The *four* species of the genus have all powerful tails, not so long as the body.

The KANGAROO, *Macropus major* (Plate LXIII), is sometimes called the Great Kangaroo. The average dimensions of an adult male are generally as follows: the head and body exceed four feet, and the tail is rather more than three feet in length; the circumference of the tail at its base is about a foot. When it sits erect after its curious tripedal fashion, supported by its hind-quarters and tail, its height is rather more than fifty inches; but when it wishes to survey the country, and stands erect upon its toes, it surpasses in height many a well-grown man. The female is very much smaller than her mate, being under six feet in total length, and the difference in size is so great that the two sexes might well be taken for different species.

The color of the animal is brown, mingled with gray, the gray predominating on the under portions of the body and the under-faces of the limbs. The fore-feet are black, as is also the tip of the tail.

The WOOLLY KANGAROO, *Macropus laniger*, has a woolly matted coat, of a rusty yellow color, changing to gray upon the head and shoulders. The sides of the mouth are white, and the toes are covered with black hairs. The tail is uncommonly large and powerful, and is covered with short hair.

GENUS HALMATURUS.

The WALLABEE, *Halmaturus walabatus*, is the most typical of the eighteen species. It is not as large as the common or the woolly kangaroo, being only four feet six inches in total length, of which measurement the tail occupies two feet. Its fur is rather long and coarse in texture. The color is a darkish-brown washed with a warm rusty hue, and obscurely pencilled with whitish-gray. The under portions of the body are of a yellowish tint, and the feet and the wrists are quite black. The dorsal third of the tail is of the same color as the back, but the remaining two-thirds change abruptly from brown to black.

The animal is an inhabitant of New South Wales, and is of tolerably frequent occurrence in the neighborhood of Port Jackson. It is sometimes known by the name of the Aroe Kangaroo.

GENUS PETROGALE.

The genus comprises seven species of Kangaroos, which have derived their generic title from their rock-loving habits.

The ROCK KANGAROO, *Petrogale penicillata*, is the most striking of the group.

The color of the animal is of a purplish-gray, warming into a rusty red upon the hind-quarters. A white band runs along the throat to the chest. The total length of a male adult is about four feet. The tail has a tuft of dark hairs about three inches in length. The body is robust, and the feet thickly covered with fur.

GENUS DENDROLOGUS.

The *two* species are peculiar to New Guinea.

The TREE KANGAROO, *Dendrologus ursinus* (Plate LXIII), is called *ursinus*, or bear-like, as the hairs of its fur are thought to bear some resemblance to those of the American black bear. The whole of the back and the upper parts of the body are a deep, glossy black, the hairs being rather coarse, and running to some length. The under parts of the body are of a yellowish hue, and the breast is washed with a richer and deeper tint of chestnut. The tail is of the same color as the body, and is of very great length, probably to aid the animal in balancing itself as it climbs among the branches of the trees on which it loves to disport itself.

GENUS LAGORCHESTES.

The KANGAROO HARE, *Lagorchestes leproides*, is found in the interior of Australia. All the *five* species in the genus have many hare-like traits, such as squatting close to the ground, "doubling" when pursued, and running with great celerity.

The total length of this animal is about two feet, the tail occupying about thirteen inches. The color of the fur is very like that of the European hare, but the different specimens vary much, some being much redder than others. The fore-legs are black, the hinder feet a brownish white. A buff-colored ring surrounds the eye, and the back of the neck is washed with yellow. The tail is of a very pale brownish-gray.

GENUS BETTONGIA.

The *six* species forming this genus have very short, broad heads. Their general color is a palish brown, pencilled with white, and their size is about that of a hare.

The BRUSH-TAILED BETTONG, *Bettongia penicillata*, is also called the Jerboa Kangaroo. It is very common over the whole of New South Wales.

It is a nocturnal animal, and its nest is a very ingenious specimen of architecture. It conveys the materials for it by twisting its prehensile

PLATE LXIII MARSUPIALIA

tail round a bundle of grass and hopping away merrily with its burden. The mother invariably drags a tuft of grass over the entrance whenever she leaves or enters her home.

GENUS HYPSIPRYMNUS.

The POTOROO or KANGAROO RAT, *Hypsiprymnus minor*, is the most typical of the *four* species.

It is but a diminutive animal, the head and body being only fifteen inches long, and the tail between ten and eleven inches. The color of the fur is brownish-black, pencilled along the back with a gray-white. The under parts of the body are white, and the fore-feet are brown. The tail is equal to the body in length, and is covered with scales, through the intervals of which sundry short, stiff, and black hairs protrude.

The family PHALANGISTIDÆ constitutes one of the most varied and interesting groups of Marsupials, being modified in various ways for an arboreal life. These variations within the range of a single family indicate the great antiquity of the Australian fauna. It comprises *eight* genera.

GENUS PHASCOLARCTUS.

The KOALA, *Phascolarctus cinereus* (Plate LXIII), is the Australian representative of the American sloth. The toes of the forefeet or paws are divided into two sets, the one composed of the two inner toes, and the other of the three outer.

The head has a very unique aspect, on account of the tufts of long hairs which decorate the ears. The muzzle is devoid of hair, but feels like cotton velvet when gently stroked with the fingers.

GENUS PHALANGISTA.

The PHALANGISTS are characterized by possessing a prehensile tail. This member, although to all appearance covered with a heavy coating of long hair, has its inferior surface for some distance from the extremity bare of fur. The genus contains *five* species.

The SOOTY PHALANGIST or TAPOA, *Phalangista fuliginosa*, is prized for its soft and beautiful fur which usually is of a deep blackish-brown color. The tail is very full, the hair being thick, long, and bushy. The ears are rather elongated, and externally naked. The underside of the body is of the same color as the upper portions.

The VULPINE PHALANGIST, *Phalangista Vulpina*, is often called simply the Opossum. It merits by its form and nature, its distinguishing epithet of Vulpine or Fox-like.

It is an extremely common animal, and is the widest diffused of all the Australian opossum-like animals. It is nocturnal, residing during the day in the hollows of decaying trees. The nature of its food is of a mixed character. If a small bird be given to a Vulpine Phalangist, the creature seizes it in his paws, and then tears it to pieces and eats it. In all probability, therefore, the creature makes no small portion of its meals on various animal substances, such as insects, reptiles, and eggs. It is a tolerably large animal, equalling a large cat in dimensions.

GENUS CUSCUS.

The WOOLLY PHALANGERS are divided into *eight* species. The tail, instead of being covered with hair, is naked, except at the base, and is thickly studded with minute tubercles. They are not found in Southern Australia; but are natives of New Guinea, the Moluccas, and Northern Australia.

The SPOTTED CUSCUS, *Cuscus maculatus*, in size is equal to a tolerably large cat, and its tail is remarkably prehensile. Its movements are slow and cautious, and its food is chiefly of a vegetable nature. Its fur is beautifully soft and silken in its texture, but the color is singularly variable. Another species, called the Ursine Cuscus, is of a uniform deep brown.

GENUS PETAURISTA.

The TAGUAN, *Petaurista taguanoides*, is the *only* species of the genus. It is provided with a parachute membrane on the flanks, which enables it to float through the air like the flying squirrels.

In color it is extremely variable; the back is of a rather deep blackish-brown, darker or lighter in different individuals, the feet and muzzle are nearly black, and the under surface of the body and membrane is white. The upper surface of the parachute membrane is rather grizzled, on account of the variegated tints of black and gray with which the hairs are annulated. Many varieties of color, however, exist in the animal.

GENUS BELIDEUS.

The *five* species of Flying Opossums are found in New Guinea, as well as in Australia.

The GREAT FLYING PHALANGER, *Belideus Australis* (Plate LXIII), is an inhabitant of New Holland. Its color is variable, but generally may be described as brown tinged with gray, with a darker line along the spine. The abdomen and under portions of the flying membrane are white, very perceptibly worked with yellow. The feet are blackish-brown, the tail is covered with long and soft fur of a brown tint. The total length of the animal is a little over three feet. The native name for the creature is the HEPOONA ROO.

The SUGAR SQUIRREL, *Belideus sciureus*, is also called the NORFOLK ISLAND FLYING SQUIRREL, and the SQUIRREL PETAURUS. Its fur is very beautiful, being of a nearly uniform brownish-gray, of a peculiarly delicate hue, and remarkably soft in its texture. The parachute membrane is gray above, but is edged with a rich brown band, and a bold stripe of blackish-brown is drawn along the curve of the spine, reaching from the point of the nose to the root of the tail. The head is somewhat darker than the rest of the body. The under parts of the body are nearly white. Its long and bushy tail is covered with a profusion of very long, full, soft hair, grayish-brown above, and of a beautiful white underneath.

GENUS ACROBATA.

The OPOSSUM MOUSE, *Acrobata pygmæus*, is also called the Flying Mouse. It is about the size of our common mouse, and when it is resting upon a branch, with its parachute, or umbrella of skin, drawn close to the body by its own elasticity, it looks very like it.

Its color is the well-known mouse-tint; but on the abdomen, and under portions of the skin-parachute, the fur is beautifully white.

The remaining genera DROMICIA and TARRIPES call for only slight mention. The former contains *five* species of beautiful dormouse-like creatures. The latter is formed of *one* species which is a true honey-sucker, with an extensile tongue, and of the size of a mouse.

The family PHASCOLOMGIDÆ contains only *one* genus of *three* species.

The animals comprised in it are tailless, terrestrial burrowing animals, about the size of a badger, but feeding on roots and grass.

GENUS PHASCOLOMYS.

The WOMBAT, *Phascolomys ursinus* (Plate LXIII), is popularly styled the Australian Badger. In all its exterior appearance it is a rodent; in its internal anatomy it approximates to the beaver.

Its fur is warm, long, and very harsh to the touch, and its color is gray, mottled with black and white. The under parts of the body are grayish-white, and the feet are black. The muzzle is very broad and thick. The length of the animal is about three feet, the head measuring seven inches.

It is nocturnal in its habits, living during the day in the depths of a capacious burrow.

Its teeth present a curious resemblance to those of the rodent animals, and its feet are broad, and provided with very strong claws, that are formed for digging in the earth. There are five toes to each foot, but the thumb of the hinder feet is extremely small, and devoid of a claw. This animal is remarkable for possessing fifteen pairs of ribs—in one case sixteen pairs of ribs were found—only six pairs of which reach the breast-bone.

ORDER XIII.
MONOTREMATA.

FAMILIES.

83. ORNITHORHYNCIDÆ . . . . . DUCKBILLS.
84. ECHIDNIDÆ - - - - - - - - - - ECHIDNÆ.

MONOTREMATA

THE DUCK MOLE AND AUSTRALIAN HEDGE-HOG.

THE MONOTREMATA—THE FAMILY ORNITHORHYNCHIDÆ—THE DUCK MOLE—THE FAMILY ECHIDNIDÆ—THE NATIVE HEDGEHOG—THE TASMANIAN SPECIES—CONCLUSION.

THE strange animals belonging to the order MONOTREMATA appear to be a link between Mammals, Birds, and Reptiles. That they are Mammals, is beyond dispute, for the females have lacteal glands, and secrete genuine milk; but their interior anatomy is that which is familiar to us in birds; externally, and as regards the skeleton, they are the very opposite of birds. They are small mammals, with short limbs, beak-like jaws covered with a dry skin, small eyes, a short flat tail, feet with five long toes and powerful claws. They have no external ear.

THE DUCK-BILLS.

The family ORNITHORHYNCHIDÆ is found in East and South Australia and Tasmania. It is represented by a *solitary* species of a *single* genus, an animal of so strange an appearance, that when the first stuffed specimen was brought to England, it was taken for the contrivance of some swindler.

GENUS ORNITHORHYNCHUS.

The DUCK MOLE, *Ornithorhynchus paradoxus* (Plate LXIV), is known also as the MULLINGONG. The name *Ornithorhynchus* signifies "bird's beak," and describes the duck-like bill the creature possesses. It is an animal duly organized for aquatic life. Its feet have five toes, terminated by stout nails. The front feet are completely webbed. The tail is broad, of middling length, and flattened on its lower surface. The beak is flattened, and is not much unlike that of a swan or duck. Two great horny excrescences, placed on each jaw, supply the place of molars. Its coat is pretty thick, and is of a brown color, more or less tinged with russet, changing to light-fawn on the abdomen.

In the males, the heels of the posterior members are each armed with a spur or dew claw, pierced with a hole at its extremity. This spur allows to escape, at the will of the animal, a liquid, secreted by a gland which is situated on the thigh, and with which the spur communicates by a broad subcutaneous conduit. This liquid has nothing venomous about it. The organ in question, very much developed in the males, is quite rudimentary in the females, and, as she ages, disappears entirely. Dr. Bennett allowed himself frequently to be wounded with this spur, and experienced no ill effects.

The Duck Mole can run on land and swim in water with equal case. Its fore-feet are used for digging as well as swimming, and are armed with powerful claws. The animal has been seen to make a burrow two feet in length through hard gravelly soil in less than ten minutes. It uses its beak as well as its feet when digging. The burrow in which the Mullingong lives is generally from twenty to forty feet in length, and always bends upward toward a sort of chamber in which the nest is made. This nest is of the rudest description, consisting of a bundle of dried weeds thrown carelessly together. The burrow has a very evil odor, which is unpleasantly adherent to the hand that has been placed within it.

Owing to the extremely loose skin of the Mullingong, it can push its way through a very small aperture, and is not easily retained in the grasp, wriggling without much difficulty from the gripe of the fingers. The loose skin and thick fur are also preventives against injury, as the discharge of a gun, which would blow any other animal nearly to pieces, seems to take but little external effect upon the Duck Bill. The animal is, moreover, so tenacious of life, that one of these creatures which had received the two charges of a double-barreled gun, was able, after it had recovered from the shock, to run about for twenty minutes after it had been wounded.

The food of the Mullingong consists of worms, water insects, and

PORCUPINE ANT EATER
DUCK MOLE

little molluscs, which it gathers in its cheek-pouches as long as it is engaged in its search for food, and then eats quietly when it rests from its labors. The teeth, if teeth they may be called, of this animal are very peculiar, consisting of four horny, channeled plates, two in each jaw, which serve to crush the fragile shells and coverings of the animals on which it feeds. It seems seldom to feed during the day, or in the depth of night, preferring for that purpose the first dusk of evening or the dawn of morning. During the rest of the day it is generally asleep. While sleeping, it curls itself into a round ball, the tail shutting down over the head and serving to protect it.

The young Mullingongs are curious little creatures, with soft, short flexible beaks, naked skins, and almost unrecognizable as the children of their long-nosed parents. When they attain to the honor of their first coat, they are most playful little things, knocking each other about like kittens, and rolling on the ground in the exuberance of their mirth. Their little twinkling eyes are not well adapted for daylight, nor from their position can they see spots directly in their front, so that a pair of these little creatures that were kept by Dr. Bennett used to bump themselves against the chairs, tables, or any other object that might be in their way. They bear a farther similitude to the cat in their scrupulous cleanliness, and the continual washing and pecking of their fur.

THE AUSTRALIAN HEDGEHOG.

The family ECHIDNIDÆ contains one genus, which is divided into two closely allied species. These animals, although quite as extraordinary as the Duck Bill in their internal structure, are not so peculiar in their external appearance, having very much the aspect of a hedgehog or spiny ant-eater. One of the species inhabits East and South Australia, the other, Tasmania.

GENUS ECHIDNA.

Both species have squat, thick-set bodies, short legs, very short tails; the beak and tongue narrow and elongated; the toes armed with digging claws; the back covered with spines. The males, like the Duck Mole, have a spur on the heel. They inhabit sandy places, where they dig out burrows, and live on ants, which they catch by projecting their tongue covered with a viscous humor into the dwellings of these insects.

The AUSTRALIAN HEDGEHOG, *Echidna aculeata* (Plate LXIV) is found on the mainland of Australia. The head is elongated like that of the proper ant-eaters, and contains in the jaws no teeth of any kind. The spines or prickles begin at the back of the head, and form a thick covering upon the upper part of the animal. The head, limbs and abdomen are covered with hair of a dark-brown color and very stiff and bristly. The prickles are dirty yellow, with black tips.

If this animal is attacked, it rolls itself into a ball, and is then difficult to hold, as the spines can give severe wounds. It is almost impossible to drag it from its burrow, as it spreads its spines out against the walls of the tunnel. It is probable that it passes the dry season in a state of sleep, for it is, during that period, seldom seen abroad. Cold also has a great influence on it, and a slight decline of temperature seems to place it in a state of hibernation.

It is called NICOBEJAN, JANOKAMBINE, and COGERA, by the natives who roast it in its skin as European gypsies roast the hedgehog. Some of them have been kept in captivity; they were quiet, and liked to be caressed. They were fed on sugared liquids.

The second species, ECHIDNA SETOSA, is found in Tasmania. It resembles the above in every respect except color, which is of a darker brown instead of the black and yellow which decorates the spines of the common species. The prickles are comparatively few, and there is much close fur between them.

CONCLUSION

JOHN JAMES AUDUBON.

JOHN JAMES AUDUBON, the greatest artist in his own walk that ever lived, was born in New Orleans on the 4th of May, 1780. The family was of French origin. His father, who began his career as a sailor, acquired considerable property in Saint Domingo, and held a commission in the French navy. Like his son, he was handsome, with a hardy constitution, great agility, and a restless spirit. He frequently visited the North American Continent and purchased land not only in the French colony of Louisiana, but in Virginia and Pennsylvania, in which latter State he became the possessor of Mill Grove near Schuylkill Falls. The revolt of the negroes in St. Domingo compelled the family to flee and take up their residence on their Louisiana plantation, where the future naturalist first saw the light; and with that fertile land his earliest recollections were associated. To quote his own words, written in 1831: "I received light and life in the New World. When I had hardly yet learned to walk, and to articulate those first words always so endearing to parents, the productions of nature that lay spread all around were constantly pointed out to me. They soon became my playmates, and before my ideas were sufficiently formed to enable me to estimate the difference between the azure tints of the sky and the emerald hue of the bright foliage, I felt that an intimacy with them, not consisting of friendship merely, but bordering on frenzy, must accompany my steps through life; and now, more than ever, am I persuaded of the power of those early impressions. They laid such hold upon me that, when removed from the woods, the prairies and the brooks, or shut up from the view of the wide Atlantic, I experienced none of those pleasures most congenial to my mind. None but aerial companions suited my fancy. No roof seemed so secure to me as that formed of the dense foliage under which the feathered tribes were seen to resort, or the caves and fissures of the mossy rocks to which the dark-winged cormorant and the curlew retired to rest, or to protect themselves from the fury of the tempest. My father generally accompanied my steps, procured birds and flowers for me with great eagerness, pointed out the elegant movements of the former, the beauty and softness of their plumage, the manifestations of their pleasure or sense of danger, and the always perfect forms and splendid attire of the latter. My valued preceptor would then speak of the departure and return of birds with the seasons, and would describe their haunts, and, more wonderful than all, their change of livery; thus exciting me to study them, and to raise my mind toward their great Creator."

He left Louisiana, however, while still a child, and after a brief residence in St. Domingo was taken to France. His father was desirous he should become a sailor like himself, and among the branches of education prescribed for him were music, drawing, and dancing. He played skilfully on the violin, flute, and guitar, and his terpsichorean skill served him well at the crisis of his fate, for his voyage to England to publish his great work was rendered possible by gaining two thousand dollars by dancing lessons. His drawing master was the great chief of the classical school,—David, the celebrated painter. At his home at Nantes he indulged in his nest-hunting propensities. He began to collect and try to preserve the specimens. But the preparation of the birds, after death, was onerous, and required constant care, and was subject to decay as the beauty of the plumage vanished. "I wished," was the longing of the boy naturalist, "all the productions of nature, but I wished life with them." The sequel is told in rather a dramatic way: "I turned to my father and made known to him my disappointment and anxiety. He produced a book of illustrations. A new life ran in my veins. I turned over the leaves with avidity; and although what I saw was not what I longed for, it gave me a desire to copy nature." To copy nature—it is in three words the story of his future life. He worked so assiduously that he completed drawings of no less than two hundred French birds. His pencil, he says, "gave birth to a family of cripples. So maimed were most of them that they resembled the mangled corpses on a field of battle, compared with the integrity of living men." One expression strikes us as of singular force and felicity: "The worse my drawings were, the more beautiful did the originals seem."

His father, although desirous to see his son in his country's service, wisely recognized the bent of his genius, and sent him to superintend his property in America. He landed in New York, and it is strange to read that he caught the yellow fever by walking to what is now Greenwich Street in that city. On his recovery he took possession of his Mill Grove farm. "A blessed spot," as he describes it, "where hunting, fishing, and drawing occupied my every moment." In his autobiographical sketch he tells in unaffected language his first meeting with his wife, Lucy Bakewell, the daughter of an English neighbor. She taught him English, but to the last he retained his French accent, and with his intimate friends preserved the graceful French use of the second person singular; and in his letters and memoranda French words and constructions are of common occurrence. His life at Mill Grove was in every way agreeable. He had ample means; was gay and fond of dress; ridiculously fond, he says, for he used to hunt in black satin breeches, wear pumps when shooting, and dress in the finest ruffled shirts he could import. Yet with all this his mode of life was abstemious. "I ate no butcher's meat, lived chiefly on fruits, vegetables, and fish, and never drank a glass of spirits or wine until my wedding day. To this I attribute my continued good health, endurance, and an iron constitution. All this time I was fair and rosy, and active and agile as a buck."

The misconduct of a partner, Da Costa, who had been sent out by the elder Audubon, compelled him to hurriedly leave this "happy spot" and hasten to France. He remained there four years, and after a term of service in the Imperial navy as midshipman, sailed again for the western continent, where he arrived safely, after his ship had been overhauled by a British privateer. Once more at Mill Grove, he resumed his hunting, drawing, and lovemaking with Lucy Bakewell. His house became a museum, the walls were festooned with bird's eggs, the mantel-piece was covered with stuffed squirrels, raccoons, and opossums, the shelves laden with fishes, frogs, and reptiles, and the walls hung with paintings of birds. Patiently and with industry he applied himself to study, determined to make his sketches represent life. "David," he writes, "had guided my hand in tracing objects of large size; eyes and noses of giants and heads of horses from the antique were my models. These subjects I laid aside; I returned to the woods of the new world with fresh ardor, and commenced a collection of drawings." The idea of an "Ornithological Biography" already took possession of his mind, and he labored to produce lifelike pictures, giving the form, plumage, attitude, and characteristic marks of his feathered favorites. He acquired, too, great skill in stuffing and preserving birds. But these artistic instincts found little favor in the eyes of Mr. Bakewell, who insisted on his obtaining some knowledge of commercial pursuits. To this end he proceeded to New York, where he lost his money, and continued as far as possible his favorite pursuits. His rooms in the city became a laboratory, and the odor of his bird-flaying and bird-stuffing achievements produced a visit from a constable who came to abate the nuisance. Some of the specimens prepared by Audubon at this period were made for Dr. Samuel Mitchel, one of the founders of the Lyceum of Natural History, and were finally, it is believed, deposited in the New York Museum. Unsuccessful in his commercial attempts, Audubon resolved to "go West." He sold Mill Grove, married Lucy Bakewell, and set out for Louisville, Kentucky.

At this time of his life his appearance was fascinating. "I measured," he naively writes, "five feet ten and a half, was of a fair mien and quite a handsome figure, large, dark, and rather sunken eyes, aquiline nose, and a fine set of teeth; hair, fine texture and luxuriant, divided and passing down behind each ear in luxuriant ringlets as far as the shoulders." He was an admirable marksman, an expert swimmer, a clever writer, possessing great activity, prodigious strength, and was notable for the careful attention to the care of his dress. "When I first knew Mr. Audubon," said a lady, "people used to ask who was this gay young Frenchman who danced with all the girls." Such was Audubon in 1808.

The life of the Apprentice was over; the life of the Journeyman was beginning. "For a period of twenty years," writes Audubon, "my life was a succession of vicissitudes. I tried various branches of commerce, but they all proved unprofitable, doubtless because my whole mind was ever filled with my passion for rambling and admiring those objects of nature from which alone I received the purest gratification. I had to struggle against the will of all who at that period called themselves my friends. I must here, however, except my wife and children. The remarks of my other friends irritated me beyond endurance, and breaking through all bonds, I gave myself entirely up to my pursuits. Any one unacquainted with the extraordinary desire which I then felt of seeing and judging for myself, would doubtless have pronounced me callous to every sense of duty, and regardless of every interest. I undertook long and tedious journeys, ransacked the woods, the lakes, the prairies, and the shores of the Atlantic. Years were spent away from my family. Yet, reader, will you believe it, I had no other object in view than simply to enjoy the sight of nature."

It was while he was in business in Louisville that Audubon met Wilson, the celebrated author of the "American Ornithology," in the year 1810. Wilson called on him to solicit his patronage for his book. "Do not subscribe, my dear Audubon," said a French friend, "your own drawings are certainly far better." Neither at the moment appears to have had any previous knowledge of the pursuits of the other. Audubon examined the engravings of Wilson with interest, and the latter was still more surprised to witness such drawings of birds in the portfolio of a Western storekeeper. Wilson asked if it was his intention to publish, and appeared still more perplexed when he learned that so patient a student had no such object. He borrowed the drawings to examine during his stay in the town, and was introduced to birds new to him in the neighborhood, in hunting with his chance acquaintance. Audubon, who as yet had not "taken unto the height the measure of himself,' placed all his drawings at the disposal of his visitor, with the result of his researches, proposing a correspondence, and stipulating only in return for an acknowledgment, in the published work, of what came from his pencil. The offer was not accepted, and Wilson left Louisville, declaring he had not got one subscriber or one new bird, or received one act of civility. "Science and literature has not one friend in this place." A most unjust remark, had he but known the man before him. They were alike pupils in the great school of nature, taking their lessons in the wilderness, encountering, Wilson particularly, more enemies in the indifference of the world than the "winter and rough weather" to which they voluntarily subjected themselves.

Audubon found business bad at Louisville and removed to Henderson. He found it bad at Henderson and removed to St. Genevieve. The population here was French Canadian, to which Audubon had all the French Creole's antipathy, and he speedily returned to Henderson. It was on the way to St. Genevieve, an old town forty miles below St. Louis, that he first saw the great eagle that he named after Washington. All these wanderings of Audubon were training him for his great work, and during the whole period his soul was given to his favorite study rather than to mercantile adventures. The country he traversed was unbroken wilderness—no towns, no villages, no roads, no steamboats. When he left one of his temporary dwelling-places he embarked his stock in trade—usually, it seems, whisky—his provisions, his arms and powder, and pushed into the stream. The highest speed with the current was five miles an hour; days were passed in going a few miles, but then these miles were miles of wild fowl, on the water and on the land, and herds of wild deer and countless bears and raccoons. Encampments of Indians were visited, and he joined the red men in their hunt. Some of his vivid descriptions are of the adventures encountered in the company of these Sons of the Forest. At times the boat had to be towed against the stream, and every man had to haul. "While I was tugging with my back to the cordella," he writes, "I kept my eyes on the forest or the ground, *looking for birds.*" Then came the camp on shore; the deep slumber, the early start, the same round of tugging and pulling, the same keen lookout for birds. Often there was not a white man's cabin within twenty miles; the Osage and Shawnee Indians became his guides and comrades as he investigated the habits of deer, bears, cougars, raccoons, and turkeys. "I drew more or less by the side of the camp-fire every day." On another occasion he tells how with his knapsack, his gun, and his dog, and well moccasined, he moved slowly over the prairies, attracted by the brilliancy of the flowers and the gambols of the fawns. At night the night-hawks skimmed around him, and alone the distant howling of the wolves gave him hope that he would soon arrive at some woodland where settlers might be found. At one settler's hut he would have been murdered by his hostess in his sleep had not a friendly Indian put him on his guard. His hairbreadth escapes by flood and field would fill a volume, and would be more entertaining than any of Cooper's tales.

While he was at Hendersonville his father died, leaving him his property in France and $17,000 in money. The former he never claimed till his sons were grown up; the latter was lost by the bankruptcy of the merchant in whose hands the money had been deposited, yet he never lost heart. For a time he seemed to settle; he refused a colonel's commission in an expedition to South America; he bought a house and some negroes, and began to prosper. Then, in an unlucky moment, he erected a steam-mill at Hendersonville, which ruined all concerned. The naturalist speaks with bitterness of the "infernal mill," and with equal fierceness of a steamer that he and his mill-partners had purchased and sold without ever being paid. Difficulties now increased, bills became due, he handed over to his creditors all he possessed, and left Hendersonville with his sick wife, his gun, his dog, and his drawings. The family once more pitched their tent in Louisville, where Audubon turned his artistic skill to practical use, and started as a portrait draughtsman. In a week or two he had as much work as he could do. His business spread, and he enjoyed an extended reputation for the success with which he portrayed the features of the dead. One success brought on another, and as a portrait painter he seemed to have got a new start in life. Then he became a taxidermist to the Museum at Cincinnati, and was liberally remunerated, and at the same time he opened a drawing-school. But this gleam of affluence was short; the work at the Museum was soon done, his sitters would not pay, and his friends spoke of his wandering habits. In truth, during his whole life in Kentucky and Ohio he was rambling through the woods in search of specimens. His gun supplied all his wants, a maple-sugar camp was a pleasant refuge in the woods, and a glade of the primeval forest, near some spring or salt-lick where game was plentiful, his chosen bivouac. His chance acquaintances were all kinds of wandering adventurers, who were wending westward with their wagons and household goods, or drifting down from the upper waters of the Ohio in a family ark; he describes graphically his meetings with Daniel Boone, the Kentucky hero, and with the eccentric Rafinesque, like himself a naturalist, but in another branch, that of botany. Audubon was now known as a student of nature, for Rafinesque came with a letter of introduction to him, and when he left Cincinnati to descend the river to New Orleans, he received letters of recommendation from Henry Clay and General Harrison to the Governor of Arkansas and other persons likely to be of service to him. His design on this expedition was to collect specimens in Mississippi, Alabama, and Florida, retracing his steps to New Orleans, up the Red River, down the Arkansas, and then homewards. He was determined in any case to complete one hundred drawings before he returned to Cincinnati, and he fulfilled his resolve.

On the 12th October, 1820, the naturalist left Cincinnati in company with an officer of engineers, who had been commissioned to make a survey of the Mississippi. For fourteen days they drifted down the Ohio in the flat-boat which conveyed the scientific expedition. Audubon had a relative in the old French town of Natchez, and made his first stoppage on his way down the Father of Waters at that place. "On a clear frosty morning in December," writes Audubon in his journal, "I arrived at Natchez, and found the levee lined with various sorts of boats full of Western produce. The crowd was immense, and the market appeared to be a sort of fair. Scrambling up to the cliffs on which the city is built, I found flocks of vultures flying along the ground, with outspread wings, in pursuit of food. Large pines and superb magnolias crowned the bluff, and their evergreen foliage showed with magnificent effect. I was delighted with the spectacle of white-headed eagles pursuing fishing-hawks, and surveyed the river scenery, sparkling in bright sunlight, with a new pleasure."

After some stay in Natchez, Audubon left for New Orleans, with his friend Berthoud, in a keel-boat, which was taken in tow of the steamer. Not long after leaving, he discovered that one of his portfolios, containing some drawings of birds he prized highly, was missing. On the voyage he remarked the thousands of swallows in their winter quarters, the soft notes of the doves, the glowing crests of the grosbeak, and the perfume of the oranges across the stream. On arriving at New Orleans, Audubon was relieved to find that the lost portfolio had been found, and was located safely in the office of the *Mississippi Republican* newspaper. He, however, found no work to do, and had to live for some days in the boat he came with. The money he had—not much—was stolen from him, and he had not even as much as would pay a lodging he took in advance. Amid all his difficulties he still kept wandering to the woods, made additions to his specimens, and filled his portfolio with new drawings. Meeting an Italian painter, Audubon explained his anxiety to have work. The Italian introduced him to the director of the theatre, who offered the naturalist one hundred dollars per month to draw for him, but a fixed engagement could not be entered upon.

At New Orleans Audubon made the acquaintance of Jarvis, the painter. Jarvis told him that he did not know how to paint birds, but perhaps might be useful in the studio for filling-in backgrounds. He returned again to the painter, willing to undertake this task, but Jarvis received him coolly, had no use for him, and in Audubon's opinion seemed to fear his rivalry. He again sought portraits to paint. Time passed sadly in seeking ineffectually for employment, but a lucky hit with the likeness of a prominent citizen brought a few orders that relieved his immediate necessities. He describes himself as spending his time in "vain endeavors to obtain a sight of Alexander Wilson's 'Ornithology,' which is very high-priced," and was successful in procuring some new birds. He writes in March: "Of late I have been unable to make many entries in my journal. Near our lodgings, on the south angle of a neighboring chimney-top, a mocking-bird regularly resorts, and pleases us with the sweetest notes from the rising of the moon until about midnight, and every morning from about eight o'clock until eleven, when he flies away to the convent gardens to feed. I have noticed that bird always in the same spot and same position, and have been particularly pleased at hearing him imitate the watchman's cry of 'All's well!' which comes from the fort, about three squares distant; and so well has he sometimes mocked it that I should have been deceived, if he had not repeated it too often, sometimes several times in ten minutes."

But his fortunes did not mend in New Orleans; his unconquerable restlessness stood in his path. His spirits never flagged. In his sorest penury and direst distress he resumed his gaiety when he had got a few dollars for the likeness of some patron. He had some pupils also, and accepted a situation in the family of Mrs. Perrie, at Bayou Sara, where he had to teach drawing during the summer months to Mrs. Perrie's daughter. His salary was sixty dollars a month, and board and lodging for himself and his friend Mason. The arrangement, in fact, seems to have been proposed by the lady to give an opportunity to carry on his favorite pursuits. The lessons were short, and the rest of his time was free for hunting. Here he remained from June to October, in a country the aspect of which was entirely new to him. "Since I left Cincinnati, Oct. 12, 1820," he writes, "I have finished sixty-two drawings of birds and plants, three quadrupeds, two snakes, fifty portraits, and have subsisted by my humble talents, not having had a dollar when I started." In the month of October he returned to New Orleans, rented a house, and began life there "with forty-two dollars, health and much anxiety to pursue his plan of collecting all the birds in America." He speaks with childish glee of the delight a new suit of clothes gave him. "My long flowing hair and loose yellow nankeen dress and the unfortunate cut of my features attracted much attention, and made me desire to be dressed like other people." Now begins, perhaps, the hardest portion of Audubon's life. He could get no work, no pupils, and between December 8, 1821, when his wife arrived at New Orleans, to March 7, 1822, he wrote no journal, because he could not buy a book to write in. The one obtained at last is of thin, poor paper, and the records entered in its pages are in keeping with his financial troubles. His health began to suffer from depression; he resolved to return to Natchez, and obtained his passage by the steamer in return for a crayon portrait of the captain and his wife. Mrs. Audubon was left behind, as she had undertaken the charge of the children of a Mr. Brand. At Natchez he got an appointment to teach drawing at Wellington College, but complained that the work interfered with his ornithological pursuits; and he was full of despair, fearing that his hopes of becoming known to Europe as a naturalist were destined to be forever blasted. Nor were his spirits raised by the remarks of an Englishman who examined his drawings, and, after advising him to take them to England, added that he would have to spend years in making himself known. In December a portrait-painter named Stein arrived in Natchez, and from him Audubon took his first lesson in oil-painting. His wife urged him to go to Europe and perfect himself in this branch of art, and with this view she entered into an engagement to educate the children of a Mrs. Percy, of Bayou Sara, along with her own and a limited number of pupils. On her departure for this sphere of employment, Audubon and Stein resolved to start out as wandering artists, painting portraits for a livelihood. "I had finally determined to break through all bonds and pursue my ornithological pursuits. My best friends solemnly regarded me as a madman, and my wife and family alone gave me encouragement. My wife determined that my genius should prevail, and that my final success as an ornithologist should be triumphant."

So far Audubon's career had been marked by few gleams of prosperity, but his desperate venture of coming East to obtain help to complete his work proved to be the stepping-stone of fame. He reached Philadelphia April 5, 1824, and was welcomed by the artists there, receiving great kindness from Mr. Sully. An old friend, Dr. Mease, introduced him to Charles Bonaparte, who was engaged on his own book of "American Birds." Bonaparte examined his drawings, and praised them highly. He introduced him to Peale, the artist to the Academy of Arts, and pronounced his birds superb and worthy of a pupil of David. Mr. Lawson, the engraver of Wilson's plates, said the drawings were too soft, and objected to engrave them. Another engraver, Mr. Fairman, strongly advised him to go to England. Mr. Murtrie, the conchologist, gave the same counsel; and all strengthened him in his resolution to go to Europe with his treasures, assuring him that nothing so fine in the way of ornithological representations existed. He kept working away at oils under Sully's instruction, "who gave me," to quote his own words, "all the possible encouragement which his affectionate heart could dictate." He mentions the disinterested generosity of Mr. Edward Harris, who at parting squeezed a hundred-dollar bill into his hand, a gift returned by Audubon insisting on his friend receiving the drawings of all his French birds. He paid a flying visit to Mill Grove, the old home where his father had resided, and where he himself had been married, and then fortified with letters from Mr. Sully to Gilbert Stuart, Washington Alston, and Colonel Trumbull, left for New York, "free of debt, and free from anxiety about the future." In New York he made inquiries about the publication of his drawings, but the project met with no favor. "I feel depressed," he writes. "I fear I shall die unknown. I am strange to all but the birds of America." Money was all this time scarce; so scarce that he could not visit Boston. He went to Niagara, however. "What a scene!" he exclaims. "My blood shudders still at the grandeur of the Creator's power." At Buffalo he met the chief Red Jacket, a noble-looking man, and ate a good dinner for twelve cents. "Went to bed thinking of Franklin eating his roll in the streets of Philadelphia, of Goldsmith travelling by the help of his musical powers, and fell asleep, hoping by persevering industry to make a name for myself among my countrymen." He next spent a month at Pittsburgh, scouring the country for birds, and finally reached Cincinnati again. He took a deck passage to Louisville, so low were his finances, and slept on some shavings, yet on the page recording this, he continues: "The spirit of contentment I now feel borders on the sublime. Enthusiast or lunatic, as some of my relatives will have me, I am glad to possess it." Audubon's temperament did not, indeed, attract any but artists like himself. "My friends think only of my apparel," he tells us, "and those upon whom I have conferred kindnesses prefer to remind me of my errors." Disgusted with such friends, he once more sought Bayou Sara, where his wife was still engaged. She was receiving what he describes as a large income, three thousand dollars a year, which she offered him to help forward the publication of his drawings. Numerous pupils desired lessons from him; a special invitation to teach dancing was received, and a class of sixty was soon formed. He gives an amusing description of the scene: and adds that the result of the speculation was two thousand dollars, and that with this sum and what his wife could furnish, he now saw more clearly the prospect of completing the task on which he had been working for five-and-twenty years.

On May 19, 1826, Audubon left New Orleans on the ship *Delos*, and July 20th, landed at Liverpool. He exhibited his drawings there, and after a brief visit in Manchester, went to Edinburgh. Here he met a warm welcome. Sir William Jardine, in the midst of his ornithological publications, spent hours beside him as he worked, and men like Scott, Jeffrey, and Patrick Neill did their best to encourage him. Lizars, the engraver, on seeing his drawings exclaimed, "My God, I have never seen anything like these before!" and offered to bring out a first number of his "BIRDS OF AMERICA." The book was to be published in numbers, containing four birds in each, the size of life, at two guineas ($10) a number. "One hundred subscribers for my book," he writes, "will pay all expenses." In the following year he issued, on March 17th, the "Prospectus" of his great work. He had now assumed a more civilized garb, dressed twice a day, and wore silk stockings, but still for a time let his hair grow as long as usual. "It does as well for me as my paintings," he exclaimed; but at last the importunities of his friends prevailed, and he sacrificed his cherished locks to the Caledonian taste of Edinburgh.

After a provincial canvass for subscribers he reached London, and removed the publication of his work from Lizars to Robert Havell. In the autumn he visited Paris, where Cuvier made a report on his work, describing it "as the most magnificent monument which has as yet been erected to ornithology." Subscriptions in Paris came in slowly, four only in seven weeks, and after a visit of two months he returned to London and thence to America, where he landed on May 5th. He returned to England with his wife in 1830, and at the close of this year the first volume of his great work was completed. He was back in America in 1831, and made several excursions. He procured letters from the Department at Washington to the military outposts, explored the Carolinas and Florida, and following the birds in their migrations, proceeded northward to Maine and Labrador, everywhere enriching his portfolio with the results of his explorations.

Audubon thus passed nearly three years of "travel and research" in America before he returned to England, where he was greeted by his completed second volume, one-half of his projected work. A third appeared in due time, and the fourth and last was finished in 1838.

After this the author was at liberty to make his permanent home in

America, and consequently in 1839 returned to the United States and became the purchaser of a country-seat in the immediate vicinity of New York, on the banks of the Hudson, in the upper portion of the island on which the city is situated. To bring the results of his great work within the reach of a larger number of the public, he employed himself upon its reduction. This was published in New York in seven octavo volumes, between 1840 and 1844. Nor had the author meantime relinquished his active habits of exploration. In company with his sons Victor Gifford and John Woodhouse he traversed the remoter regions of the country, collecting materials for a new work on which he now became engaged, on the "Viviparous Quadrupeds of North America." Besides the aid of his sons, he had the assistance in this work of his friend, the accomplished naturalist, the Rev. John Bachman, of South Carolina. It was in size similar to the "Ornithology," and was completed in three volumes in 1848. This was the last publishing enterprise which the author lived to see completed, a smaller edition of the work having appeared since his death.

With his fame established, in the society of his wife, who had so ably aided him, and of his sons, who grew up to work with him, Audubon spent his declining years in his rural home. Some time before his death his mind gave way till he peacefully passed from earth on the 27th January, 1851.

Rufus W. Griswold, who visited Audubon in 1846, gives us the following picture of his home: "Several graceful fawns, and a noble elk, were stalking in the shade of the trees, apparently unconscious of the presence of a few dogs, and not caring for the numerous turkeys, geese, and other domestic animals that gobbled and screamed around them. Nor did my own approach startle the wild, beautiful creatures, that seemed as docile as any of their tame companions." His house became the dwelling of animals of various kinds, some very far from attractive. At one time he aroused the household in the night to aid him to catch a number of white mice which had escaped from their cage; at another time a polecat was a resident of his painting-room; at another a California buzzard in the last stages of decomposition lay for days on the piazza. Whenever he could do so, he drew from life; bears, wolves, foxes, deer, moose, elk, and many smaller quadrupeds were kept in inclosures near the house, and sketched as they moved about. Animals that could not be sketched from life were painted as soon after death as possible before the muscles had relaxed or the coloring had lost the gloss and brilliancy of life. Every motion and attitude in each creature's native home had been carefully studied.

Another friend writes:

"The unconscious greatness of the man seemed only equalled by his childlike tenderness. The sweet unity between his wife and himself, as they turned over the original drawings of his birds, and recalled the circumstances of the drawings, some of which had been made when she was with him; her quickness of perception, and their mutual enthusiasm regarding these works of his heart and hand, and the tenderness with which they unconsciously treated each other, remain impressed upon my memory. Ever since, I have been convinced that Audubon owed more to his wife than the world knew, or ever would know. That she was always a reliance, often a help, and ever a sympathizing sister-soul to her noble husband, was fully apparent to me."

"His enthusiasm was sometimes intense; he would rise before daylight and walk about eagerly, waiting for the dawn that he might begin work, and once at work would steadily and earnestly continue to paint all day. Sunset found him at his picture full of vigor and energy, but with no interest in anything else. He would pursue this course till the fever left him, when he would lay his brushes aside and roam through woods and fields." At no time did he lose sight of his work; it was ever on his mind, as might be seen from the questions addressed to those he was with, and the quick glances of his eagle eye. His quick, nervous temperament caused strange contradictions in his character. "One can hardly believe," writes his granddaughter, "that the man who, for three weeks, spent every day and all day long, lying on his back under a tree watching two little birds build their nest could be so impatient when the effect he desired could not be produced, that he would throw canvas, easel, paints and brushes from him, and rush from the house to find consolation in his beloved woods."

He was a tall, thin man, with a high, arched and serene forehead, and a bright, penetrating gray eye; his white locks fell in clusters upon his shoulders, but were the only signs of age, for his form was erect, and his step as light as that of a deer. The expression of his face was sharp, but noble and commanding, and there was something in it, partly derived from the aquiline nose, and partly from the shutting of the mouth, which made you think of the imperial eagle. A traveller who met him on his return from one of his hunting trips describes him: "I was at a fashionable hotel at Niagara when an elderly man arrived whose appearance excited much comment. He seemed to have sprung from the woods; his dress, which was of leather and heavy cloth, was dreadfully dilapidated; a worn-out blanket was strapped to his shoulders, a large knife hung at one side, a rusty tin box on the other, and his hair and beard were so long and thick that they alone would have rendered him remarkable."

Prof. Wilson, the "Christopher North" of *Blackwood's Magazine*, gives an equally striking picture of Audubon as he appeared in the literary circles of Scotland: "When, some five years ago, we first set eyes on him, in a party of literati, in 'stately Edinborough throned on crags,' he was such an American woodsman as took the shine out of us modern Athenians. Though dressed, of course, somewhat after the fashion of ourselves, his long raven locks hung curling over his shoulders, yet unshorn from the wilderness. They were shaded across his open forehead with a simple elegance, such as a civilized Christian might be supposed to give his 'fell of hair,' when practising 'every man his own perruquier' in some liquid mirror in the forest glade, employing, perhaps, for a comb, the claw of the bald eagle. His sallow, fine-featured face bespoke a sort of wild independence; and then such an eye, keen as that of the falcon! His foreign accent and broken English speech—for he is of French descent—removed him still further out of the commonplace circle of this every-day world of ours; and his whole demeanor—it might be with us partly imagination—was colored to our thought by a character of conscious freedom and dignity, which he had habitually acquired in his long and lonely wanderings among the woods, where he had lived in the uncompanied love and delight of nature, and in the studious observation of all the ways of her winged children, that forever fluttered over his paths and roosted on the tree at whose feet he lay at night, beholding them still the sole images that haunted his dreams."

An unfinished portrait represents him in his woodland dress, taken on his return from the Rocky Mountains. It is a half-length, life-size, the head a little thrown back, the keen eyes undimmed, though the face shows deep lines of age and thought. He holds his gun in his hand, and his backwoodsman's coat of green baize with fur collar and cuffs is but roughly painted. It was never completed by his son, as the old man never retained his woodland garb longer than he could help, and one of his first steps on returning to civilized life was to remove the greater part of his luxuriant hair and beard.

We have recounted at greater length the early years of struggle than the later years of fame and success. It is in the narrative of his days of wandering, of poverty, of neglect, that the true lesson of Audubon's life is taught. He had one object before him, and nothing deterred him from his quest; neither hardship, nor difficulty, nor misconstruction, nor the opinion of the world. Great as his works may be, much as they may do to foster a love of nature and of nature's works, Audubon's greatest, most endearing, most effective work is the example he has set, and the lesson of unselfish perseverance which his life teaches.

LOUIS AGASSIZ.

LOUIS JOHN RODOLPH AGASSIZ, one of the most distinguished naturalists and scientific explorers of the present day, was born in the parish of Mottier, between the lake of Neufchâtel and the lake of Morat, in Switzerland, on the 28th of May, 1807. Of Huguenot race his father was a village pastor, as for six generations in lineal descent his ancestors had been before him. The pastor's wife, a woman of rare worth and intelligence, was the daughter of a Swiss physician.

At the age of eleven, Louis entered the gymnasium of Bienne, whence he was removed, in 1822, as a reward for his attainments in his scientific studies, to the Academy of Lausanne. Two years later he engaged in the study of medicine at the school at Zurich, and subsequently pursued the scientific and philosophical courses at the Universities of Heidelberg and Munich, receiving his degree as doctor of medicine at the latter. The bent of his mind was already shown at these latter institutions, in his devotion to the study of botany and comparative anatomy.

His father designed him for a commercial life, and was impatient at his devotion to frogs, snakes, and fishes. He came to London with letters to Sir Roderick Murchison. The great English geologist took the lad, the same evening, to a meeting of the Royal Society, and at the close said: "I have a young friend here from Switzerland, who thinks he knows something about fishes. There is, under this cloth," pointing to a heap on the table, "the skeleton of a fish which existed long before man." He then gave the precise locality where it was found, with one or two other facts concerning it. "Can you sketch for me on the blackboard your idea of the fish?" he asked in conclusion.

Agassiz took the chalk and rapidly sketched the skeleton. The portrait was correct in every bone and line. "Sir," said Agassiz, when he told the story, "that was the proudest moment of my life."

In 1828, at the age of twenty-one, Agassiz began his public career as a naturalist by the description of two new fishes in the "Isis" and "Linnæa," two foreign periodicals occupied with natural history. The following year he was selected to assist the eminent German naturalist, Von Martius, in his report of the scientific results of his expedition to Brazil, undertaken under the auspices of the Austrian and Bavarian governments. The portion of the work entrusted to his charge was the preparation of an account of the genera and species of the fish collected by the naturalist Von Spix in the expedition. The successful accomplishment of this work gave him reputation as an ichthyologist. His labors were noticed with approval, and brought before a Berlin meeting of German naturalists by the eminent transcendental anatomist, Oken. Encouraged by this success he pursued his ichthyological studies with great perseverance, recording the results from time to time in the natural history publications of the day. His labors also secured him the friendship of Humboldt and Cuvier in a visit to Paris, where he was enabled to pursue his researches by the friendly pecuniary assistance of a clergyman and friend of his father, Mr. Christinat.

In 1832, he was appointed Professor of Zoology at Neufchâtel. In 1834, he published a paper on the "Fossil Fish of Scotland," in the "Transactions of the British Association for the Advancement of Science," and others subsequently on the classification of fossil fishes in various foreign journals. He devoted seven years to this subject, completing the publication of his great work on "Fossil Fishes," in five volumes, in 1844. Associated with these studies and results, was the preparation of his important work on Star-Fishes, or Echinodermata, published in parts from 1837 to 1842, under the title "Monographies d'Echinodermes Vivans et Fossiles." He had also, during this period, completed another leading work, a "Natural History of the Fresh-water Fishes of Europe," which was published in 1839.

"The researches of Agassiz upon fossil animals," says a writer in the "English Cyclopædia," "would naturally draw his attention to the circumstances by which they have been placed in their present position. The geologist has been developed as the result of natural history studies. Surrounded by the ice-covered mountains of Switzerland, his mind was naturally led to the study of the phenomena which they presented. The moving glaciers and their resulting moraines, furnished him with facts which seemed to supply the theory of a large number of phenomena in the past history of the world. He saw in other parts of the world, whence glaciers have long since retired, proofs of their existence in the parallel roads and terraces, at the bases of hills and mountains, and in the scratched, polished, and striated surface of rocks. Although this theory has been applied much more extensively than is consistent with all the facts of particular cases by his disciples, there is no question in the minds of the most competent geologists of the present day, that Agassiz has, by his researches on this subject, pointed out the cause of a large series of geological phenomena. His papers on this subject are numerous, and will be found in the 'Transactions of the British Association' for 1840, in the 3d volume of the 'Proceedings of the Geological Society,' in the 18th volume of the 'Philosophical Magazine,' third series; and in the 6th volume of the 'Annals and Magazine of Natural History.'"

In 1846, Agassiz came to the United States to continue his explorations and to fulfil an engagement to deliver a course of Lectures on the Animal Kingdom before the Lowell Institute at Boston. The lectures excited much interest and were followed in successive seasons by three other courses on Natural History before the same institution. While these were in progress he had, at the close of 1847, accepted the appointment of Professor of Zoology and Botany in the scientific school founded by Mr. Abbott Lawrence, in connection with Harvard University at Cambridge. He became a master of English composition, and spoke the language not only with fluency, but with a voluble eloquence which was peculiarly his own. He studied the modes of thought among the people, and learned to know in what they differed from the European. His family ties, his household, his associates were of the country; and yet, after all, he was unchanged. A genius like his could put itself in communication with many and different people; it could grow also, but it could not change. In the following year he was engaged, with some of his pupils, in a scientific exploration of the shores of Lake Superior, the results of which were published in a volume written by Mr. Elliott Cabot and others, entitled "Lake Superior." In conjunction with Dr. A. A. Gould, of Boston, Professor Agassiz published in the same year a work on "The Principles of Zoology." Devoting himself to an assiduous practical study of the natural history of the country, he visited its most important portions in the Atlantic and Gulf States, the valley of the Mississippi, and the regions of the Rocky Mountains. In 1850, he spent a winter upon the reefs of Florida, in the service of the United States Coast Survey; and subsequently, during the winter of 1852-53, was Professor of Comparative Anatomy in the Medical College of Charleston, S. C., which afforded him the opportunity of making other scientific researches in the southern region and seaboard. The results of his investigations in these various journeys have been given to the world in a series of volumes in quarto entitled "Contributions to the Natural History of the United States," a work for which an extraordinary popular subscription was obtained.

On the death of Mr. F. C. Gray, in 1858, it was found that he had left fifty thousand dollars to establish a Museum of Comparative Zoology; and his nephew, Mr. William Gray, scrupulously following his uncle's inclinations, selected Harvard College as the proper institution. During the following year, a committee of gentlemen raised more than seventy thousand dollars, and the State gave one hundred thousand. Of this institution Agassiz became the presiding spirit.

In the summer of 1865, Professor Agassiz extended his American researches to the Southern Continent, in an expedition at the head of a chosen party of assistants, in an exploration of Brazil, where he devoted eighteen months to a thorough survey of the valley of the Amazon and other portions of the country. An account of this tour, in a volume entitled "A Journey to Brazil," from the pen of Mrs. Agassiz, a devoted companion to her husband in his scientific studies, was published in 1867. He subsequently was engaged in a like exhaustive study of the regions of the United States bordering on the Pacific; and, in 1872, in a voyage of scientific observation on the western shores of South America. In these expeditions he was accompanied by a corps of pupils devoted to natural history, and vast materials were gathered by him, to be added to the collections of animals, plants, and fossils, of which he undertook the classification and preservation, as curator of the Museum of Comparative Zoology, in connection with his Professorship at Cambridge.

The Brazilian expedition of 1865 brought home barrels and cases by the hundred, and so did the Hassler expedition of 1871. Nor were these half, for the incessant eagerness of the director sought original collections from all parts of the world, some by exchange and some by purchase. In 1870 the building was increased to double its former capacity, but it does not afford room to-day for the arrangement of the collections stored in it. When the first section of the edifice was finished, fire-proof and fairly fitted with shelves and cases, a grand moving took place, and the motley boxes and bottles were carried or carted in all haste to the new quarters.

While pursuing his career of original study at Cambridge and giving to the public the result of his observations in his series of philosophical lectures before the Museum of Comparative Zoology, involving original and elaborate *constructions* of animal life, the sphere of his investigations was enlarged in the summer of 1873 by the gift of Mr. John Anderson, a gentleman of Massachusetts, of Penikese Island, off the coast of New England. This piece of land, valued at one hundred thousand dollars, was presented to him for the establishment of a school of Investigation in Natural History, with an additional gift in money of fifty thousand dollars to carry out the design.

In the prosecution of this liberal plan, Prof. Agassiz became at once engaged in the effective organization of the school or college, endeavoring to "extend the range of its usefulness in the application of science to the practical art of modern civilization," his object being particularly "to combine physical and chemical experiment with the instruction and work of research to be carried on upon the island—physiological experiments being at the very foundation of the exhaustive study of zoology."

"He is," says the accomplished critic, Mr. Whipple, in the course of an able review of his "Essay on Classification" in the first volume of the "Contributions to the Natural History of North America," "not merely a scientific thinker, he is a scientific force; and no small portion of the immense influence he exerts is due to the energy, intensity, and geniality which distinguish the nature of the man. In personal intercourse he inspires as well as informs; communicates not only knowledge, but the love of knowledge, and makes, for the time, everything appear of small account in comparison with the subject which has possession of his soul. To hear him speak on his favorite themes is to become inflamed with his enthusiasm. He is at once one of the most dominating and one of the most sympathetic of men, having the qualities of leader and companion combined in singular harmony. People follow him, work for him, contribute money for his objects, not only from the love inspired by his good-fellowship, but from the compulsion exercised by his force. Divorced from his congeniality, his energy would make him disliked as a dictator; divorced from his energy, his geniality would be barren of practical effects. The good-will he inspires in others quickens their active faculties as well as their benevolent feelings. They feel that, magnetized by the man, they must do something for the science impersonated in the man—that there is no way of enjoying his companionship without catching the contagion of his spirit. He consequently wields, through his social qualities, a wider personal influence over a wider variety of persons than any other scientific man of his time. At his genial instigation laborers delve and dive, students toil for specimens, merchants open their purses, legislatures pass appropriation bills."

Professor Agassiz received the most distinguished attentions from the French Academy of Sciences and other numerous scientific associations of Europe. The Emperor offered him the post of director of the *Jardin des Plantes*, at Paris, with a seat in the Senate. The position was one which a scientific man of the highest rank might well covet, and the emoluments of the office, with that of the other office associated with it, were quite large. His acceptance of the offer would have given him at Paris a rank equal to that which Cuvier occupied in his time. He respectfully declined it, on the ground that he was then engaged in original researches in the United States, which promised to be very fruitful in zoological discovery, and which would take him some years to complete. He considered that the correspondence was closed; but he was surprised by receiving another letter from the minister, renewing the offer, and informing him that the high office would be kept open for him until his American researches were completed. Agassiz justly thought that was the greatest compliment ever paid to him, but his determination to live and die in his adopted country was fixed, and his letter indicating this determination closed the correspondence.

On the death of the eminent Professor Edward Forbes, in 1854, he was invited to succeed to his chair of Natural History in the University of Edinburgh, but declined the offer in favor of his adopted home, and the field of some of his most distinguished researches in America.

At the time when he was absorbed in some minute investigations in a difficult department of zoology, he received a letter from the president of a lyceum at the West, offering him a large sum for a course of popular lectures on natural history. His answer was: "I can not afford to waste my time in making money."

The marvel of Agassiz, and a never-ceasing source of wonder and delight to his friends and companions, was the union in his individuality of solidity, breadth and depth of mind, with a joyousness of spirit, an immense overwhelming geniality of disposition, which flooded every company he entered with the wealth of his own opulent nature. Placed at the head of a table with a shoulder of mutton before him, he so carved the meat that every guest flattered himself that the host had given him the best piece. His social power exceeded that of the most brilliant conversationalists and the most delicate epicures; for he was not only fertile in thoughts, but wise in wines and infallible in matters of fish and game. It was impossible to place him in any company where he was out of place.

"I look upon him," said a pupil, "as a prophet, as an apostle of science; he has made every honest investigator his debtor; he has not only elevated in public esteem the intellectual class to which he belongs, but he has induced the moneyed class to give science the means of carrying out its purposes. Since Agassiz came into the country, you can not but have noticed that private capitalists, State Legislatures, and the Congress of the country have been liberal of aid to every good scientific enterprise. We owe a great part of this liberality to Agassiz. He it was who magnetized the people with his own scientific enthusiasm. He made science popular, because in him science was individualized in the most fascinating and persuasive of human beings. All the rest of us are more or less dominated by our special lines of investigation, or so infirm in physical health, or so unsympathetic with ignorant people, or so supercilious, or so controlled by some innate 'cussedness' of disposition, that we can not readily adapt ourselves to the ways of men of the world; but Agassiz, with his enormous physical health and vitality, and his capacity to meet all kinds of men on their own level, drew into our net hundreds of people, powerful through their wealth or their political influence, who would never have taken any interest in science if they had not first been interested in Agassiz. And these men were the men who gave us the money we needed for the extension of scientific knowledge and the promotion of scientific discovery."

Another student and admirer writes:

"Some thirty-five years ago, at a meeting of a literary and scientific club of which I happened to be a member, a discussion sprang up concerning Dr. Hitchcock's book on 'bird-tracks,' and plates were exhibited representing his geological discoveries. After much time had been consumed in describing the bird-tracks as isolated phenomena, and in lavishing compliments on Dr. Hitchcock, a man suddenly rose who in five minutes dominated the whole assembly. He was, he said, much interested in the specimens before them, and he would add that he thought highly of Dr. Hitchcock's book as far as it accurately described the curious and interesting facts he had unearthed; but, he added, the defect in Dr. Hitchcock's volume is this, that 'it is dees-*creep-teeve*, and not com-par-a-*teeve*.' The moment he contrasted 'dees-creep-teeve' with 'com-par-a-teeve,' one felt the vast gulf that yawned between mere scientific observation and scientific intelligence; between eyesight and insight; between minds that doggedly perceive and describe, and minds that instinctively compare and combine."

To illustrate the lectures on geology, he used to invite students to accompany him on excursions to neighboring towns. From boyhood an associate of students, there was no company in which he felt more at ease; and he regarded with unfeigned consternation the stiff relations that, twenty years ago, subsisted between our professors and their pupils. It was pleasant to see him, at the head of a score of us youngsters, taking his way toward the pudding-stone quarries in Roxbury. His face wore an easy smile, and, as his quick brown eye wandered over the landscape, it saw more than did all our eyes put together, for he looked, but we only stared. Near by, like a sort of lieutenant, walked Jacques Burkhardt, the lifelong friend and artist of the great professor. Though his beard was white, he never grew old; and, to the last, preferred the cheerful company of the collegians. Whenever we came to a gravel-pit, or a railway cut, the professor would stop, and would expatiate on the structure of the drift with as much interest as if he saw it for the first time. This enthusiasm, fresh and untiring over trite facts, was a source of immense power to him. It showed his French blood, for it was but an enlargement of that peculiar temper which renders the Parisian workman at once the most interesting and the most successful in the world.

Of the section of conglomerate in Roxbury he was never tired of talking; and, over and over again, to different sets of hearers, would he explain the cleavage planes of the rock, and how the cleavage had cut right through hard pebbles like a knife; then the structure of the stone itself, and the different origins of flat and of rounded pebbles; and, finally, he would climb to the top of the ledge, and earnestly show the grooves and scratches running north and south, and the surface polished by glaciers.

It is very curious that he never learned to make finished drawings—curious, because he had often been too poor to employ an artist, and because his accuracy of eye and of touch were remarkable. If there were ten hairs in the field of the microscope and the artist had put eleven in the drawing, the professor would exclaim, the moment he got his head over the eye-piece, "Those cilia are crowded; there must be too many." He would hold the dried shell of the turtle in his left hand and with a saw divide it lengthwise into precise halves, with no other guide than his eye. Although he never attempted to become an artist, his chalk outlines on the blackboard were what few artists can make. The thousands of people who have heard his lectures will always recollect the astonishing rapidity with which he drew an animal, putting in only the characteristic points. If he were saying, " The salmons have a peculiar fatty fin, called the adipose," almost with the words would appear an unmistakable outline of the fish. There was no better nor more pitiless critic of a zoological drawing. He rarely was satisfied with the finest work. Were the artist painstaking, he would encourage him with, "Try it once again; it is all wrong, but don't get out of patience." The careless or self-sufficient draughtsman got a brisk admonition. The man who never failed to please him was Sonrel, who made the plates for the " Embryology of Turtles," of which Claparède said, " I had supposed that such lithography was impossible."

It was interesting to notice his behavior in the presence of domesticated animals. The ugliest, filthiest, stupidest, most unreasonable, most obstinate creature in the barnyard is the pig, yet, with a stick in his hand, Agassiz would go up to the most unsociable. "cantakerous," misanthropic grunter, and after a few soft words and a movement of the stick over the bristles of the creature in the right direction, the pig would lift its head erect, its small eyes would glisten with a vague intelligence, it would remain almost motionless in a kind of pleased surprise, and emit a sound indicative of as much content and comfort as are indicated by the purring of a cat. The neigh of a horse to him was a more friendly neigh than any ever heard by a hostler or a jockey. He carried serpents in his hat and in his pockets with a grand unconcern, and dropped them sometimes in his bedroom, so that his wife was frequently troubled by finding them coiled up in her boots. Whenever he entered a menagerie he was eagerly welcomed by lions, tigers, wolves, hyenas, and other beasts of prey, which considered even their keepers as stupid louts, but recognized in him the one person that they could have a rational conversation with.

Year by year Agassiz strove to support the ever-increasing burdens of his task,—his vast correspondence carried on in three languages; the superintendence of numerous assistants; protracted conferences almost daily with the learned men who were at the head of the different departments; and a constant and intense study of the grand question of arrangement. In addition to this labor, especially devoted to the Museum, he exerted himself in many other ways. He gave lectures and contributed to scientific literature. He was at the disposal of every one who came to ask questions; and he found time to attend agricultural meetings, learned societies, and literary clubs. Besides all this, he undertook a task very disagreeable to him in asking aid to carry on so expensive an establishment. More than once his warm friend and admirer, Brown-Séquard, warned him that such a strain was not to be borne. Agassiz *could not* stop. He was driven by a power like that which the Greeks called mighty fate. At length, in December of 1869, his system gave way, and his brain was attacked in a manner which threatened paralysis. Nothing saved him then but his powerful constitution, seconded by the most careful treatment. Weakened by disease and with death imminent, his heroism was at once noble and pathetic. One day the tears began to roll down his cheeks, and he said: "Brown-Séquard tells me I must not think. Nobody can ever know the tortures I endure in trying to stop thinking!"

His physical health had been so great that, when he was superintending the arrangement and publication of one of his early works, he labored for a couple of months steadily at his desk at the rate of sixteen or eighteen hours a day, taking no exercise; and when the delightful task was completed, he started on an excursion among the Alps, which exacted as much labor from his limbs, as the months preceding it had exacted from his brain. In fact, he seemed, up to the period of his first attack of disease, utterly insensible to bodily as to mental fatigue.

In the midst of this active career of usefulness, after a summer of unusual exertion in the establishment of his School of Natural History, Professor Agassiz, who had already suffered some symptoms of failing health, was, in the beginning of December, 1873, suddenly stricken down by an attack of paralysis, and, after a few days of lingering illness, on the night of the 14th expired at his residence in Cambridge, Mass. His death was attended by the profoundest regret and the noblest tributes to his memory in America, and by the friends of science throughout the civilized world. We conclude with a few lines from the poem "Agassiz," by J. Russell Lowell:

> "We have not lost him all; he is not gone
> To the dumb herd of them that wholly die;
> The beauty of his better self lives on
> In minds he touched with fire, in many an eye
> He trained to Truth's exact severity;
> He was a Teacher: why be grieved for him
> Whose living word still stimulates the air?
> In endless file shall loving scholars come,
> The glow of his transmitted touch to share,
> And trace his features with an eye less dim
> Than ours whose sense familiar want makes dumb."

INDEX.

The Animals named in Capitals appear in the Illustrations.

A.

Aard Vark, 743.
" Wolf, 240.
Adder, 614.
Adjag, 266.
Aguara, 313.
Agouta, 161.
Agouti, 726.
Aguarachay, 253.
Ai, 738.
Alactaga, 697.
Almiqui, 162.
Alpaca, 528.
Angwantibo, 112.
Anoa, 597.
Ani Bear, 744.
Ant-Eater, Great, 744.
" Little, 746.
" Native, 752.
" Porcupine, 766.
Antelope, Indian, 623.
" Musk, 629.
" Sable, 615.
" Sabre, 614.
Aoudad, 646.
Apar, 741.
Ape, Bonnet, 67.
" Death's Head, 94.
" Hussar, 51.
" Javanese, 65.
" Moor, 52.
" Nun, 50.
" Red, 51.
" Satan, 90.
" Shaggy, 90.
" White-nosed, 51.
Argali, Bearded, 647.
Armadillo, 741.
" Great, 740.
Atrianha, 303.
Ashkoko, 682.
Ass, Common, 448.
" Wild, 447.
" Wild African, 448.
Atak, 340.
Atumba, 105.
Aye-Aye, 115.

B.

Babac, 715.
Babakoto, 103.
Baboon, 56.
Babyroussa, 588.
Badger, 304.
" American, 304.
" Mexican, 304.
Bandicoot, Long-nosed, 753.
" Striped, 752.
Bangsring, 154.
Banteng, 595.
Barasinga, 556.
Barrigudo, 80.
Bat, African Leaf, 135.
" Barbastelle, 139.
" Big-eared, 142.
" Bull Dog, 139.
" California, 140.
" Carolina, 141.
" Creek, 143.
" Daubenton's, 140.
" Dog Headed, 143.
" Fruit Eating, 128.
" Georgian, 141.
" Hoary, 142.
" Horseshoe, Great, 135.
" " Lesser, 134.
" " Noble, 135.
" Little Brown, 141.
" Long-eared, 142.
" Mouse-colored, 139.
" New York, 141.
" Northern, 138.
" Pale, 143.
" Red, 141.
" Serotine, 139.
" Vampire, 132.
Beagle, 277.
Bear, Black, 322.
" Brown, 320.
" Cinnamon, 323.
" Grizzly, 323.
" Polar, 318.
" Sand, 303.
" Sloth, 326.
" Spectacled, 327.

Bear, Sun, 325.
" Syrian, 321.
Beaver, American, 702.
" European, 700.
Beden, 644.
Beisa, 614.
Bettong, Brush-tailed, 756.
Big Horn, 648.
Binturong, 231.
Bison, American, 586.
Black Cat, 295.
Black Fish, 373.
Blau Bok, 616.
Bloodhound, 278.
Blue Buck, 631.
Boar, Wild, 499.
Bokambul, 107.
Bonassus, 584.
Borele, 477.
Brahmin Bull, 592.
Huanbuah, 266.
Budeng, 44.
Buffalo, American, 586.
" Mountain, 589.
" Cape, 591.
" Indian, 603.
Bulan, 159.
Bunder, 69.
Burunduk, 713.

C.

Cabramontes, 643.
Cacajao, 91.
Camel, 514.
" Bactrian, 522.
Camelopard, 568.
Campagnol, 697.
Capybara, 727.
Caracal, 218.
Carcajou, 554.
Caribou, 548.
Cat, Angola, 214.
" Chinese, 211.
" Domestic, 214.
" Egyptian, 212.
" Long-tailed, 207.
" Malay, 211.

Cat, Manx, 214.
" Marbled, 205.
" Marten, 215.
" Native, 751.
" Pampas, 207.
" Red, 223.
CATTLE, ALDERNEY, 583.
" Brahmin, 592.
" Durham, 582.
" Pampas, 577.
" Scotch, 583.
" Wild, 576.
Cavy, Patagonian, 728.
Chacma, 58.
Chameck, 82.
CHAMOIS, 633.
" Cape, 613.
Cheti, 207.
CHEETAH, 224.
Chickara, 631.
Chickaree, 696.
CHIMPANZEE, 27.
" Black, 31.
CHINCHILLA, 719.
CHIPMUCK, 712.
CHŒROPUS, 753.
CIVET, 228.
" American, 316.
Coaita, 81.
COATI, 313.
" Red, 314
" White, 314.
Coendou, 724.
Colocolo, 208.
COLUGO, 149.
COUGAR, 196.
COYOTE, 253.
COYPU, 723
CRYPTOPROCTA, 226.
Cuscus, Spotted, 758.
Cuxio, 91

D.

Dachshund, 278.
Daman, Tree, 653.
DASYURE, 751.
Dauw, 453.
DEER, AXIS, 557.
" FALLOW, 560.
" Hog, 557.
" MOOSE, 532.
" MUSK, 565.
" Pampas, 558.
" Persian, 546.
" RED or STAG, 552.
" ROE, 563.
" VIRGINIAN, 554
Dego, 721.
Desman, Pyrenean, 166.
" Russian, 166.
DEVIL, TASMANIAN, 754.
DHOLE, 205.
DINGO, 201.
Dog, BULL, 288.

Dog, Chesapeake Bay, 281.
" CHINESE, 291.
" COACH, 289.
" Colley, 283.
" DALMATIAN, 289.
" DANISH, 289.
" ESQUIMAUX, 283.
" Hare-Indian, 276.
" Labrador, 287.
" Leonberg, 287.
" Maltese, 290.
" NEWFOUNDLAND, 285.
" Poodle, 280.
" PRAIRIE, 716.
" PUG, 289.
" ST. BERNARD, 284.
" Saint John's, 287.
" SHEPHERD'S, 282.
" Spitz, 293.
" TIBET, 285.
DOLPHIN, 386.
" Bottle-nosed, 387.
" Gladiator, 390.
" White Beaked, 387.
DORMOUSE, COMMON, 699.
" Fat, 698.
Douc, 45.
Douroucouli, 92.
Drill, 63.
DROMEDARY, 514
Dseren, 622.
DUCK-MOLE, 763.
DUGONG, 400.
" Australian, 401.
Duyker Bok, 630.
Dziggetai, 445.

E.

ELAND, 607.
ELEPHANT, AFRICAN, 672.
" ASIATIC, 665.
ELK, 539.
ERMINE, 296.
" Kane's, 297.
" New York, 297.

F.

FENNEK, 260.
FERRET, 296.
" Black-footed, 298.
Fisher, Marten, 295.
FOX, ARCTIC, 262.
" BLUE, 263.
" COMMON, 256.
" Gray, 262.
" Kit, 262.
" Large-eared, 264.
" SILVER, 261.

G.

Galago, 112
" Giant, 113
GALLI, 226.

Garangang, 234.
GAUR, 594.
Gayal, 593.
GAZELLE, 617.
" Ariel, 618.
Gelada, 53.
Genet, Common, 230.
" PALE, 230.
Gibbon, Agile, 35.
" HULOCK, 35.
GIRAFFE, 569.
GNU, 633.
GOAT, ANGORA, 640.
" Bezoar, 639
" CASHMERE, 639.
" COMMON, 637.
" Dwarf, 642.
" Egyptian, 641.
" Mamber, 641.
" MOUNTAIN, 635.
Goffer, 701.
Goral, 635
GORILLA, 23.
Grampus, 390.
Greyhound, Italian, 276.
" IRISH, 274.
" Persian, 274.
" Russian, 274.
" Scotch 273.
Grison, 299.
Ground Pig, 72.
Guanaco, 525.
GUENON, 52.
Guereza, 48.
GUINEA PIG, 725.
Gundy, 722.

H.

Hamadryad, 60.
HAMSTER, 692.
Hare, Cape Leaping, 697.
" COMMON, 732.
" Irish, 733.
" Marsh, 731.
" NORTHERN, 731.
" Polar, 730.
" Sand, 732.
" Snow, 733.
" Trowbridge's, 731.
" Wood, 731.
Harrier, 277
" Welsh, 277.
Hartebeest, 632.
HEDGEHOG, AUSTRALIAN 766.
" EUROPEAN, 158.
" Long-eared, 158.
" Madagascar, 159
HEMIGALE, 230.
HIPPOPOTAMUS, 486.
" Small, 492.
HOCCLAI, 346.
HOG, BERKSHIRE, 504.
" Bush, 507.

INDEX. 793

Hog, Pencilled, 506.
" Wild, of India, 501.
Horse, 410.
" Arab, 418.
" Austrian, 437.
" Barb, 422.
" Black Dray, 444.
" Clydesdale, 443.
" French, 438.
" Galloway, 441.
" Hackney, 434.
" Holstein, 438.
" Hunter, 434.
" Italian, 439.
" Pacer, 432.
" Percheron, 443.
" Race, 427.
" Russian, 436.
" Spanish, 440.
" Trotting, 428.
Hound, Fox, 277.
" Otter, 277.
" Stag, 277.
Hulman, 42.
Huneman, 42.
Hunting dog, 264.
Huron, 294.
Hutia Conga, 720.
Hyena, Brown, 243.
" Dog, 264.
" Spotted, 244.
" Striped, 242.
Hyrax, 681.

I.
Ibex, 643.
Ichneumon, 235.
Indris, Crowned, 103.
Inia, 383.

J.
Jackal, 255.
" Black-handed, 256.
" Cape, 256.
Jaguar, 198.
Jairou, 619.
Jelerang, 708.
Jemlack, 641.
Jerboa, 697.

K.
Kaberoo, 250.
Kahau, 44.
Kakang, 109.
Kalan, 302.
Kalong, 128.
Kanchil, 532.
Kangaroo, 754.
" Rock, 755.
" Tree, 756.
" Woolly, 755.
" Hare, 756.
" Rat, 757.

Katshkar, 648.
Keitloa, 479.
Killer, 390.
" Cape, 391.
Kinkajou, 315.
Koala, 757.
Koodoo, 659.
Kuala Ayer, 462.
Kuichua, 297.
Kulan, 445.
Kuwuk, 211.

L.
Laudjak, 256.
Lapunder, 71.
Lemming, 698.
Lemur, Black-fronted, 105.
" Cat, 106.
" Diadem, 109.
" Flying, 149.
" Grey, 107.
" Ring-tailed, 106.
" Red, 106.
" Ruffed, 105.
" White-fronted, 105.
" Slow-paced, 109.
Leopard, African, 201.
" Asiatic, 202.
" Seal, 343.
Linsang, 239.
Lion, African, 183.
" Asiatic, 183.
" Maneless, 183.
Llama, 527.
Lobo, 251.
Lori, Bengal, 110.
Lynx, Bay, 222.
" Booted, 220.
" Canadian, 220.
" Common, 219.
" Marsh, 217.
" Persian, 217.
" Polar, 220.
" Southern, 220.

M.
Macaque, 65.
Magots, 72.
Markong, 254.
Maki, 104.
" Dwarf, 107.
Mammoth, 657.
Mampalon, 233.
Manatee, 399.
Mandrill, 62.
Mangabey, 52.
Mangouste, 234.
" Crab-eating, 237.
" Zebra, 237.
Mangue, 239.
Manul, 211.
Manis, Long-tailed, 739.
" Short-tailed, 740.

Mara, 728.
Maracaya, 207.
Marikina, 99.
Marimonda, 81.
Markhor, 641.
Marmoset, 96.
" Dwarf, 99.
" Leoncito, 98.
Marmot, 714.
" Leopard, 713.
" Pouched, 715.
Marquay, 206.
Marten, Beech, 294.
" Pine, 293.
Mastiff, 284.
Mastodon, 658.
Magotte, 104.
Meerkat, 238.
Meloncillo, 234.
Mexican Mopsey, 291.
Miko, 79.
Mink, 297.
Misiki, 83.
Mohogoo, 481.
Mohoh, 113.
Mole, American, 165.
" Blind, 164.
" Broad-handed, 166.
" Changeable, 163.
" Common European, 164.
" Gibb's, 168.
" Hairy-tailed, 166.
" Japanese, 167.
" Oregon, 166.
" Prairie, 166.
" Star-nosed, 165.
" Rat, 696.
Mongoose, 106.
Monkey, Bartlett's, 83.
" Bearded, 51.
" Capucin, 78.
" Diana, 54.
" Gibraltar, 72.
" Green, 51.
" Howling, Black, 86.
" " Red, 86.
" Negro, 44.
" Proboscis, 44.
" Spider, 81.
" Squirrel, 93.
Moose, 543.
Mouflon, 627.
Mouse, Barbary, 692.
" Common, 690.
" Harvest, 691.
" Jumping, 697.
" Le Conte's, 694.
" Pouched, 751.
" Wilson's Meadow, 698.
" Field, New Jersey, 696.
" " Sonora, 697.

Mouse, Field, Texan, 697.
Mule, 456.
MULLINGONG, 764.
Munga, 67.
Mungos, Banded, 239.
Musujak, 564.
Musang, 231.
MUSKOX, 653.
MUSTANG, 411.

N.

NARWHAL, 377.
Nelbander, 71.
Nicobejan, 766.
Noctule, 140.
Nyeutok, 310.
Nylghau, 611.
Nyule, 234.

O.

OCELOT, 205.
" GRAY, 206.
" Painted, 206.
Opossum, Crab-eating, 750.
" MERRIAN'S, 750.
" Mouse, 750.
" VIRGINIA, 750.
" YAPOCK, 750.
Otter, California, 302.
" Chinese, 303.
" Common, 300.
" Javanese, 304.
" NORTH AMERICAN, 301.
" SEA, 302.
OUISTITI 98.
OUNCE, 203.
OURANG-OUTAN, 32.
Ourebi, 627.
OX, 575

P.

PACA, SOOTY, 727.
Paguma, Marked, 232.
" Whiskered, 232.
" Woolly, 232.
Pallah, 624.
PANDA, 316.
PANTHER, 202.
" Black, 203.
" JAPANESE, 202.
" Sunda, 202.
Paseng, 630.
Passan, 613.
PECCARY, COLLARED, 494.
" White-lipped, 495.
Pee-shoo, 220.
PENTAIL, 155.
PHALANGER, FLYING, 759.
Phalangist, Sooty, 757.
" Valpine, 758.
PHATAGIN, 739.
Pichcrogo, 742.
Pika, Alpine, 729.

Pika, North American, 729.
" Ogotono, 729.
Pinchaque, 463.
Pinche, 98.
Pipistrelle, 138.
POINTER, 279.
POLECAT, 298.
Pony, Exmoor, 442.
" MIDLAND 441.
" Welsh, 442
PORCUPINE, BRAZILIAN, 724.
" CANADIAN, 724.
" COMMON, 725.
" TUFTED-TAILED, 726.
PORPOISE, 388.
" Right-whale, 387.
Potoroo, 757.
Potto, 111.
Pongoune, 231.
Press, 154.
PRONG HORN, 625.
PUMA, 196.

Q.

QUAGGA, 452.

R.

RABBIT, 733.
" JACKASS, 731.
" ROCK, 684.
RACCOON, 311.
" Black-footed, 313.
" California, 313.
" CRAB-EATING, 313.
Raccoon Dog, 259.
Rat, Beaver, 695.
" BROWN, 688.
" BLACK, 689.
" Cotton, 694.
" Florida, 693.
" Rocky Mountain, 693.
" Madagascar, 108.
" Musk, 695.
" POUCHED, FUR COUNTRY 702.
" Trunked, 152.
" WATER, 697.
RATEL, 306.
Rasse, 229.
REINDEER, 545.
RIEDRIVER, 281.
RHINOCEROS, INDIAN, 471.
" Javanese, 473.
" LITTLE BLACK, 477.
" Rough-eared, 476.
" Sumatran, 474.
" WHITE, 481.
Riman-dahan, 208.
Roode Bok, 691.
RORQUAL, 362.

Rorqual, Great Indian, 365.
" Southern, 367.

S.

SAGOUIN, 98.
" Silver, 99.
Saiga, 624.
Saimiri, 93.
Saki, Black, 91.
" Scarlet-faced, 91.
" WHITE-HEADED, 90.
Salamander, 701.
Sambur, 557.
SABLE, 294.
" American, 295.
Sanga, 579.
SAPAJOU, 77.
" Horned, 79.
Sasin, 623.
SEA BEAR, NORTHERN, 3--9.
SEA-COWS, 397.
" of Steller, 404.
SEA-ELEPHANT, 344.
Sea-Leopard, False, 343.
SEA-LION, 329.
" Northern, 332.
Sea-Unicorn, 377.
Seal, Bearded, 341.
" Caspian, 340.
" Crab-eating, 342.
" CRESTED, 345.
" COMMON, TRUE, 339.
" Fur, Australian, 331.
" " Cape, 331.
" " Southern, 331.
" Grey, 342.
" Hair, Australian, 332.
" " Cape, 332.
" " California, 331.
" HARP, 340.
" Richard's, 341.
" Ross's Large-eyed, 343.
" White-bellied, 342.
Serval, 215.
SETTER, 279.
" Irish, 280.
Sewellel, 718.
Sheep, CRETAN, 650.
" Fat-tail, 649.
" HIGHLAND, 653.
" Leicester, 651.
" MERINO, 652.
" Rocky Mountain, 648
" SO UTHDOWN, 654.
Shrew, Berlandier's, 170.
" Broad-nosed, 170.
" Carolina, 170.
" ELEPHANT, 151.
" Etruscan, 169.
" Forster's, 169.
" House, 169.
" Masked, 170.
" Mouse, 168.
" Navigator, 170.

INDEX. 795

Shrew, Otter, 162.
" Pentail, 155
" Short tailed, 170.
" Squirrel, 155.
" Thick tailed, 169.
" Thompson's, 170.
" Water, 169.
Siamang, 38.
Skunk, 309.
" California, 310.
" Large tailed, 310.
" Texan, 310
Sloth, Spotted, 739
" Three-toed, 738.
" Two-toed, 738.
Sosoook, 382.
Spaniel, Blenheim, 290.
" Cocker, 280.
" King Charles, 290
" Water, 250.
Spermophile, Leopard, 713.
" Line tailed, 714
Sphinx, 60
Spring Bok, 620.
Squirrel, Black, 708.
" European, 707.
" Flying, 710.
" Grey, 708.
" Ground, 712
" Hare, 709.
" Long eared, 710.
" Northern Grey, 709
" Red, 709
" Sugar, 709.
Stag, Maned, 557.
Stoat, 296.
Surilho, 308.
Syra, 207.

T.

Taguan (Squirrel), 711.
" (Marsupial), 758.
Tagnicate, 495.
Tajaca, 404.
Talapoin, 47.
Tamandua, 745.
Tamanoir, 744.
Tamarin, 99.
Tangalung, 229.
Tapir, American, 458.
" Baird's, 463.
" Malay, 462.
Tapoa, 715.
Tarpan, 410.
Tarsier, Spectre, 114.
Tatouay, 741.
Tatouhou, 740.

Tayra, 300.
Tee-Tee, 94.
" Collared, 95.
Teledu, 305.
Tendrec, 161.
Tenrec, 160.
" Banded, 160.
Terrier, Black and Tan, 288.
" Bull, 288.
" Fox, 289
" Scotch, 288.
" Skye, 288
" Yorkshire, 288.
Thresher, 391
Tiger, Bengal, 188.
" Clouded, 208.
Tikus, 159.
Tucuxi, 384.
Tukotuko, 722.
Tupaia Tana, 153.
Turnspit, 278.

U.

Uakari, 91.
Unan, 733.
Unko, 38.
Urchin, 158.

V.

Vampire, 132.
" Californian, 133.
Vicuna, 529.
Viscacha, 720
" Alpine, 719.

W.

Wah, 316.
Walawy, 168.
Wallabee, 755.
Walrus, 334
Wanderoo, 71.
Wapiti, 550.
War-Elliot, African, 510.
" Elian's, 511.
Water-buck, 628.
Wauwau, 36.
Weasel, 298.
" Bridled, 298.
" Little Nimble, 297.
" Richardson's, 298.
" Small Brown, 298.
" Small, 297.
" Tawny, 297.
" Yellow cheeked, 297.

Whale, Biscay, 358.
" Black, Howling or Cooing, 362.
" Bottle-nosed, 375.
" Bunched, 360.
" California Grey, 367.
" Cape, 358.
" Greenland or Right, 359
" Humpback, 360
" Northern Finner, 364
" Pike, 366.
" Pilot, 392.
" Right Australian, 358.
" Scrag, 358.
" Shortheaded, 374.
" Sperm, 368.
" Sulphur Bottom, 364.
" White, 303.
Wild Cat, 222.
" European, 209
" Texas, 223.
Wolf, Alpine, 266.
" Black, 251.
" Crab-eating, 254.
" Grey, 251.
" Jackal, 250.
" Maned, 254.
" Prairie, 252.
" Red, 254.
" Red of Texas, 252.
" Striped, 250.
" White, 252.
" Zebra, 752.
Wolverene, 299.
Wombat, 760.
Woodchuck (Marmot), 715.
" (Marten), 295.

X.

Xiphius, 376.

Y.

Yaguarundi, 198.
Yak, 595.
Yarke, 90.
Yapock, 750.

Z.

Zebu, 592.
Zebra, 454.
Zenick, 238.
Zibeth, 228.
Zorilla, 307.

INDEX OF SCIENTIFIC NAMES.

A.

Acrobata pygmæus, 759.
Addax nasomaculatus, 614.
Ælurus fulgens, 316.
Æpyceros melampus, 624.
Alcephalus caama, 632.
Alces Americanus, 541.
  " palmatus, 539.
Amphisorex Linneanus, 169.
Anoa depressicornis, 597.
Antechinus flavipes, 751.
Antilocapra Americana, 625.
Antilope bezoartica, 623.
Antrozous pallidus, 143.
Aonyx leptonyx, 303.
Aplocerus Americanus, 635.
Arctitis binturong, 231.
Arctocebus Calabarensis, 112.
Arctocephalus Antarcticus, 331.
  " cinereus, 331.
  " nigrescens, 331.
  " niveosus, 332.
Arctomys habac, 715.
  " marmota, 714.
  " monax, 715.
Arctomyx collaris, 303.
Arctopithecus flaccidus, 739.
Aricla taniata, 237.
Arvicola amphibius, 697.
  " arvalis, 698.
  " Pennsylvanica, 698.
Ateles bartletti, 83.
  " Beelzebub, 81.
  " paniscus, 81.
  " pentadactylos, 82.
Atherura Africana, 726.
Auchenia guanaco, 525.
  " llama, 527.
  " paco, 528.
  " vicuña, 529.
Aulacodes swinderianus, 723.

B.

Babyroussa Alfurus, 508.
Balæna Biscayensis, 358.
  " gibbosa, 352.
  " marginata, 358.
  " mystacetus, 350.
Balænoptera Australis, 367.
  " rostrata, 366.
Bassaris astuta, 316.
Belideus Australis, 759.
  " sciureus, 759.
Beluga Canadensis, 394.
  " catodon, 393.
Bettongia penicillata, 756.
Bibos frontalis, 593.
  " gaurus, 594.
  " Sondaicus, 595.
Bison Americanus, 587.
  " bonassus, 584.
Bos taurus, 575.
Brachyotus Daubentonii, 140.
Bradypus tridactylus, 738.
Brachyurus calvus, 91.
  " melanocephalus, 91.
Bubalus buffelus, 603.
  " Caffer, 599.
  " vulgaris, 605.

C.

Callicephalus caspicus, 340.
  " vitulinus, 339.
Callithrix lugens, 95.
  " personata, 94.
  " torquata, 95.
Callorhinus ursinus, 329.
Camelopardalis girafa, 569.
Camelus Arabicus, 514.
  " Bactrianus, 522.
Canis adustus, 250.
  " Alpinus, 206.

Canis aureus, 254.
  " Azaræ, 253.
  " cancrivorus, 254.
  " dingo, 291.
  " domesticus, 266.
  " Dukhunensis, 26
  " griseus, 262.
  " jubatus, 254.
  " lagopus, 262.
  " fuliginosus, 263.
  " latrans, 252.
  " lupaster, 250.
  " occidentalis, 251.
  " " var. albus, 252.
  " " " ater, 251.
  " " " gigas, 251.
  " " " rufus, 252.
  " ochropus, 253.
  " pallipes, 256.
  " Primævus, 266.
  " procyonides, 259.
  " simensis, 250.
  " Sumatrensis, 266.
  " velox, 262.
Capra ægagrus, 639.
  " Ægyptiana, 641.
  " angoriensis, 640.
  " Argali, 647.
  " ibex, 643.
  " Jemlaica, 641.
  " laniger, 634.
  " mambrica, 641.
  " megaceros, 641.
  " montana, 648.
  " musimon, 647.
  " polii, 648.
  " Pyrenaica, 643.
  " reversa, 642.
  " Syriaca, 644.
  " tragelaphus, 636.
Capreolus vulgaris, 563.
Capromys pilorides, 720.
Castor Canadensis, 702.
  " fiber 701.

INDEX OF SCIENTIFIC NAMES. 797

Catoblepas gnu, 633.
Catodon macrocephalus, 368.
Cavia cobaya, 728.
Cebus apella, 79.
" capucinus, 78.
" fatuellus, 79.
" hypoleucus, 78.
" leucogenys, 78.
" olivaceus, 78.
Centetes ecaudatus, 160.
" variegatus, 160.
Cephalophus mergens, 630.
" natalensis, 631.
Cercocebus collaris, 52.
" fuliginosus, 52
Cercolabes prehensilis, 724.
Cercoleptes caudivolvulus, 315.
Cercopithecus Diana, 51.
" Mona, 51.
" petaurista, 51.
" ruber, 51.
" Sabæus, 51.
Cervicapra atchi, 627.
Cervulus muntjak, 564.
Cervus Aristotelis, 557.
" axis, 557.
" barasinga, 557.
" campestris, 558.
" Canadensis, 559.
" Elephas, 552.
" hippelaphus, 557.
" hyelaphus, 558.
" Virginianus, 554.
" Wallechii, 556.
Cheirogaleus milii, 108.
" murinus, 108.
Cheiromys Madagascariensis, 115.
Cheironectes vapock, 750.
Chinchilla laniger, 719.
Chlamydophorus truncatus, 742.
Chæropus castanotis, 753.
Cholæpus didactylus, 738.
Chrysochloris holoserica, 163.
Cœlogenys paca, 727.
Colobus guereza, 45.
" Satanas, 46.
" ursinus, 46.
Condylura cristata, 165.
Cricetus frumentarius, 692.
Crossarchus obscurus, 239.
Cryptoprocta ferox, 226.
Ctenodactylus massoni, 722.
Ctenomys Magellanicus, 722.
Cuscus maculatus, 758.
Cyclothurus didactylus, 746.
Cynælurus jubatus, 224.
Cynictis Levaillantii, 238.
Cynocephalus babuin, 56.
" hamadryas, 60.
" Maimon, 62.
" Mormon, 63.
" porcarius, 58.

Cynocephalus sphinx, 60.
Cynogale Bennettii, 233.
Cynomys ludovicianus, 716.
Cynopithecus niger, 74.
Cystophora cristata, 345.

D.

Dama vulgaris, 563.
Dasyprocta agouti, 726.
Dasypus sexcinctus, 741
Dasyurus viverrinus, 754.
Delphinapterus Peronii, 387.
Delphinus Delphis, 386.
Dendrologus ursinus, 756.
Desmodus rufus, 133.
Didelphys cancrivora, 750.
" dorsigera, 750.
" Virginiana, 750.
Dioplodon Schellensis, 377.
Dipus Ægyptus, 697.
" alactaga, 697.
Dolichotis Patagonicus, 728.
Dysopes nasutus, 144.

E.

Echidna aculeata, 766.
Elasmognathus Bairdii, 463.
Elephas Africanus, 672.
" Indicus, 666.
" mastodon, 658.
" primigenius, 657.
Enhydris marina, 302.
Equus Asinus, 448.
" Burchelli, 453.
" caballus, 410.
" hemionus, 445.
" onager, 447.
" Quagga, 452.
" tæniopus, 448.
" Zebra, 454.
Erinaceus auritus, 158.
" Europæus, 158.
Eriodes hypoxanthus, 83.
Erethizon dorsatum, 724.
Eubalæna Australis, 358.
Eumetopias, Stelleri, 332.

F.

Felis catus, 209.
" chati, 207.
" concolor, 196.
" domesticus, 214.
" " angolensis, 214.
" " ecaudatus, 214.
" ferox, 208.
" griseus, 206.
" Japonicus, 202.

Felis Javanensis, 211.
" leo var. Barbarus, 183.
" " var. Goojratensis, 183.
" " var. Persicus, 183.
" Leopardus, 202.
" macrocelis, 208.
" macrourus, 207.
" maniculatus, 212.
" manul, 211.
" marmoratus, 205.
" melas, 203.
" pajeros, 207.
" pardalis, 205.
" pardus, 201.
" pictus, 206.
" onca, 199.
" serval, 215.
" syra, 207.
" tigrinus, 206.
" tigris, 188.
" uncia, 203.
" undatus, 211.
" variegatus, 202.
" viverrinus, 215.
" yaguarundi, 198.
Fiber zibethicus, 695.

G.

Galago agisymbanus, 113.
" crassicaudatus, 114.
" Moholi, 113.
Galeopithecus volans, 149.
Galictis barbara, 300.
" vittata, 299.
Galidictis vittata, 233.
Gazella dorcas, 617.
" euchore, 620.
Genetta Senegalensis, 230.
" vulgaris, 230.
Geomys bursarius, 700.
" pinetus, 701.
Globiocephalus deductor, 392.
Grampus Cuvierii, 396.
Gulo luscus, 299.
Gymnura Rafflesii, 159.

H.

Halichœrus gryphus, 342.
Halicore Dugong, 400.
" tabernaculi, 401.
Halicyon Richardsii, 341.
Halmaturus ualabatus, 755.
Hapale Jacchus, 98.
" Œdipus, 97.
" Pygmæus, 99.
Haplodon rufus, 718.
Helarctos Malayanus, 325.
Helictis Nepaulensis, 310.
Hemicentetes speciosus, 161.

Hemigale Hardwickii, 239.
Herpestes griseus, 234.
" Ichneumon, 235.
" Javanicus, 234.
" Nyula, 234.
" Widdringtonii, 234.
Hesperomys campestris, 696.
" Sonoriensis, 697.
" Texana, 697.
Hippopotamus amphibius, 486.
" liberiensis, 492.
Hippotragus leucophoeus, 616.
" niger, 615.
Hapalemur griseus, 107.
Hyaena brunnea, 243.
" crocuta, 244.
" striata, 244.
Hydrochoerus capybara, 727.
Hydromys chrysogaster, 695.
Hylobates agilis, 36.
" Hulock, 35.
" lar, 35.
" Raffle-ii, 35.
Hylomys suillus, 155.
Hyperoodon bidens, 375.
Hypsiprymnus minor, 757.
Hyrax Abyssinicus, 682.
" arboreus, 683.
" capensis, 683.
Hystrix cristata, 725.

I.

Ictonyx capensis, 307.
Indris brevicaudatus, 103.
" mitratus, 103.
Inia Amazonica, 383.

J.

Jaculus Hudsonius, 698.

K.

Kobus ellipsoprymnus, 628.

L.

Lagidium Cuvieri, 719.
Lagomys Alpinus, 729.
" ogotona, 729.
" princeps, 729.
Lagorchestes leporoides, 758.
Lagostomus trichodactylus, 720.
Lagothrix Humboldtii, 80.
Lasiurus cinereus, 142.
" Noveboracensis, 141.
Lemur catta, 106.
" leucomystax, 105.
" macaco, 105.
" mongos, 106.

Lemur ruber, 106.
" varius, 105.
Lepilemur tarcifer, 108.
Leptonyx Weddellii, 343.
Lepus Æthiopicus, 732.
" Alpinus, 733.
" Americanus, 731.
" callotis, 731.
" cuniculus, 733.
" Hibernicus, 733.
" palustris, 731.
" Sylvaticus, 731.
" Texianus, 731.
" timidus, 730.
" Trowbridgei, 731.
" vulgaris, 732.
Linsang gracilis, 230.
Lobodon carcinophaga, 342.
Lontra Brasiliensis, 303.
Loris gracilis, 110.
Lutra Californica, 302.
" Canadensis, 301.
" vulgaris, 300.
Lycaon pictus, 264.
Lynx caligatus, 220.
" Canadensis, 220.
" chaus, 217.
" fasciatus, 223.
" maculatus, 223.
" melanotis, 218.
" pardinus, 220.
" rufus, 222.
" vulgaris, 219.

M.

Macacus Cynomolgus, 65.
" erythraeus, 71.
" Inuus, 72.
" nemestrinus, 71.
" pileatus, 67.
" rhesus, 69.
" silenus, 71.
" sinicus, 67.
Macropus laniger, 755.
" major, 754.
Macroscelides proboscideus, 181.
" Rozetti, 182.
Macrotus Californicus, 133.
Manatus Australis, 399.
" latirostris, 400.
Manis brevicaudatus, 740.
" longicaudatus, 739.
Martes abietum, 293.
" Americana, 295.
" Canadensis, 295.
" foina, 294.
" Pennantii, 295.
" zibellina, 294.
Megaderma frons, 135.
Megalotis Lalandii, 264.
Megaptera longimana, 360.

Meles taxus, 304.
Mellivora capensis, 306.
Mephitis mesoleuca, 310.
Mephitis mephitica, 308.
" mesoleuca, 310.
" occidentalis, 310.
" suffocans, 308.
Microcebus myoxinus, 107.
Midas argentatus, 99.
" rosalia, 99.
" ursulus, 99.
Molossus nasutus, 141.
Monodon monoceros, 377.
Moschus moschiferus, 565.
Mungos fasciatus, 239.
Mus barbarus, 692.
" minutus, 691.
" musculus, 689.
" rattus, 688.
Mustela vulgaris, 298.
Mycetes Caraya, 86.
" seniculus, 85.

Myogale moschata, 166.
" pyrenaica, 166.

Mystacoceus sulcatus, 744.

N.

Nanotragus Hemprichii, 630.
Nasua leucorhyncha, 314.
" nasica, 313.
" rufa, 314.
Nemorhedus goral, 635.
Neotoma Drummondii, 693.
" Floridana, 693.
Neotragus moschatus, 629.
Noctilio Americanus, 144.
" leporinus, 144.
Nycticetus tardigradus, 109.
Nycticetus crepuscalensis, 143.
Nyctinomus nasutus, 144.
" obscurus, 144.
Nyctipithecus trivirgatus, 92.
Nycteris, 135.

O.

Octodon Cummingii, 721.
Ommatophoca Rossii, 343.
Orca capensis, 391.
" gladiator, 390.
Orcas canna, 607.
Ornithorhynchus paradoxus, 763.

INDEX OF SCIENTIFIC NAMES.

Orycteropus capensis, 743.
Oryx beisa, 614.
" capensis. 613.
" leucoryx, 614.
Otaria jubata, 329.
Otolicnus, 113.
Ovibos moschatus, 653.

P.

Pagomys fœtidus, 340.
Paguma larvata, 232.
Paradoxurus fasciatus. 231.
" typus, 231
Pedetes capensis, 697
Pelagius albiventer, 342.
Perameles fasciata, 752.
" nasuta, 753.
Perodicticus Potto, 111.
Petaurista taguanoides, 758.
Petrogale penicillata. 755.
Phacochærus Æliani, 511.
" Æthiopicus, 510.
Phalangista fuliginosa, 757.
" vulpina, 758.
Phascolarctus cinereus, 757.
Phascolomys ursinus, 760.
Phoca Barbata, 341.
Phocæna communis, 388.
Physostoma spectrum, 132.
Physalus antiquarum, 363.
" gibbaldii, 364.
Physalus Indicus, 365.
" sulphureus, 364.
Physeter tursio, 373.
Pithecia hirsuta, 90.
" leucocephala, 90.
" satanas, 91.
Platanista gangetica, 382.
Plecotus auritus, 142.
Poephagus grunniens, 505.
Portax pictus, 611.
Potamochærus Africanus, 507.
" pictus, 506.
Potamogale velox, 162.
Primodontes gigas, 740.
Procapra gutturosa, 622.
Prochilus labiatus, 326.
Procyon cancrivorus, 313.
" Hernandezii, 313.
" lotor, 321.
" psora, 313.
Propithecus diadema, 109.
Proteles cristatus, 240.
Pteromys petaurista, 711.
Pteropus edulis, 138.
" Edwardsii, 138.
Ptilocercus Lowii, 155.
Putorius agilis, 297.
" cicognantii, 298.
" ermineus, 296.
" fœtidus, 295.
" frenatus, 295.

Putorius furo, 296.
" fuscus, 297.
" Kaneii, 297.
" Nigripes, 298.
" Noveboracensis, 297.
" pusillus, 297.
" Richardsonii, 298.
" vison, 297.
" xanthogenys, 297.

R.

Reithrodon Le Contes, 694.
Rhinoceros bicornis, 477.
" indicus, 471
" Keitloa, 479.
" lasiotis, 476.
" simus, 481.
" sondaicus, 473.
" Sumatranus, 474
Rhinolophus ferrum equinum, 135.
" hipposideros, 134.
" nobilis, 135.
Rhinopoma microphyllum, 136
Rhyncocyon cirnei, 152.
Rhytina Stelleri, 404.
Rupicapra tragus, 633.

S.

Saiga tartarica, 624.
Saimiris sciureus, 94
Sarcophilus ursinus. 751.
Scalops aquaticus, 165.
" argentatus, 166.
" Brewerii, 166.
" latimanus, 166.
" Townsendii, 166.
Sciropterus volucella, 710.
Sciurus Carolinianus, 708.
" Europeus, 708.
" Hudsonius, 709.
" Javanicus, 708.
" leporinus, 708.
" macrotus, 710.]
" niger, 708.
Scotophilus Carolinensis, 141.
" Georgianus, 141.
Semnopithecus entellus, 42.
" maurus, 44.
" nasica, 44.
" nemœus, 45.
Siamanga syndactyla, 38.
Sigmodon hispidum, 694.
Simia Satyrus, 32.
Solenodon Cubanus, 161.
" paradoxus, 162.
Sorex araneus, 169
" Bulandieri. 170.

Sorex brevicaudatus, 170.
" Carolinensis, 170.
" Etruscus, 169.
" fodicus, 169
" Forsteri, 169.
" longirostris, 170.
" navigator, 170.
" pachyurus, 169.
" personatus, 170.
" platyrrhinus, 170.
" talpoides, 170.
" Thompsoni, 170.
" vulgaris, 168.
Spalax typhlus, 696.
Spermophilus grammatius, 714.
" Hoodii, 713.
Steno Tucuxi, 384.
Suricata zenick, 238.
Sus aper, 499.
" cristatus, 501.
Synotus macrotis, 142.
" Townsendii, 142.

T.

Talpa cæca, 164.
" Europea, 164.
Tamandua tetradactyla, 745.
Tamias lysteri, 712.
" striatus, 713.
Taphozous, 138.
Tapirus indicus, 462.
" terrestris, 458.
" villosus, 463.
Tarandus caribou, 548.
" rangifer, 545.
Tarsius spectrum, 114.
Tartaria septemcinctus, 740.
Taxidea Berlandieri, 304.
" Labradoria, 304.
Tetraceros quadricornis, 631.
Thalassarctos ursinus, 318.
Theropithecus Gelada, 53.
Thylocenus cynocephalus, 752.
Tolypeutes tricinctus, 741.
Tragelaphus kudu, 609.
Tragulus pygmœus, 532.
Tremarctos ornatus, 327.
Trichechus Rosmarus, 334
Troglolytes Gorilla, 23.
" Niger, 27.
" calvus, 31.
Tapaia ferruginea, 154.
" Javanica, 154.
" Tana, 153.
Tursio erebennus, 387.

U.

Urotrichus Gibbsii, 168.
" talpoides, 167.

INDEX OF SCIENTIFIC NAMES.

Ursus Americanus, 322.
" Arctos, 320.
" cinnamoneus, 323.
" ferox, 323.
" Isabellinus, 321.
Urva cancrivora, 237.

V.

Vespertilio barbastellus, 139.
" Daubentonii, 140.

Vespertilio murinus, 139.
" nitidus, 140.
" noctula, 140.
" pipistrellus, 138.
" serotinus, 139.
" subulatus, 141.
Vesperugo Nilsonii, 138.
Viverricula malayensis, 229.
Viverra civetta, 228.
" tangarunga, 229.
" zibetha, 228.

X.

Xenurus unicinctus, 741.
Xiphius Sowerbiensis, 376.

Z.

Zalophus Gillespii, 331.
" lobatus, 332.

www.ingramcontent.com/pod-product-compliance
Lightning Source LLC
Chambersburg PA
CBHW020833020526
44114CB00040B/701